Results and Problems in Cell Differentiation

51

Series Editors
Dietmar Richter, Henri Tiedge

Roland Martin • Andreas Lutterotti (eds.)

Molecular Basis of Multiple Sclerosis

The Immune System

 Springer

Editors
Roland Martin
Universitätsklinikum Hamburg-Eppendorf
Institut für Neuroimmunologie und
Klinische Multiple-Sklerose-Forschung
Falkenried 94
20251 Hamburg
Germany

Andreas Lutterotti
Clinical Department of Neurology
Innsbruck Medical University
Anichstrasse 35
A-6020 Innsbruck
Austria
andreas.lutterotti@i-med.ac.at

Series Editors
Dietmar Richter
Center for Molecular Neurobiology
University Medical Center Hamburg-
Eppendorf (UKE)
University of Hamburg
Martinistrasse 52
20246 Hamburg
Germany
richter@uke.uni-hamburg.de

Henri Tiedge
The Robert F. Furchgott Center for Neural
and Behavioral Science
Department of Physiology and
Pharmacology
Department of Neurology
SUNY Health Science Center at Brooklyn
Brooklyn, New York 11203
USA
htiedge@downstate.edu

ISBN 978-3-642-14152-2   e-ISBN 978-3-642-14153-9
DOI 10.1007/978-3-642-14153-9
Springer Heidelberg Dordrecht London New York

Results and Problems in Cell Differentiation ISSN 0080-1844

Library of Congress Control Number: 2010930623

Cover design: WMXDesign GmbH, Heidelberg

Printed on acid-free paper

Springer is part of Springer Science+Business Media (www.springer.com)

Preface

Despite major efforts by the scientific community over the years, our understanding of the pathogenesis or the mechanisms of injury of multiple sclerosis is still limited. Consequently, the current strategies for treatment and management of patients, which are sometimes based on unproven hypotheses, are limited in their efficacy. The mechanisms of tissue protection and repair are probably even less understood. Further, the knowledge of disease pathogenesis is often derived from animal models, and it is often not clear whether animal data apply to the human condition; e.g. therapy for multiple sclerosis has failed many times. One reason for these limitations is the enormous complexity of the disease and every facet of its pathogenesis, the mechanisms of tissue injury, the diagnostic procedures and finally the efficacy of treatments and their side effects. For individual researchers and clinicians it is more and more difficult to oversee all aspects of the disease and to regularly review the advances in different fields of research.

In this volume of the Springer book series "Results and Problems in Cell Differentiation" we have tried to pay tribute to this high complexity by choosing several topics from basic and clinical immunology, genetics, diagnostics and therapy, which are discussed by outstanding experts in the field. It was the aim of all authors to review the latest advances in their speciality, but foremost to highlight the big challenges for the future.

We are aware that this series cannot include all the relevant aspects, but we hope to have reached our aim of bridging different fields such as neurobiology, immunology, genetics and therapy. We would like to thank all authors for their contributions.

Hamburg
June 2010

Roland Martin
Andreas Lutterotti

Contents

The Genetics of Multiple Sclerosis

Jan Hillert

Abstract After many years of slow progress in the genetic analysis of complex diseases, including multiple sclerosis (MS), the past few years have eventually seen the first steps of turning a frustrating quest into a success story with the identification of a rapidly increasing number of confirmed disease promoting or protecting genetic variants. The principal reason for the change in situation is numbers: number of individuals assessed in the studies and number of genetic variations available for genotyping. In short, we now know that it takes thousands of patients to identify new genes but also to pinpoint the complex interplay of several genetic factors within the strongest genetic locus in MS, the HLA gene complex. However, the mission is not completed until the gene identification has been translated into useful understanding of disease mechanisms, and this is likely to necessitate further methodological progress as well as hard work in functional studies.

Obviously, the reason for gene identification is never going to be genetic counseling or prenatal diagnosis as not even full knowledge of the genetic background on MS will allow even 30% accuracy in disease prediction, i.e., the concordance rate in female monozygotic twins. Therefore, it is only in the way that risk genes reveal details of disease triggering events, or that severity genes influence mechanisms driving the tissue damage, that genetic information is likely to be useful.

Abbreviations

CNS	Central nervous system
EAE	Experimental autoimmune encephalomyelitis
GWAS	Genome wide association study
HLA	Human leukocyte antigen
LD	Linkage disequilibrium

J. Hillert
Department of Clinical Neuroscience, Karolinska Institute, Stockholm, Sweden
e-mail: Jan.Hillert@ki.se

Results Probl Cell Differ, DOI 10.1007/400_2009_13

MHC   Major histocompatibility complex
MS   Multiple sclerosis
QTL   Quantitative trait locus

1 Genetic Epidemiology

For a complex disease, multiple sclerosis (MS) is unusually well studied in terms of risk of disease in relatives of MS patients. A series of seminal papers based on the Canadian MS registry clearly shows that the risk in monozygotic (identical) twins (28.4%) exceeds that of dizygotic (fraternal) twins (5.6%), thereby arguing strongly for the importance of genes in determining the risk of MS (Ebers et al. 1986; Sadovnick et al. 1993). Subsequent papers on the risk of relatives by adoption and half-siblings clearly show that the immediate environment plays no or little role in determining who will get MS in a family, whereas genes do (Ebers et al. 1995, 2000; Willer et al. 2003).

On the other hand, it has been less clear whether genetic factors also influence the severity or other characteristics of the course – although some studies indicate that there may be similarities in disease course within a family at a statistical level (Brassat et al. 1999; Chataway et al. 2001; Oturai et al. 2004; Hensiek et al. 2007), the general impression has been that family members are often strikingly different in this respect (Hensiek et al. 2007). Thus, it is not uncommon to see a relative pair of a parent with mild disease and a child with severe course, or a sibling pair where one has a relapsing course and the other a primary progressive. Considering that systematic bias is more likely to result in similarities rather than differences, this indicates that course is likely to be largely influenced by nongenetic mechanisms.

A further interesting aspect of MS genetics is the possibility that inheritance of several autoimmune diseases may share genetic background. Also here, studies are partly conflicting, and whereas some indicate that diseases such as inflammatory bowel disease are more common among relatives of MS patients (Barcellos et al.2006a,b; Nielsen et al. 2008; Hemminki et al. 2009), other studies fail to confirm this impression (Ramagopalan et al. 2007a,b). Indeed, as will be described below, there are now strong indications that there is a set of "autoimmune genes" of relevance in multiple autoimmune conditions. Then, it may well be that the picture has been obscured by different associations with specific alleles of the human leukocyte antigen (HLA) genes, class I as well as class II, since MS differs from most other diseases in this respect.

2 MS as a Homogeneous Entity

The great spread in the clinical picture of MS during the first 10–20 years of disease course has traditionally been seen as a strong evidence for heterogeneity within the condition we so far see as MS. The four proposed pathological subtypes of MS proposed by Lucchinetti, have revitalized this discussion (Lucchinetti et al. 2000).

Indeed, the identification of neuromyelitis optica (NMO) as a clearly distinguishable disease entity with a unique immunopathogenesis (Lennon et al. 2004), and the description of an MS-indistinguishable disease in the presence of mitochondrial DNA mutations typical of Leber's hereditary opticoneuropathy (Kellar-Wood et al. 1994), have served as early examples to what may be a common development – the MS entity falling apart by revealing its "subphenotypes."

However, if MS were a heterogeneous disease, then it would be a poor subject for genetic analysis since separate conditions would be likely to have their own risk genes. Thus, the recent success in gene identification indicates that the situation is actually more favorable than usually presumed.

In fact, there are several reasons to doubt the dogma of heterogeneity in MS. First, the weak genetic background for disease course and severity cited above (Hensiek et al. 2007), exemplified by strikingly different courses in the same family, speaks in this direction. Second, the uniform response to disease modifying drugs, particularly in terms of MRI parameters, supports the same concept. Third, the uniform distribution of the strongest genetic risk factor, HLA-DR15, as well as the second strongest, HLA-A*02, similarly frequent in familial and nonfamilial MS, benign or severe MS, primary progressive MS and relapsing remitting MS, may well serve to indicate similarity (Masterman et al. 2000; Brynedal et al. 2007; Smestad et al. 2007). Last, although the short-term course in MS may differ strikingly, the long-term prognosis gives a more coherent picture (Weinshenker 1989; Confavreux 2006; Sayao 2007). Thus, it may be argued that MS is indeed a condition well suited for gene identification, which is indeed supported by the current progress in such efforts.

3 A Short History of MS Genetics (Including Methodology Outline)

After the pioneering discoveries, starting in the first years of the 1970s, of the HLA molecules and genes being important for the risk of MS and several other autoimmune diseases in, the rate of success was very slow. For many years, linkage analysis in MS families appeared as the logical way forward in complex diseases such as MS as it had been for typically inherited Mendelian diseases. However, findings were typically weak and varied from study to study, not even giving formal evidence for the already confirmed importance of HLA (Ebers et al. 1996; Haines et al. 1996; Sawcer et al. 1996; Kuokkanen et al. 1997). Eventually, Sawcer and coworkers collected a substantial number of MS families and while corroborating the importance of the HLA locus, clearly demonstrated the absence of other strong loci, intriguingly concluding that other factors would have odds ratios below 1.3 (Sawcer et al. 2005). Indeed, the first two non-HLA MS risk genes, the IL7R and IL2R, display odds ratios at this level (Lundmark et al. 2007; Gregory et al. 2007; Hafler et al. 2007). The failure, in complex diseases, of linkage analysis to identify non-HLA genes, also suggested that the alternative approach of association analysis should be sought, but to be successful very large case–control materials would be required due to the minute effect of each gene.

In fact, after many years of frustrating search for genes, methodological progress has eventually allowed a radical change in our ability to discover risk genes in complex diseases. The success of the microarray based genome wide association screen has since early 2007 resulted in a rapidly growing list of genes with an undisputed influence on the risk of complex diseases. In particular, this has been productive for the group of supposedly autoimmune diseases. At this moment, March 2009, this has resulted in reports presenting convincing evidence for a handful of new MS genes originating from the genome wide association screening (GWAS) strategy (see below), but there is convincing data for a handful of new genes in the process of being reported.

4   HLA Complex Genes in MS

The HLA gene complex, the human major histocompatibility complex, is located on the short arm of the sixth chromosome (6p21) and spans some four million base pairs of DNA sequence, including around 250 known genes, many with immune functions. Figure 1 represents the physical map of the HLA complex. The HLA class II region contains the classical HLA class II genes DR, DQ, and DP. Class II molecules are heterodimers of one alpha and one beta chain, respectively, coded for by an A and a B gene, each of which makes up two of the globular domains characteristic

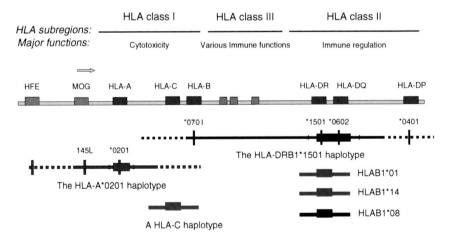

Fig. 1 Schematic representation of the HLA gene complex on the short arm of chromosome 6. The MS-associated alleles are depicted below as haplotypes (risk haplotypes in *black*, protective haplotypes in *blue*), signifying the extensive linkage disequilibrium of this region, whereby the exact location of the truly associated variant in each haplotype is bound to be uncertain. The figure illustrates the fact that several association with MS are likely to exist in combination with one or several HLA class I haplotypes, so far represented by association with HLA-A and HLA-C alleles

of the immunoglobulin super-gene family of molecules. Class II molecules are primarily expressed not only on professional antigen presenting cells such as dendritic cells, but also on B-cells and monocyte-macrophages. In case of inflammation, class II molecules can be induced temporarily in a number of cell types, including astrocytes. Class II molecules present antigens (i.e., serve as "restriction elements") to CD4 positive T-cells in the event of immune recognition at the initiation of the immune response. HLA class I molecules, of which the classical are HLA-A, -B and -C, are made up by single three-domain chains coded for by their respective genes. The fourth domain of the class I molecules is made up of the beta-2 microglobulin protein, coded elsewhere. Whereas the expression of class II molecules is highly regulated, class I molecules can be expressed by all cells and present antigen to mostly cytotoxic T-cells.

The original reports from 1972 to 1973 of an increased frequency in MS patients of the HLA specificities HLA-A3 (by serology) and HLA-Dw2 (by a cellular assay of lymphocyte proliferation) documented the importance of the HLA gene complex in MS long before these genes were accessible for direct study (Naito et al. 1972; Jersild and Fog 1972; Jersild et al. 1973). During the following decade, these reports were corroborated in several studies. Although HLA-A3 is a HLA class I molecule, it was soon recognized that the strongest effect resided with HLA-Dw2, shortly afterwards renamed HLA-DR2 when serological reagents became available (Jersild et al. 1973). The development of more specific sera allowed the refinement of this association to the DR2 subtype, DR15. In addition, it was noted that the MS-associated allele DR15 almost always occurs together with HLA-DQ6, an allele of the neighboring HLA-DQ genes (see below), which signifies that the genes coding for these molecules are carried as a haplotype on the same chromosome. During the 1990s, serologic HLA-typing was complemented by DNA-based genotyping. This gave access to the full underlying sequence variation, i.e., further alleles could be identified compared with serology and therefore made a new, DNA-based, nomenclature necessary. Since then, DNA-typing has largely replaced serology for clinical use and has become the only used method in genetic studies. The classic MS-associated haplotype is, therefore, now referred to as HLA-DRB1*1501,DRB5*0101,DQA1*0102, DQB1*0602 (for a review, see Hillert 1994). This signifies a haplotype of specific alleles, besides the HLA-DRB1 gene, also of DRB5, a second isotypic HLA-DR gene carried on DR15-carrying chromosomes, as well as the DQA1 and DQB1 genes that encode the two chains (α and β) of the HLA-DQ molecule (Fogdell et al. 1995). The haplotype also carries a HLA-DRA gene, but this gene is virtually nonpolymorphic, i.e., nonallelic. The HLA-DR15, DQ6 haplotype is also known for being strongly associated with the risk of cataplectic narcolepsy but to be virtually absent in individuals with type 1 diabetes (T1D). On the other hand, it is not associated with an increased risk for other autoimmune diseases.

During the years, efforts have been made to identify which component of this haplotype is responsible for the risk of MS, and views have varied. The difficulties depend on the strong linkage disequilibrium between its components – the separate alleles only occasionally occur in combination with other alleles, i.e., on other haplotypes. To find other haplotypes, one has to go to non-European populations where large sample sizes are difficult to obtain. Presently, most believe and tend to support the importance of the DRB part of the haplotype, based on a study on US African-Americans (Oksenberg et al. 2004). However, an opposing view has been proposed based on observations on a group of Brazilian MS patients (Alves-Leon et al. 2007).

Apart from the classical MS-associated haplotype, by convenience here referred to as DR15, there have been reports of other associated haplotypes in MS. However, reports have been inconsistent until the past few years when investigators have started to scale up studies to around 1,000 patients/families or more. Thus, with several large materials now available, it is apparent that a number of additional haplotypes may increase or decrease the risk of MS. Table 1 lists the findings of three such studies (Dyment et al. 2005; Barcellos et al. 2006a,b; Brynedal et al. 2007). Thus, HLA-DR3, -D8, -DR7, and -DR13 all seem to increase the risk whereas -DR1 and -DR14 seem to diminish the risk. This creates a hierarchy of risks, from DR15 to DR1, much resembling what is now thought to be present in T1D.

In addition, interesting data suggest that these haplotypes indeed interact with each other in various ways. For instance, it appears that HLA-DR1 exerts its protecting effect only in individuals that carry HLA-DR15 on their other chromosome (Dyment et al. 2005). In contrast, DR8 seems to increase the risk of MS only in the absence of DR15 (Dyment et al. 2005). This indicates that the most relevant level of analysis would be that of genotypes (both chromosomes considered together) rather than carriage. The notion of carriage as the most relevant level of analysis stems from transplantation, where the principle of codominant expression (all HLA gene products are expressed, at a roughly similar level) is critical from a practical point of view. Accordingly, to assess the importance of genotypes rather than carriage would require dramatically increased numbers of individuals, i.e., many thousands.

As described above, the original HLA association in MS was actually to a class I specificity, HLA-A3, now referred to as HLA-A*03 (Jersild and Fog 1972; Naito et al. 1972). However, it was soon noted that the association with HLA-Dw2 was stronger and this was, therefore, considered to be of primary importance (Jersild et al. 1973). Although no formal test was performed, it was not until the late 1990s when a set of MS patients was analyzed for both HLA class I and class II and an attempt was made to seek associations at both loci. Thus, Fogdell-Hahn reported an association with HLA-A*02 which appeared independent from DR15 (Fogdell-Hahn et al. 2000). Reciprocally, an increase of HLA-A*03 was observed. Similar findings were somewhat later reported from Norway (Harbo et al. 2004). Apart from occasional reports in small studies in other populations, no large study addressed the possible association in MS with HLA class I genes until two studies were reported in 2007. Thus, Brynedal reported a confirmation of the HLA-A*02 association in an independent case-control material from Sweden of 1,200 MS patients and 1,300 controls, all typed for HLA-DRB as well as for HLA-A and analyzed by stepwise logistic regression (Brynedal et al. 2007). In parallel, Yeo reported in a British-American study of concomitant typing of HLA-DRB, HLA-A, HLA-B, and HLA-C that there was indeed evidence for an independent HLA class I factor in MS, best represented by an association with HLA-C*05, but not of any HLA-A allele (Yeo et al. 2007). However, the statistical approach was different from that of Brynedal, using a

Table 1 HLA DRB1 association in MS in three large data sets. An increasingly consistent pattern is emerging where DRB1*15 and DRB1*8 stand out as predisposing whereas DRB1*01 and DRB1*14 appear to be protective (Dyment et al. 2005; Barcellos et al. 2006a,b; Brynedal et al. 2007 with additional data to be published by J Link)

Gene marker	Chromosomal location	Other name	Function	Pooled P	Estimated odds ratio	Patients	Controls
IL2R	10p15.1		T-cells				
rs2104286				10^{-27}	1.25	15,714	15,486
rs12722489				10^{-18}	1.24	16,907	15,486
IL7R	5p13		T-cells				
rs6897932				10^{-20}	1.20	18,117	16,720
KIAA0350	16p13	CLEC16A	Immune system				
rs6498169				10^{-8}	1.14	4,115	4,252
rs12708716				6.00×10^{-12}	1.20	8,106	10,296
KIF1b	1p36.2		Axonal transport of mitochondria				
rs10492972				2.00×10^{-8}	1.35	2,106	2,930
EV15	1p22		Retroviral integration?				
rs10735781				10^{-7}	1.11	1,364	1,702
rs6680578				10^{-7}	1.11	1,364	1,702
RPL5	1p		Ribosomal protein				
rs6604026				10^{-6}	1.15	4,065	4,252
CD58	1p13	LFA3	Immune system				
rs12044852				10^{-6}	1.24	4,065	4,252
CD226	18q22.3	DNAM1	Immune system				
rs763361				2.00×10^{-6}	1.13	8,106	10,296
SH2B3	12q24	LNK	T-cell activation				
rs3184504				7.00×10^{-5}	1.11	8,106	10,296

stepwise exclusion of patients carrying the best, second, third, etc., associated alleles. This technique may offer some advantages over that of logistic regression, but results in a rapidly decreasing number of remaining individuals/chromosomes.

The existence of independent class I factors in typically class-II-associated disorders has until recently not been widely accepted, but recent reports of a class I effect in T1D serves to support this concept (Nejentsev et al. 2007; Howson et al. 2009). Here, there seems to be factors both at the HLA-B and HLA-A loci, although class II remains by far the most important. In these studies, the investigators used a combination of a method referred to as recursive partitioning and logistic regression. In a similar analysis, unpublished data from our group, based on over 1,600 MS patients and 1,700 controls, all typed for HLA-DR, HLA-A, and HLA-C show that the main class I effect among these genes is represented by HLA-A and A*02 but that a small contribution of HLA-C is also demonstrated (Link, to be published). Clearly, the ultimate analysis will require typing also of HLA-B. Interestingly, in a family-based study by Ramagopalan et al., however, no influence was seen of the HLA-class I part of the transmitted haplotypes (Chao et al. 2008). It remains to be determined whether this is a correct conclusion or if the failure to see a class I signal depends on the family-based nature of the study where risk alleles are likely to be enriched, thus decreasing the power to detect protective influences.

4.1 HLA on Disease Characteristics

Although HLA genes confer risk for MS, it is much less clear whether disease characteristics are influenced. Although some early studies reported a likelihood of a severe or benign course as well as differences between relapsing remitting MS and primary progressive MS, later and larger studies have failed to support this (Olerup et al. 1989; Olerup 1991; Masterman 2000). Thus, in a recent survey Smestad and coworkers failed to find any effect on clinical characteristics by any alleles of the HLA-DR or HLA-A genes, with the exception of an effect on age at onset by DR15 – carriers of this variant at an average develop MS 1.5 years earlier, which on the other hand is an expected effect of a reasonably powerful risk gene (Smestad 2007). An observation which has yet to be independently confirmed is the report by DeLuca that DRB1*01 exerts a beneficial influence on MS prognosis, but this effect is only visible by comparing the 5% patients with the worst outcome to the 10% with the best (DeLuca et al. 2007).

As we have seen, whereas an importance of one, or more likely several HLA complex genes in MS is uncontroversial, the mechanisms by which these genes play their role is less clear. The best clues are likely to be found in genetic studies of the MS-like experimental model experimental autoimmune encephalomyelitis (EAE) in rodents. Here, the identification of associated class II molecules has allowed experiments with specific peptides which interfere with the EAE-critical antigen presentation (reviewed by Olsson and Hillert 2008).

Even more interesting from an MS perspective, combined breeding of transgenically manipulated mice carrying human HLA class II genes as well as human T-cell receptor genes has led to a spontaneous MS-like disease (Madsen et al. 1999; Gregersen et al. 2006; Friese et al. 2008). Thus, although far from formally proven, EAE studies support the functional importance of the HLA genes themselves also in MS.

In conclusion, the HLA complex harbors by far the strongest genetic factors in MS, which includes a complex interaction of several genetic variants both in the class II and class I regions. A first attempt to study the combined effect of class II and class I genes in combination suggests that the risk between genotypes may vary substantially, over 20 times, which exceeds the risks conferred by class II genotypes in isolation (Brynedal et al. 2007). Thus, it is possible that the HLA locus explains more of the total risk of MS than hitherto recognized.

5 The Candidate Gene and Linkage Screen Approaches in MS Genetics

When polymorphic markers became available for individual genes, first in the form of restriction fragment length polymorphisms (RFLP), later as microsatellites and subsequently as single nucleotide polymorphisms (SNPs), candidate gene studies became an evident possibility to identify genes in complex genetics. However, for many years, in hindsight, methods were not adequate to allow success, markers were too few in each gene and far too few individuals were investigated. In addition, assumptions were incorrect regarding the impact of the expected genes. Thus, almost 20 years of case-control association studies of candidate genes (HLA genes excluded) resulted in practically no useful information. A number of genes yielded suggestive associations but in the end, no consensus was established for any of these genes.

Whereas most of the studied candidate genes were selected based on supposed function in the pathogenesis of MS, some were selected based on their chromosomal location, i.e., under the suggestive peaks in previous linkage studies.

The first HLA linkage studies in MS families were published in 1996 (Ebers et al. 1996; Haines et al. 1996; Sawcer et al. 1996). Also, here, confusion was great. Inconsistent signals were obtained even from the HLA gene locus and other suggestive peaks were very modest and seemed to vary between the studies. As for case-control studies, the problem was later shown to be an overestimated expectation of the importance of each factor, resulting in insufficient clinical materials being studied. Thus, it was an important step forward when Sawcer in 2005 published a study of sufficient size to demonstrate that the HLA complex is by far the dominating genetic factor and that all other chromosomal regions scored below the level of significance (Sawcer et al. 2005). Thus, non-HLA genes could hardly be expected to be very powerful.

The only non-HLA linkage peak of some consistency, on the long arm of chromosome 17 (17q24) may, in fact, still be viewed as promising. Here, promising data have been reported for the PRKCA gene and some of its neighbors (Saarela et al. 2006), but so far no confirmation of this locus has been obtained from the recent GWA screen reports in autoimmune disorders, MS included.

6   The Interleukin Receptor Gene (IL7R)

However, there is one exception from the rule of failure of candidate genes in association studies in MS. After a few less-recognized studies since 2003 (Teutsch et al. 2003; Booth et al. 2005; Zhang et al. 2005), in 2007, two independent candidate gene studies reported an identical association with a nonsynonymous base exchange in the sixth exon of the IL7R gene (Lundmark et al. 2007; Gregory et al. 2007). As will be described below, these findings have subsequently been confirmed in several studies.

The IL7R gene codes for the alpha-chain of the interleukin 7 receptor and is vital for IL7 signaling. In turn, IL7 is a nonredundant cytokine of critical importance for the development of the immune system. Thus, knockout mice lacking an IL7 receptor are left with a severe combined immunodeficiency and are virtually void of T-cells and a similar condition has been reported in humans with IL7R mutations (He and Malek 1996; Lai et al. 1997). All in all, IL7 is a vital factor for T-cell survival, and it appears that the expression of IL7R is critical rather than the level of IL7, which seems to be less regulated (for a review, see Mazzucchelli and Durum 2007).

The background of the selection of IL7R as a candidate is interesting. Apart from an assumed functional importance of IL7 signaling, its location in one of the earliest suggestive linkage peaks was another argument for selection. In addition, IL7R is one of a handful of genes in the EAE locus eae2, identified already in 1995 (Sundvall et al. 1995).

The reports by Lundmark and Gregory were similar in the way by which early signals for separate SNP markers were followed by the identification of the same best marker, rs6798932. In addition, Lundmark added some functional data in observing an increased expression of IL7R and its ligand in the cerebrospinal fluid (CSF) cells in MS patients (Lundmark et al. 2007). Gregory interestingly tested the exon 6 amino acid substitution in an expression system and found that presence of the risk increasing allele increased the expression of a transcript lacking exon 6 (Gregory et al. 2007). As IL7R exon 6 codes for the trans-membrane part of the protein, this is likely to increase the expression of soluble receptor, which may be hypothesized to decrease IL7 signaling by reducing the amount of IL7 available for receptor binding. How this should be translated into an increased risk for MS remains to be explained, and efforts are ongoing to increase the understanding of these mechanisms.

In summary, although made at the verge of the GWAS era, the discovery of the MS IL7R association demonstrated that the concept of candidate genes in a complex disease such as MS was not altogether wrong or hopeless, but only frustratingly difficult.

7   The New MS Genes: Identification Through
Genome Wide Association Screen

The technique that eventually allowed efficient identification of non-HLA genes in MS, like in other complex disorders, was the large-scale SNP genotyping by microarray techniques, pioneered by Affymetrix and soon also offered by Illumina.

The seemingly simple principle was to genotype patients and controls for large numbers of SNPs representing most parts of our 23 chromosomal pairs, thus justifying the eponym, GWAS. In fact, already in 2002, the completion of a first GWAS in MS was reported. However, instead of using SNPs, still not available at a large enough scale, the investigators applied approximately 5,000 microsatellite markers (Sawcer et al. 2002). As a consequence of an insufficient density of markers (i.e., too great average distance between neighboring markers), only a fraction of the genome was covered and no true MS risk gene was identified. A few years later, in 2004, researchers at Serono Genetics Institute reported the completion of an MS GWA using some 90,000 SNPO markers, and the subsequent identification of a list of MS risk genes. However, so far these findings have remained unpublished.

In June 2007, in a seminal paper, the Wellcome Trust Case–Control Consortium published the completion of a GWA using half a million SNP markers genotyped on the Affymetrix platform in 14,000 patients representing seven sets of 2,000 patients with different complex diseases; each set was compared to a common set of 3,000 healthy controls (WTCCC 2007). Three of the disorders, T1D, rheumatoid arthritis (RA), and Crohn's disease (CD) were of supposedly autoimmune origin, in contrast to the others, acute myocardial infarction, hypertension, type 2 diabetes, and bipolar disease. In short, markers representing 12 genes scored positively at the genome-wide level, i.e., $P < 5 \times 10^{-7}$, a level of significance low enough to compensate for the multiple comparisons performed in any kind of high-throughput gene-mapping study. Of these, six were previously known, and thus served as positive controls. The other six were new and might therefore be expected to have a risk of being false positives. However, five of these six were already confirmed by subsequent independent studies at the time of publication of the WTCCC paper (WTCCC 2007). Thus, the methodology proved to be possibly better than expected. On the one hand, the risk of false signals, due for instance to poor comparability between patients and controls, did not appear to be overwhelming. On the other hand, the finding of several confirmed signals also proved that the studied diseases, and particularly CD and T1D, were homogeneous enough to allow the identification of genes. In fact, previous expectations of a widespread heterogeneity within complex disease entities has been a theoretical argument against the chance of success of GWAS even at a large scale. All in all, the paper demonstrated effectively that GWAS technology was now mature enough to provide genes for complex diseases. A long series of papers from various research groups or consortia have in the past year corroborated these notions, and lists of newly identified genes are being assembled for many complex diseases, including MS. Thus, the past years have seen a success rate in complex genetics previously unheard of.

For MS, the era of GWAS is still only in its beginning. A few materials have so far been screened with high numbers of SNPs, but to date only one study has been presented, that of the International MS genetics Consortium (IMSGC), presented in July 2007 (Hafler et al. 2007). In parallel, a number of scans are being analyzed and will be presented soon. In addition, a true second generation GWAS is currently being performed by the IMSGC under the auspices of the Wellcome Trust, targeting to screen 11,000 MS patients. However, already at the present stage, success has been substantial also for MS and progress is likely to be fast. Here, the emphasis will be on the first examples of GWAS-generated MS genes, with the assumption that these will be followed by several others and possibly more exciting ones already when this chapter is being printed.

The first MS GWAS was based on a screen of MS 931 trios (patients with both parents) (Hafler et al. 2007). From a primary list of results, 100 SNP markers were selected for a confirmation test based on a combination of screen results and previous information from earlier studies or supposed importance in MS. These markers were then tested for importance in 609 further trios and 2,322 sporadic patients and 2,987 controls.

The resulting list of combined data presented intriguing information. The strongest signal was, as expected, obtained for an HLA class II SNP, representing the HLA-DRA gene. In addition, two genes obtained support at the genome-wide level, IL2R, represented by two markers, and IL7R. The latter was expected due to the great overlap in material with the previously obtained data by the IMSGC on IL7R (see above), although eventually published simultaneously with the GWAS paper. In addition, a number of genes scored below the genome-wide cutoff and were thus of unclear relevance, but all the same interesting. As will be seen below, several of these have already received additional support to approach, and are likely to exceed the genome wide significance level of 5×10^{-7}. In summary, the IL7R gene, with the independent support provided by the Lundmark study, was widely accepted as the first confirmed non-HLA gene in MS.

Evidence confirming an importance of IL2R gene was soon afterwards provided in an independent study by the Ebers group (Ramagopalan et al. 2007a,b) and subsequently in at least two more studies from Australia (Rubio et al. 2008) and Europe (Weber et al. 2008).

Sawcer recently coordinated a report by the IMSGC and presented pooled data on the top IL7R and IL2R markers from several populations, together mounting to the impressive numbers of almost 14,000 patients and as many controls (IMSGC et al. 2009). The accumulated relative risk for the IL7R marker rs6987932 was 1.20 at an impressive significance level of $P = 10^{-17}$. The corresponding figures for the two IL2R markers rs2104286 and rs12722489 were 1.25 ($P = 10^{-23}$) and 1.23 ($P = 10^{-15}$). Thus, evidence for an importance in MS is indeed impressive for these two genes in terms of individuals investigated and statistical certainty of an association. On the functional side, in addition to what was initially presented for IL7R (Lundmark et al. 2007; Gregory et al. 2007), recent studies suggest a similar mechanism for the IL2R (Maier et al. 2009a,b).

KIAA0350, also referred to as CLEC16A, stands out as the third most convincing non-HLA MS risk gene. Originally scoring nicely in the GWAS study for marker rs6498169 (OR = $1.14/P = 4 \times 10^{-6}$), this gene received attention by giving a convincing signal also in T1D, although with another marker rs12708716 (WTCC 2007). In a very recent report, Booth reported an impressive OR = $1.20/P = 6 \times 10^{-12}$ for this marker in 8,106 patients and 10,296 controls (Booth et al. 2009). KIA0350 codes for a C-type lectin, a group of molecules thought to be of importance for cell-cell interaction within the immune system. In the same report, Booth reported promising data for two other genes selected due to their identification as T1D genes, CD226, and

SH2B3, of which CD226 has recently received further support, both in MS and other autoimmune diseases (Hafler et al. 2009).

As expected, several investigators have followed up on the list of candidate genes provided by the original MS GWAS (EV15, RPL5, CD58). The EV15 gene, of which two markers were among the top 15 scorers in the GWAS received further support for a role in MS by a subsequent report by Hoppenbrouwers, adding up to a potential score of OR = $1.11/P = 10^{-8}$ for its best marker (Hoppenbrouwers et al. 2008). The functional importance of this gene and its product is less clear, but EV15 is a known site for retroviral integration, pointing to other intriguing pathogenetic mechanisms in MS. Further support was found for CD58, which codes for LFA3, a well-known adhesion molecule. Subsequently, a recent study corroborates the candidacy of these genes even further (De Jager et al. 2009). A final example of a promising GWAS gene is RPL5 which codes for a ribosomal protein (Hoppenbrouwers et al. 2008).

A different story is the identification of KIF1B as an exciting and likely new MS gene. Performing a GWAS on the limited number of 49 Dutch MS patients, Hintzen and coworkers identified a signal in several markers of the KIF1B gene (Aulchenko et al. 2008). Somewhat surprisingly, this finding was substantiated in three other case–control populations from The Netherlands, Canada, and Sweden and a resulting score of OR = $1.35/P = 2 \times 10^{-8}$ was provided; i.e., passing the level of genome-wide significance after investigating a mere 2,106 patients and 2,930 controls. Although confirmation in independent studies is still pending, this finding is especially intriguing since the KIF1B gene product is thought to be of importance not within the immune system but for the transport of mitochondria in axons, i.e., within the nervous system itself. This suggests that not only the properties of the immune system, but also of the target organ itself, the CNS, influence the risk of MS, a principally important observation.

7.1   Genetic Control of Disease Characteristics in MS

As outlined above, while it has been evident from family studies that risk of MS is inherited, it is much less so for disease characteristics such as severity or course. Thus, although a number of studies have claimed the importance of certain genes for severity of MS, no coherent pattern has yet developed. However, this question has not so far been addressed by the GWAS technology, as would the question of genes influencing the course of disease. In addition, it is not clear what methodology should be ideally applied in assessing the severity, i.e., the studies yet performed have had an unknown but probably low statistical power. An illustration of this is a study by our group (Lundström to be published) in which the influence of the first five non-HLA MS risk genes were studied in conjunction for an importance on severity: no such risk could be observed, neither by comparison of MSSS scores, nor by survival analysis to EDSS milestones. In fact, this raises the possibility that many genes of importance for triggering of disease may have little to do with the pathogenetic mechanisms driving disease activity at later stages. Therefore, the identification of severity genes should be a prioritized task.

7.2 Coinheritance with Other Autoimmune Diseases

A number of epidemiological studies have indicated that the risk of autoimmune diseases may be increased in relatives of MS patients, if not in the patients themselves. This is highly compatible with the increasing list of GWAS-generated susceptibility genes. Even before the GWAS era, not only the HLA complex, but also the CTLA4 and PTPN22 were confirmed to be of importance for more than one autoimmune disease. As has been mentioned above, it now seems certain that this is paralleled by IL2R, IL7R, KIAA0350, CD226, and SH2B3 and possibly several others (International Multiple Sclerosis Genetics Consortium (IMSGC) (2009)). The reason why co-inheritance between, for instance, MS and T1D has been less obvious may be the fact that the strongest factor in each disease are "opposing" HLA-DR alleles: HLA-DR15 which has an OR for MS of approximately 3, is strongly protective against T1D, something that efficiently counteracts co-occurrence in an individual, and diminished co-inheritance within a family.

7.3 Discussion

In summary, the first generation MS GWAS has provided several leads that have been productive in follow-up already at this early stage. A few GWAS studies of similar size are ongoing, one of which was recently reported (Baranzini et al. 2009) and may add novel genes. The ongoing second generation GWAS which applies up to one million SNP markers, will encompass up to 20,000 patients thanks to generous grants provided by various bodies, foremost the Wellcome Trust and national MS societies in various countries such as Canada and Australia, and occasionally by industry. The prospects for further rapid progress are bright. Within a reasonably short period we are likely to have a substantial list of MS risk genes each reflecting a pathogenetic bottleneck in the precipitation of MS. As seen above, so far most of the presently available genes point towards an importance of the immune system, i.e., giving support for the autoimmune hypothesis.

It may be worthwhile to emphasize our short perspective on GWAS data in complex diseases and that methodological surprises may still appear. Markers of limited risk, such as all the genes in Table 2 are potentially sensitive to biases, such as population stratification, a problem which is likely to be dealt with increased sample sizes. More importantly, factors influencing the expressivity of MS, for instance, affecting the chance of subclinical disease may masquerade as risk factors even if they are essentially course modifiers. As argued above, such genes are likely to be of even greater importance as drug development targets than mere risk genes. In conclusion, we have not proven the case for an associated gene until its mechanisms have been unraveled, and we may be in for more surprises with increased experience of GWAS data in complex diseases.

Table 2 Emerging non-HLA MS risk genes. These genes have all been identified as predisposing en at least three ethnic group and collective *P*-values are the genome wide significance level in published or soon to be published material. The given *P*-values are estimations of published *P*-values without formal meta-analysis having been performed. The most likely odds ratios from published material are given

Allele	Dyment et al. (2005)		Barcellos et al. (2006a,b)		Brynedal et al. (2007)	
	Effect	*P*-value	Effect	*P*-value	Effect	*P*-value
DRB1*01	Protection	10^{-2}	Protection	10^{-6}	Protection	10^{-3}
DRB1*15	Risk	10^{-14}	Risk	10^{-33}	Risk	10^{-16}
DRB1*16						
DRB1*17	Risk	10^{-3}				
DRB1*04				10^{-2}		
DRB1*011						
DRB1*12						
DRB1*13	Risk	10^{-2}				
DRB1*14	Protection	10^{-5}	Protection	10^{-5}	Protection	10^{-2}
DRB1*07					Protection	10^{-4}
DRB1*08	Risk	10^{-2}	Risk	10^{-4}		
DRB1*09						
DRB1*10						

References

Alves-Leon SV, Papais-Alvarenga R, Magalhães M, Alvarenga M, Thuler LC, Fernández y Fernandez O (2007) Ethnicity-dependent association of HLA DRB1-DQA1-DQB1 alleles in Brazilian multiple sclerosis patients. Acta Neurol Scand 115(5):306–311 PMID: 17489940

Aulchenko YS, Hoppenbrouwers IA, Ramagopalan SV, Broer L, Jafari N, Hillert J, Link J, Lundström W, Greiner E, Dessa Sadovnick A, Goossens D, Broeckhoven CV, Del-Favero J, Ebers GC, Oostra BA, van Duijn CM, Hintzen RQ (2008) Genetic variation in the KIF1B locus influences susceptibility to multiple sclerosis. Nat Genet 40(12):1402–1403

Baranzini SE, Wang J, Gibson RA, Galwey N, Naegelin Y, Barkhof F, Radue EW, Lindberg RL, Uitdehaag BM, Johnson MR, Angelakopoulou A, Hall L, Richardson JC, Prinjha RK, Gass A, Geurts JJ, Kragt J, Sombekke M, Vrenken H, Qualley P, Lincoln RR, Gomez R, Caillier SJ, George MF, Mousavi H, Guerrero R, Okuda DT, Cree BA, Green AJ, Waubant E, Goodin DS, Pelletier D, Matthews PM, Hauser SL, Kappos L, Polman CH, Oksenberg JR (2009) Genome-wide association analysis of susceptibility and clinical phenotype in multiple sclerosis. Hum Mol Genet 18(4):767–778 Epub 2008. PMID: 19010793

Barcellos LF, Sawcer S, Ramsay PP, Baranzini SE, Thomson G, Briggs F, Cree BC, Begovich AB, Villoslada P, Montalban X, et al (2006a) Heterogeneity at the HLA-DRB1 locus and risk for multiple sclerosis. Hum Mol Genet 15:2813–2824

Barcellos LF, Kamdar BB, Ramsay PP, DeLoa C, Lincoln RR, Caillier S, Schmidt S, Haines JL, Pericak-Vance MA, Oksenberg JR, Hauser SL (2006b) Clustering of autoimmune diseases in families with a high-risk for multiple sclerosis: a descriptive study. Lancet Neurol 5(11):924–931

Booth DR, Arthur AT, Teutsch SM, Bye C, Rubio J, Armati PJ, Pollard JD, Heard RN, Stewart GJ (2005) Gene expression and genotyping studies implicate the interleukin 7 receptor in the pathogenesis of primary progressive multiple sclerosis. J Mol Med 83:822–830

Booth DR, Heard RN, Stewart GJ, Goris A, Dobosi R, Dubois B, Lorentzen AR, Celius EG, Harbo HF, Spurkland A, Olsson T, Kockum I, Link J, Hillert J, Ban M, Baker A, Sawcer S, Compston A, Mihalova T, Strange R, Hawkins C, Ingram G, Robertson NP, De Jager PL,

Hafler DA, Barcellos LF, Ivinson AJ, Pericak-Vance M, Oksenberg JR, Hauser SL, McCauley JL, Sexton D, Haines J (2009) The expanding genetic overlap between multiple sclerosis and type I diabetes. Genes Immun 10(1):11–14

Brassat D, Azais-Vuillemin C, Yaouanq J, Semana G, Reboul J, Cournu I, Mertens C, Edan G, Lyon-Caen O, Clanet M, Fontaine B (1999) Familial factors influence disability in MS multiplex families. French Multiple Sclerosis Genetics Group. Neurology 52(8): 1632–1636

Brynedal B, Duvefelt K, Jonasdottir G, Roos IM, Akesson E, Palmgren J, Hillert J (2007) HLA-A confers an HLA-DRB1 independent influence on the risk of multiple sclerosis. PLoS ONE 2(7):e664

Chao MJ, Barnardo MCNM, Lincoln MR, Ramagopalan SV, Herrera BM, Dyment DA, Montpetit A, Sadovnick AD, Knight JC, Ebers GC (2008) HLA class I alleles tag HLA-DRB1*1501 haplotypes for differential risk in multiple sclerosis susceptibility. Proc Natl Acad Sci U S A 105(35):13069–13074

Chataway J, Mander A, Robertson N, Sawcer S, Deans J, Fraser M, Broadley S, Clayton D, Compston A (2001) Multiple sclerosis in sibling pairs: an analysis of 250 families. J Neurol Neurosurg Psychiatry 71(6):757–761

De Jager PL, Baecher-Allan C, Maier LM, Arthur AT, Ottoboni L, Barcellos L, McCauley JL, Sawcer S, Goris A, Saarela J, Yelensky R, Price A, Leppa V, Patterson N, de Bakker PI, Tran D, Aubin C, Pobywajlo S, Rossin E, Hu X, Ashley CW, Choy E, Rioux JD, Pericak-Vance MA, Ivinson A, Booth DR, Stewart GJ, Palotie A, Peltonen L, Dubois B, Haines JL, Weiner HL, Compston A, Hauser SL, Daly MJ, Reich D, Oksenberg JR, Hafler DA (2009) The role of the CD58 locus in multiple sclerosis. Proc Natl Acad Sci U S A 106(13):5264–5269

DeLuca GC, Ramagopalan SV, Herrera BM, Dyment DA, Lincoln MR, Montpetit A, Pugliatti M, Barnardo MC, Risch NJ, Sadovnick AD, Chao M, Sotgiu S, Hudson TJ, Ebers GC (2007) An extremes of outcome strategy provides evidence that multiple sclerosis severity is determined by alleles at the HLA-DRB1 locus. Proc Natl Acad Sci U S A 104:20896–20901

Dyment DA, Herrera BM, Cader MZ, Willer CJ, Lincoln MR, et al (2005) Complex interactions among MHC haplotypes in multiple sclerosis: susceptibility and resistance. Hum Mol Genet 14:2019–2026

Ebers GC et al (1986) A population-based study of multiple sclerosis in twins. N Engl J Med 315(26):1638–1642

Ebers GC, Sadovnick AD, Risch NJ (1995) A genetic basis for familial aggregation in multiple sclerosis. Canadian Collaborative Study Group. Nature 377(6545):150–151

Ebers GC et al (1996) A full genome search in multiple sclerosis. Nat Genet 13(4):472–476

Ebers GC et al (2000) Conjugal multiple sclerosis: population-based prevalence and recurrence risks in offspring. Canadian Collaborative Study Group. Ann Neurol 48(6):927–931

Fogdell A, Hillert J, Sachs C, Olerup O (1995) HLA-DRB5 alleles in multiple sclerosis and narcolepsy. Tissue Antigens 46:333–336

Fogdell-Hahn A, Ligers A, Gronning M, Hillert J, Olerup O (2000) Multiple sclerosis: a modifying influence of HLA class I genes in an HLA class II associated autoimmune disease. Tissue Antigens 55:140–148

Friese MA, Jakobsen KB, Friis L, Etzensperger R, Craner MJ, McMahon RM, Jensen LT, Huygelen V, Jones EY, Bell JI, Fugger L (2008) Opposing effects of HLA class I molecules in tuning autoreactive CD8+ T cells in multiple sclerosis. Nat Med 14(11):1227–1235

Gregersen JW, Kranc KR, Ke X, Svendsen P, Madsen LS, Thomsen AR, Cardon LR, Bell JI, Fugger L (2006) Functional epistasis on a common MHC haplotype associated with multiple sclerosis. Nature 443:574–577

Gregory SG, Schmidt S, Seth P, Oksenberg JR, Hart J, Prokop A, Caillier SJ, Ban M, Goris A, Barcellos LF, et al (2007) Interleukin 7 receptor alpha chain (IL7R) shows allelic and functional association with multiple sclerosis. Nat Genet 39:1083–1091

Hafler DA, Compston A, Sawcer S, Lander ES, Daly MJ, De Jager PL, de Bakker PI, Gabriel SB, Mirel DB, Ivinson AJ, et al (2007) Risk alleles for multiple sclerosis identified by a genomewide study. N Engl J Med 357:851–862

Hafler JP, Maier LM, Cooper JD, Plagnol V, Hinks A, Simmonds MJ, Stevens HE, Walker NM, Healy B, Howson JM, Maisuria M, Duley S, Coleman G, Gough SC, International Multiple Sclerosis Genetics Consortium (IMSGC), Worthington J, Kuchroo VK, Wicker LS, Todd JA (2009) CD226 Gly307Ser association with multiple autoimmune diseases. Genes Immun 10(1):5–10 PMID: 18971939

Haines JL et al (1996) A complete genomic screen for multiple sclerosis underscores a role for the major histocompatability complex. The Multiple Sclerosis Genetics Group. Nat Genet 13(4):469–471

Harbo HF, Lie BA, Sawcer S, Celius EG, Dai KZ, Oturai A, Hillert J, Lorentzen AR, Laaksonen M, Myhr KM, et al (2004) Genes in the HLA class I region may contribute to the HLA class II-associated genetic susceptibility to multiple sclerosis. Tissue Antigens 63:237–247

He YW, Malek TR (1996) Interleukin-7 receptor alpha is essential for the development of gamma delta+ T cells, but not natural killer cells. J Exp Med 184(1):289–293

Hemminki K, Li X, Sundquist J, Hillert J, Sundquist K (2009) Risk for multiple sclerosis in relatives and spouses of patients diagnosed with autoimmune and related conditions. Neurogenetics 10(1):5–11

Hensiek AE, Seaman SR, Barcellos LF, Oturai A, Eraksoi M, Cocco E, Vecsei L, Stewart G, Dubois B, Bellman-Strobl J, Leone M, Andersen O, Bencsik K, Booth D, Celius EG, Harbo HF, Hauser SL, Heard R, Hillert J, Myhr KM, Marrosu MG, Oksenberg JR, Rajda C, Sawcer SJ, Sørensen PS, Zipp F, Compston DA (2007) Familial effects on the clinical course of multiple sclerosis. Neurology 68(5):376–383

Hillert J (1994) Human leukocyte antigen studies in multiple sclerosis. Ann Neurol 36(Suppl): S15–S17

Hoppenbrouwers IA, Aulchenko YS, Ebers GC, Ramagopalan SV, Oostra BA, et al (2008) EVI5 is a risk gene for multiple sclerosis. Genes Immun 9:334–337

Howson JMM, Walker NM, Clayton D, Todd JA, The Type 1 Diabetes Genetics Consortium (2009) Confirmation of HLA class II independent type 1 diabetes associations in the major histocompatibility complex including HLA-B and HLA-A. Diabetes Obes Metab 11(Suppl 1):31–45

International Multiple Sclerosis Genetics Consortium (IMSGC) (2009) The expanding genetic overlap between multiple sclerosis and type I diabete. Genes Immun 10:11–14

Jersild C, Fog T (1972) Histocompatibility (HL-A) antigens associated with multiple sclerosis. Acta Neurol Scand Suppl 51:377

Jersild C, Fog T, Hansen GS, Thomsen M, Svejgaard A, et al (1973) Histocompatibility determinants in multiple sclerosis, with special reference to clinical course. Lancet 2:1221–1225

Kellar-Wood H, Robertson N, Govan GG, Compston DA, Harding AE (1994) Leber's hereditary optic neuropathy mitochondrial DNA mutations in multiple sclerosis. Ann Neurol 36(1):109–112

Kuokkanen S et al (1997) Genomewide scan of multiple sclerosis in Finnish multiplex families. Am J Hum Genet 61(6):1379–1387

Lai SY, Molden J, Goldsmith MA (1997) Shared gamma(c) subunit within the human interleukin-7 receptor complex. A molecular basis for the pathogenesis of X-linked severe combined immunodeficiency. J Clin Invest 99(2):169–177

Lennon VA, Wingerchuk DM, Kryzer TJ, Pittock SJ, Lucchinetti CF, Fujihara K, Nakashima I, Weinshenker BG (2004) A serum autoantibody marker of neuromyelitis optica: distinction from multiple sclerosis. Lancet 364(9451):2106–2112

Link J, Lorentzen AR, Kockum I, Duvefelt K, Mero I-L, Celius EG, Harbo HF, Hillert J, Brynedal B (submitted) HLA class I genes in multiple sclerosis - two independent risk factors mapping to HLA-A and HLA-C

Lucchinetti C et al (2000) Heterogeneity of multiple sclerosis lesions: implications for the pathogenesis of demyelination. Ann Neurol 47(6):707–717

Lundmark F, Duvefelt K, Iacobaeus E, Kockum I, Wallstrom E, Khademi M, Oturai A, Ryder LP, Saarela J, Harbo HF, et al (2007) Variation in interleukin 7 receptor alpha chain (IL7R) influences risk of multiple sclerosis. Nat Genet 39:1108–1113

Lundström W, Greiner E, Lundmark F, Harbo HF, Lorentzen Å, Masterman T, Hillert J (submitted) Effects of non-HLA susceptibility genes on disease progress in Multiple Sclerosis

Maier LM, Lowe CE, Cooper J, Downes K, Anderson DE, Severson C, Clark PM, Healy B, Walker N, Aubin C, Oksenberg JR, Hauser SL, Compston A, Sawcer S, International Multiple Sclerosis Genetics Consortium, De Jager PL, Wicker LS, Todd JA, Hafler DA (2009a) IL2RA genetic heterogeneity in multiple sclerosis and type 1 diabetes susceptibility and soluble interleukin-2 receptor production. PLoS Genet 5(1):e1000322 PMID: 19119414

Maier LM, Anderson DE, Severson CA, Baecher-Allan C, Healy B, Liu DV, Wittrup KD, De Jager PL, Hafler DA (2009b) Soluble IL-2RA levels in multiple sclerosis subjects and the effect of soluble IL-2RA on immune responses. J Immunol 182(3):1541–1547 PMID: 19155502

Masterman T, Ligers A, Olerup O, Hillert J (2000) HLA-DRB1 alleles do not influence course or outcome in multiple sclerosis, but DR15 is associated with lower age of onset. Ann Neurol 48:211–219

Mazzucchelli R, Durum SK (2007) Interleukin-7 receptor expression: intelligent design. Nat Rev Immunol 7(2):144–154

Naito S, Namerow N, Mickey MR, Terasaki PI (1972) Multiple sclerosis: association with HL-A3. Tissue Antigens 2:1–4

Nejentsev S, Howson JMM, Walker NM, et al (2007) Localization of type 1 diabetes susceptibility to the MHC class I genes HLA-B and HLA-A. Nature 450:887–892

Nielsen NM, Frisch M, Rostgaard K, Wohlfahrt J, Hjalgrim H, Koch-Henriksen N, Melbye M, Westergaard T (2008) Autoimmune diseases in patients with multiple sclerosis and their first-degree relatives: a nationwide cohort study in Denmark. Mult Scler 14(9):1288–1289; author reply 1290–1

Oksenberg JR, Barcellos LF, Cree BA, Baranzini SE, Bugawan TL, Khan O, Lincoln RR, Swerdlin A, Mignot E, Lin L, Goodin D, Erlich HA, Schmidt S, Thomson G, Reich DE, Pericak-Vance MA, Haines JL, Hauser SL (2004) Mapping multiple sclerosis susceptibility to the HLA-DR locus in African Americans. Am J Hum Genet 74(1):160–167 PMID: 14669136

Olerup O, Hillert J, Fredrikson S, Olsson T, Kam-Hansen S, Möller E, Carlsson B, Wallin J (1989) Chronic progressive and relapsing/remitting multiple sclerosis – two distinct disease entities? Proc Natl Acad Sci U S A 86:7113–7117

Olerup O, Hillert J (1991) HLA class II-associated genetic susceptibility in multiple sclerosis: A critical evaluation. Tissue Antigens 38:1–15

Olsson T, Hillert J (2008) The genetics of multiple sclerosis and its experimental models. Curr Opin Neurol 21(3):255–260 Review. PMID: 18451707

Oturai AB, Ryder LP, Fredrikson S, Myhr KM, Celius EG, Harbo HF, Andersen O, Akesson E, Hillert J, Madsen HO, Nyland H, Spurkland A, Datta P, Svejgaard A, Sorensen PS (2004) Concordance for disease course and age of onset in Scandinavian multiple sclerosis coaffected sib pairs. Mult Scler 10(1):5–8

Ramagopalan SV, Dyment DA, Valdar W, Herrera BM, Criscuoli M, Yee IM, Sadovnick AD, Ebers GC, Canadian Collaborative Study Group (2007a) Autoimmune disease in families with multiple sclerosis: a population-based study. Lancet Neurol 6(7):604–610 PMID: 17560172

Ramagopalan SV, Anderson C, Sadovnick AD, Ebers GC (2007b) Genomewide study of multiple sclerosis. N Engl J Med 357(21):2199–2200 author reply 2200–1. PMID: 18032773

Rubio JP, Stankovich J, Field J, Tubridy N, Marriott M, et al (2008) Replication of KIAA0350, IL2RA, RPL5 and CD58 as multiple sclerosis susceptibility genes in Australians. Genes Immun 9(7):624–630

Saarela J et al (2006) PRKCA and multiple sclerosis: association in two independent populations. PLoS Genet 2(3):e42

Sadovnick AD et al (1993) A population-based study of multiple sclerosis in twins: update. Ann Neurol 33(3):281–285

Sawcer S et al (1996) A genome screen in multiple sclerosis reveals susceptibility loci on chromosome 6p21 and 17q22. Nat Genet 13(4):464–468

Sawcer S, Maranian M, Setakis E, Curwen V, Akesson E, Hensiek A, Coraddu F, Roxburgh R, Sawcer D, Gray J, Deans J, Goodfellow PN, Walker N, Clayton D, Compston A (2002)

A whole genome screen for linkage disequilibrium in multiple sclerosis confirms disease associations with regions reviously linked to susceptibility. Brain 125(Pt 6):1337–1347 PMID: 12023322

Sawcer S, Ban M, Maranian M, Yeo TW, Compston A, Kirby A, Daly MJ, De Jager PL, Walsh E, Lander ES, et al (2005) A high-density screen for linkage in multiple sclerosis. Am J Hum Genet 77:454–467

Smestad C, Brynedal B, Jonasdottir G, Lorentzen AR, Masterman T, Akesson E, Spurkland A, Lie BA, Palmgren J, Celius EG, Hillert J, Harbo HF (2007) The impact of HLA-A and -DRB1 on age at onset, disease course and severity in Scandinavian multiple sclerosis patients. Eur J Neurol 14(8):835–840. PMID: 17662002

Sundvall M, Jirholt J, Yang HT, Jansson L, Engstrom A, Pettersson U, Holmdahl R (1995) Identification of murine loci associated with susceptibility to chronic experimental autoimmune encephalomyelitis. Nat Genet 10:313–317

Teutsch SM, Booth DR, Bennetts BH, Heard RN, Stewart GJ (2003) Identification of 11 novel and common single nucleotide polymorphisms in the interleukin-7 receptor-alpha gene and their associations with multiple sclerosis. Eur J Hum Genet 11:509–515

Weber F, Fontaine B, Cournu-Rebeix I, Kroner A, Knop M, et al (2008) IL2RA and IL7RA genes confer susceptibility for multiple sclerosis in two independent European populations. Genes Immun 9:259–263

Willer CJ et al (2003) Twin concordance and sibling recurrence rates in multiple sclerosis. Proc Natl Acad Sci U S A 100(22):12877–12882

WTCCC (2007) Genome-wide association study of 14,000 cases of seven common diseases and 3,000 shared controls. Nature 447:661–678

Yeo TW, De Jager PL, Gregory SG, Barcellos LF, Walton A, Goris A, Fenoglio C, Ban M, Taylor CJ, Goodman RS, et al (2007) A second major histocompatibility complex susceptibility locus for multiple sclerosis. Ann Neurol 61:228–236

Zhang Z, Duvefelt K, Svensson F, Masterman T, Jonasdottir G, Salter H, Emahazion T, Hellgren D, Falk G, Olsson T, et al (2005) Two genes encoding immune-regulatory molecules (LAG3 and IL7R) confer susceptibility to multiple sclerosis. Genes Immun 6:145–152

Potential Triggers of MS

Jane E. Libbey and Robert S. Fujinami

Abstract MS is an immune mediated disease of the central nervous system (CNS) characterized by demyelination, axonal damage and neurologic disability. The primary cause of this CNS disease remains elusive. Here we will address our current understanding of the role of viruses as potential environmental triggers for MS. Virus infections can act peripherally (outside the CNS) or within the CNS. The association of viral infections with demyelinating disease, in both animals and humans, will be discussed, as will the potential contributions of peripheral infection with Torque Teno virus, infection outside of and/or within the CNS with Epstein–Barr virus and infection within the CNS with Human Herpesvirus 6 to MS. An experimental animal model, Theiler's murine encephalomyelitis virus infection of susceptible strains of mice is an example of viral infections of the CNS as a prerequisite for demyelination. Finally, the proposition that multiple virus infections are required, which first prime the immune system and then trigger the disease, as a model where infections outside of the CNS lead to inflammatory changes within the CNS, for the development of a MS-like disease is explored.

1 Introduction

MS is the most common inflammatory demyelinating disease of the central nervous system (CNS) in humans. Between 50 and 100 per 100,000 Caucasians (lower in other ethnic groups) are afflicted with this disease, and women are afflicted more

J.E. Libbey
Department of Pathology, University of Utah School of Medicine,
30 North 1900 East, RM 3R330, Salt Lake City, Utah 84132 USA
e-mail: Jane.Libbey@utah.edu

R.S. Fujinami(⊠)
Department of Pathology, University of Utah School of Medicine,
30 N 1900 East, Room 3R330, Salt Lake City, UT 84132 USA
e-mail: Robert.Fujinami@hsc.utah.edu

Results Probl Cell Differ, doi:10.1007/400_2008_12

than men at a ratio of about 2:1. Onset usually occurs between the ages of 20 and 40 years. Clinical features of MS include alterations in vision, sensory and motor disturbances and cognitive impairment. The clinical course of the disease may be progressive [primary progressive (PP) or secondary progressive (SP)] in nature or include relapses and remissions [relapsing-remitting (RR)]. RR–MS occurs more frequently (85–90%) but with time (10–15 years) can convert to SP–MS. The present general consensus is that MS is a CD4[+] T helper (Th) 1-mediated autoimmune disease; however, CD8[+] T cells may be involved in the pathogenesis as they are the predominant lymphocyte found in MS lesions. The cells that form myelin in the CNS (oligodendrocytes) are thought to be the main target for attack in MS, and the inflammatory demyelinating lesions characteristic of MS can occur in the optic nerve, brainstem, spinal cord and periventricular white matter. Axonal loss, another important component of MS pathology, occurs from disease onset and correlates, early in disease, with the extent of inflammation within lesions [reviewed in (Bjartmar et al. 2003; Dutta and Trapp 2007)]. The effects of axonal loss may not be apparent for years (clinically silent) due to the compensatory capacity of the CNS. With time, a threshold of axonal loss is reached which surpasses the compensatory capacity of the CNS and irreversible neurological disability develops. Once the irreversible neurological disability develops, it correlates with the degree of axonal degeneration [reviewed in (Bjartmar et al. 2003; Dutta and Trapp 2007)].

The etiology of MS remains elusive despite decades of research. The disease develops in genetically susceptible individuals, but probably, additional environmental triggers are required. A recent large-scale genomewide association study identified alleles of the interleukin (IL)-2 receptor α gene (*IL2RA*, CD25), alleles of the IL-7 receptor α gene (*IL7RA*, CD127), both of which are genes related to the regulation of the immune response, and the human leukocyte antigen (HLA)-DR locus as heritable risk factors for MS (Hafler et al. 2007). A more detailed discussion of the genetic contribution to MS is beyond the scope of this chapter and the reader is referred to recent reviews of the subject (Kantarci 2008; McElroy and Oksenberg 2008; Niino et al. 2007).

A role for environmental factors is predicted by the relatively low concordance rate (approximately 25%) for MS in identical twins (Sospedra and Martin 2005), by concordance studies, examining age of the patient at onset, calendar year at onset, presenting symptoms, disability at the time of assessment, disease course, gender and familial recurrence rate, carried out on nontwin sibling pairs (Robertson et al. 1996), as well as by the genome association study, described above, identifying genes related to the regulation of the immune response (Hafler et al. 2007). Environmental factors may include behavioral and lifestyle influences as well as infectious agents (bacterial and viral). For recent discussions of the behavioral and lifestyle influences that may play a role in MS, please see the reviews by Coo and Aronson (Coo and Aronson 2004) and Ascherio and Munger (Ascherio and Munger 2007b). A role for infectious agents is supported by epidemiological studies [(Kurtzke 1993); reviewed in (Ascherio and Munger 2007a; Fujinami 2001b; Libbey and Fujinami 2002, 2003; Oleszak et al. 2004)]. The equatorial region of the earth is a low risk area for MS and the risk increases as you move farther north or south from the equator. Infectious agents present in these areas most likely play a role in this phenomenon. Migration studies suggest that it is exposure to an agent or repeated infections before puberty that either contributes to the risk of developing MS or protects from MS. Migration studies showed that, if migration from an area of high MS risk to an area of low MS risk or vice versa occurred prior to 15 years of age, then the migrant acquired the risk of the area to which they moved; whereas, if migration occurred after 15 years of age, then the migrant maintained the risk of the area from which they moved [(Kurtzke 1993); reviewed in (Ascherio and Munger 2007a; Fujinami 2001b; Libbey and Fujinami 2002, 2003; Oleszak et al. 2004).

One proposed hypothesis for how an infectious agent could explain the epidemiology of MS is the "hygiene hypothesis" [(Leibowitz et al. 1966; Poskanzer et al. 1963a, b, 1966); reviewed in (Ascherio and Munger 2007a)]. This hypothesis proposes that exposure to multiple widespread infectious agents, which are individually relatively harmless, is protective against MS if acquired early in life, whereas exposure at an increasing age increases the risk of MS. This hypothesis could explain the latitude gradient for MS risk if the occurrence of potentially protective infectious agents is high near the equator, thus increasing the likelihood of early childhood infection and protection, and the occurrence of potentially protective infectious agents decreases with distance north and south from the equator, thus decreasing the likelihood of early childhood infection and protection and effectively increasing the likelihood of exposure later in life. This hypothesis could also explain how migration in childhood results in acquiring the risk of the area to which the migrant moves. Migration at a young age, from an area of high risk to an area of low risk, provides the opportunity for infection with potentially protective agents present in the low risk area, but lacking in the high risk area, while still in childhood. Migration at a young age, from an area of low risk to an area of high risk, would deprive the migrant of infection with potentially protective agents present in the low risk area, but lacking in the high risk area, while still in childhood, thus effectively increasing the likelihood of exposure later in life. This hypothesis could also explain how migration in late adolescence and adulthood results in maintaining the risk of the area from which the migrant moves. Adult migrants from low risk area would be protected by their plethora of childhood infections and thus would maintain their low risk, while adult migrants from high risk areas, who lack protective childhood infections, would maintain their high risk upon migration later in life to a low risk area where the occurrence of infectious agents is high. MS, according to this hypothesis, is an immune-mediated reaction triggered in response to multiple infections, such that there is no one specific etiologic agent, in susceptible individuals. It is unlikely that all infectious agents equally predispose to MS [reviewed in (Ascherio and Munger 2007a)].

Viruses have long been thought to play a role in the etiology and pathogenesis of MS. Studies with isolated populations demonstrating that populations which had no reported cases of MS experienced epidemics of MS after contact with North Americans or Europeans is one point in favor of viral involvement (Kurtzke 1993, 1997; Kurtzke and Heltberg 2001). A second point in favor of viral involvement is the finding that CD8$^+$ T cells, which function to clear viral infections, dominate the cellular infiltrate in active MS lesions (Babbe et al. 2000). As a third point, exacerbations of MS can be correlated with viral infections (Álvarez-Lafuente et al. 2006; Andersen et al. 1993; Berti et al. 2002; Buljevac et al. 2002; Christensen 2007; Correale et al. 2006; De Keyser et al. 1998; Edwards et al. 1998; Gilden 2002; Granieri et al. 2001; Kriesel et al. 2004; Kriesel and Sibley 2005; Marrie et al. 2000; Metz et al. 1998; Panitch 1994; Sibley et al. 1985; Sospedra and Martin 2006; Wandinger et al. 2000). However, to date, no immunization or vaccination has ever been demonstrated to induce exacerbations in MS [(Rutschmann et al. 2002); reviewed in (Fujinami 2001a)]. A fourth point in favor of viral involvement is the finding that antibodies against various viruses and/or the viruses themselves have been directly isolated from MS patients. The list includes such diverse viruses as rabies virus, herpes simplex virus (types 1 and 2), parainfluenza virus (type 1), measles virus, coronavirus, human T-cell lymphotrophic virus (type 1), and human herpesvirus 6 (HHV-6) [reviewed in (Johnson 1998; Libbey and Fujinami 2003)]. It should be noted that viruses most commonly associated with MS are viruses that replicate extensively within the CNS (von Herrath et al. 2003); however, no one single virus has ever been unequivocally determined to be the causative agent of MS.

Viral infections, either within the CNS or in the periphery, could induce the autoimmune reactions that result in CNS inflammatory disease through either of the two mechanisms: molecular mimicry and/or bystander activation. The two mechanisms are not mutually exclusive. Molecular mimicry occurs when viral antigens have either sequential or structural similarities to self-antigens [reviewed in (Fujinami 1996,1998, 2000, 2001b; Fujinami et al. 2006; Libbey et al. 2007; McCoy et al. 2006; Peterson and Fujinami 2006; Sospedra and Martin 2005; von Herrath et al. 2003; Whitton and Fujinami 1999)]. Individual viruses may harbor molecular mimics to different CNS antigens and these different mimicries may be mediated through HLA class I- or class II-restricted T cells or antibodies [reviewed in (Fujinami et al. 2006; Libbey et al. 2007; McCoy et al. 2006)]. Therefore, the possibility of multiple mimics and multiple mimicry pathways could explain the inability to isolate one single causative agent of MS and could explain the pheno-typic diversity of the disease, respectively. Viruses that encode molecular mimics could induce autoimmunity through either direct infection of the CNS or by peripheral infection.

Bystander activation is the nonspecific activation of autoreactive cells due to inflammatory events caused by infection [reviewed in (Fujinami et al. 2006; McCoy et al. 2006; Sospedra and Martin 2005; von Herrath et al. 2003; Whitton and Fujinami 1999)]. T cell receptor (TCR)-independent bystander activation of autoreactive T cells can occur through the production of inflammatory cytokines, the presence of superantigens or the recognition of molecular patterns by Toll-like receptors, and thus does not require direct viral infection of the CNS. TCR-dependent bystander activation of autoreactive T cells depends on the unveiling of host antigens as a result of direct viral tissue damage, followed by the presentation of host antigens by antigen-presenting cells in conjunction with an adjuvant effect of the infectious agent, which ultimately results in epitope spreading [reviewed in (Sospedra and Martin 2005; von Herrath et al. 2003)].

2 Viral Association with Demyelination

Studies using transgenic mice support the hypothesis that exposure to infectious agents (bacterial and viral) can trigger demyelination [(Goverman et al. 1993); reviewed in (Sospedra and Martin 2005)]. Transgenic mice were constructed that expressed genes encoding a rearranged TCR specific for an encephalitogenic myelin basic protein (MBP) peptide, MBP_{1-11}, in a mouse strain that is genetically susceptible to experimental autoimmune encephalomyelitis (EAE) (Goverman et al. 1993). EAE is a nonviral, autoimmune animal model for MS in which demyelinating disease is induced by the injection of whole protein or encephalitogenic peptides from CNS antigens in adjuvant, or by the adoptive transfer of CNS antigen-sensitized lymphocytes [reviewed in (Tsunoda and Fujinami 1996)]. These mice were found to develop spontaneous demyelination in the white matter of the spinal cord when housed in nonpathogen-free conditions, but remained disease free when housed in sterile, pathogen-free conditions (Goverman et al. 1993). Thus, environmental factors such as viral or bacterial infections may initiate the autoimmune response in these mice, possibly through molecular mimicry, bystander activation or damage to the blood–brain barrier which would allow access of the CNS-specific T cells to the target organ (Goverman et al. 1993).

From studies with experimental animals, certain viral infections are known to cause demyelination [reviewed in (Libbey and Fujinami 2003)]. In some cases the virally-induced demyelination closely resembles MS lesions in humans. Canine distemper virus (CDV) causes a multifocal demyelinating disease in dogs [reviewed in (Vandevelde and Zurbriggen 2005)]. Demyelination occurs at a very high frequency upon CDV infection and the demyelinating lesions generally occur within the white matter of the cerebellum, the periventricular white matter, the optic pathways and the spinal cord; all sites commonly affected in MS. The initial acute demyelinating lesions are associated with viral replication within glial cells of the white matter with extensive infection of astrocytes and restricted infection (viral transcription without translation) of oligodendrocytes, resulting in a dramatic impairment of myelin metabolism. Progression of the chronic demyelinating lesions may be due to both the inflammatory immune reaction, in which the antiviral immune response destroys oligodendrocytes by means of a bystander mechanism, and the persistence of the virus (noncytolytic) in white matter areas outside of the lesions [reviewed in (Vandevelde and Zurbriggen 2005)]. The induction of demyelination in mice can be accomplished with viruses from such diverse viral families as *Coronaviridae*, *Picornaviridae*, *Rhabdoviridae*, and *Togaviridae* [reviewed in (Libbey and Fujinami 2003)]. Various mechanisms of demyelination, which are not mutually exclusive, include: the direct lysis of infected oligodendrocytes by the virus; direct lysis of infected oligodendrocytes by the immune response; lysis of uninfected oligodendrocytes by an autoimmune response triggered by the virus; and lysis of oligodendrocytes and other cells by a nonspecific bystander immune response in the vicinity of the viral infection. Viral-induced demyelination can vary in its presentation from small focal areas to large plaques. Some similarities between MS and these viral-induced demyelinating diseases in mice include similarities in neuropathology, similarities in disease course which can be either progressive or RR and the role that genetic susceptibility plays in the development of the disease [reviewed in (Libbey and Fujinami 2003)].

Certain viral infections of humans are also known to cause demyelinating disease [reviewed in (Asher 1991; Evans et al. 1991; Sospedra and Martin 2005; Whitley and Schlitt 1991)]. These include the human polyomavirus JC, which causes progressive multifocal leukoencephalopathy (in immunosuppressed subjects); measles virus, which causes postinfectious encephalomyelitis (in approximately 1 in 1,000 cases of acute measles virus infection) and subacute sclerosing panencephalitis (in approximately 8.5 in 1 million cases of acute measles virus infection); herpes simplex virus, which causes encephalitis; and human immunodeficiency virus (HIV), which causes HIV encephalopathy [(Fujinami 2001b); reviewed in (Asher 1991; Evans et al. 1991; Sospedra and Martin 2005; Whitley and Schlitt 1991)].

3   Torque Teno Virus and MS

Torque Teno virus (TTV) is a ubiquitous, nonenveloped, nonpathogenic, small, DNA (single-stranded antisense) virus that only encodes 4–5 gene sequences (Simmonds 2002). TTV is unique in that infection of a wide range of tissues is characterized by a persistent viremia. Bone marrow, lung and lymphoid tissue harbor the highest viral loads. TTV has also been found to have high genetic diversity, with more than 28 distinct genotypes, and multiple genotypes may infect an individual at any one time (Simmonds 2002).

In the context of MS, an early study was unable to detect the presence of TTV DNA in the cerebrospinal fluid (CSF) of MS patients although the serum was shown to be positive for the virus (Maggi et al. 2001). This indicates that the CNS does not represent a common site of TTV replication and persistence and that the virus does not easily permeate an intact blood–brain barrier (Maggi et al. 2001). However, despite this finding, TTV has been implicated in the pathogenesis of MS by another study that isolated CD4$^+$ T cell clones, which were clonally expanded in vivo in the CSF during disease exacerbation in an RR–MS patient, that were found to be specific for the conserved arginine-rich region (74 amino acids) of TTV (Sospedra et al. 2005). This arginine-rich domain was found to be evolutionarily conserved among common viruses, prokaryotes and eukaryotes as part of DNA-binding regions, nuclear localization signals and other functional sequences. The T cell clones could have been activated in the periphery, perhaps in response to TTV, before migrating to the CSF, which would explain the findings of Maggi et al (Maggi et al. 2001). Thus, it has been suggested that repeated infections with common viruses, such as TTV, could induce the expansion of T cells specific for conserved protein domains. This cross-reactivity (molecular mimicry) between pathogens and autoantigens could ultimately lead to the initiation of autoimmune disease (MS) in genetically susceptible individuals (Sospedra et al. 2005). However, although an association between TTV infection and MS has been suggested, this study did not determine whether the initial activator of these T cells was TTV or some autoantigen; it may be that TTV does not trigger MS, but may play a role in perpetuating an already existing autoimmune reaction.

4 Epstein–Barr Virus and MS

Herpes viruses, to include Epstein–Barr virus (EBV), are ubiquitous, enveloped, large DNA (double-stranded) viruses that commonly produce latent, recurrent infections [reviewed in (Christensen 2007)]. EBV is a lymphotropic γ-herpesvirus for which reactivation of latent infections has been correlated with exacerbations of MS [(Wandinger et al. 2000); reviewed in (Christensen 2007)], though this finding remains controversial as other studies have been unable to find an association between EBV reactivation and exacerbations in RR–MS (Buljevac et al. 2005; Torkildsen et al. 2008). EBV seroprevalence is 90%; this seroprevalence increases to 100% in adults with MS [(Ascherio and Munch 2000); reviewed in (Pender and Greer 2007; Sospedra and Martin 2005)]. The EBV seroprevalence in children with MS (disease onset before age 16) approaches 100% and this is in comparison to a seroprevalence of only 72% in age-matched controls [(Pohl et al. 2006); reviewed in (Pender and Greer 2007)]. This is contradictory to the hygiene hypothesis that states that early infection is protective; in this case, seronegative individuals, at any age, have a very low risk of developing MS [reviewed in (Ascherio and Munger 2007a).

MS patients often have had infectious mononucleosis, for which EBV is the known etiologic agent. Primary infection with EBV, resulting in infectious mononucleosis, in adolescence or young adulthood has been correlated with an increased risk (more than twofold) of developing MS [(Thacker et al. 2006); reviewed in (Pender and Greer 2007)].

An increased humoral immune response to EBV has been found in MS patients. In examining the specificity of the oligoclonal immunoglobulin (Ig) bands (produced by intrathecal Ig synthesis) commonly observed in the CSF of MS patients, antibodies specific for two EBV proteins, EBV nuclear antigen (EBNA)-1 and BRRF2, were reported to be significantly higher in the CSF and serum of MS patients compared to controls, and these two proteins were specifically bound by the oligoclonal CSF IgG from MS patients [(Cepok et al. 2005); reviewed in (Pender and Greer 2007)]. In order to differentiate between the increased immune responsiveness to EBV seen in MS patients as being a cause or just an effect of MS, a large prospective study was performed to determine whether antibodies to EBV are elevated before the onset of MS (Levin et al. 2005). It was determined that there is an age-dependent relationship between EBV infection and the development of MS in that, prior to MS onset, serum levels of IgG specific for the EBNA complex, which were equivalent for subjects who developed MS and controls prior to age 20, were two- to threefold higher (statistically significant) in subjects who developed MS compared to controls at age ≥25 years [(Levin et al. 2005); reviewed in (Pender and Greer 2007)]. This age related increase in EBV-specific IgG could be due to alteration of the immune response to EBV by infection with other microorganisms, infection with a different strain of EBV or the development of the autoimmune reaction that leads to MS (Ascherio and Munger 2007a).

An increased cell-mediated immune response specific for EBV has also been demonstrated in MS patients. $CD4^+$ memory T cells with specificity for the C-terminal domain of EBNA-1 were increased in frequency and had increased proliferative capacity and enhanced interferon (IFN)-γ production, without a concomitant increase in viral loads, when isolated from the peripheral blood mononuclear cells (PBMC) of MS patients, compared to healthy EBV carriers as controls (Lunemann et al. 2006). The increased frequency and reactivity of these T cells did not correlate with the extent of clinical disability or disease duration. In addition, the enhanced T cell response was associated with a broadened epitope recognition pattern, rather than being focused on a distinct immunodominant region, in MS patients (Lunemann et al. 2006). Thus, studies of both the humoral immune response and the cell-mediated immune response suggest that there may be a distinct role for EBNA-1-specific immune responses in MS.

In addition to the above evidence for an increased EBV-specific humoral and cell-mediated immune response in MS patients, direct evidence of EBV infection in postmortem brain specimens from MS patients has been demonstrated (Serafini et al. 2007). EBV was found to persist and reactivate in ectopic B cell follicles forming in the cerebral meninges. This abnormal accumulation of EBV-infected B cells/plasma cells in the brain was found to be a common feature of MS which was not linked to disease course, gender, age at onset, disease duration or immunosuppressive therapy. Also, this accumulation of EBV-infected B cells/plasma cells in the brain was found to be specific to MS and not just a general occurrence driven by inflammation. At sites of EBV-infected B cell/plasma cell accumulation, $CD8^+$ T cell infiltration and activation with cytotoxicity towards EBV-infected B cells/plasma cells were found suggesting that an antiviral immune response occurs in the MS brain (Serafini et al. 2007).

EBV could act as a trigger for MS in genetically susceptible individuals through its ability to latently infect memory B cells [reviewed in (Lunemann et al. 2007; Lunemann and Munz 2007; Pender and Greer 2007)]. EBV-infected B cells, which could include CNS-reactive B cells, would not only produce pathogenic autoantibodies, but could also localize to the CNS where they could act as antigen-presenting cells (APC). These EBV-infected APC could present CNS antigens to $CD4^+$ T cells, which have been activated by common systemic infections in the periphery, and which may be cross-reactive for CNS antigens, and which may be reactivated by the APC as they traffic through the CNS, thus promoting the survival of CNS-reactive T cells within the CNS. The surviving CNS-reactive T cells could then organize an immune attack on the CNS [reviewed in (Pender and Greer 2007)]. Additionally, reactivation of a latent EBV infection, either in the CNS or in the periphery, would activate EBV-reactive $CD4^+$ T cells, some of which, isolated from the CSF of MS patients, have been shown to cross-react with the CNS protein, MBP (molecular mimicry) (Holmoy et al. 2004). These resultant myelin-specific $CD4^+$ T cells could be the mediators of MS in genetically susceptible individuals.

All of the associations, between EBV infection and MS, described above, suggest that EBV may be the etiologic agent for MS. Nevertheless, infection with EBV cannot explain the decline in risk observed among subjects who migrate from an area of high MS risk to an area of low MS risk in childhood (Ascherio and Munger 2007a). In order to explain this decline in risk, either other factors contribute to determine the risk or modify the immune response to EBV infection, or EBV strains in the low risk areas are less prone to causing MS.

5 HHV-6 and MS

HHV-6 is a lymphotropic and neurotropic β-herpesvirus whose seroprevalence is > 80% by the age of 2 years, thus making it impossible to assess the risk of MS in seronegative individuals. There are two distinct subtypes of HHV-6: HHV-6A and HHV-6B [reviewed in (Christensen 2007; Fotheringham and Jacobson 2005)]. Early reports failed to distinguish between the two subtypes. A recent report distinguishing antibodies specific for HHV-6A from antibodies specific for HHV-6B has demonstrated that the seroprevalence to HHV-6A (100%) is significantly higher in MS patients compared to controls (69.2%), while no difference was found for the seroprevalence to HHV-6B (Virtanen et al. 2007).

In comparing MS brains to control brains, it was found that HHV-6 genome was present in both while HHV-6 antigen was only detected in MS brains, specifically in oligodendrocytes in MS lessions [(Challoner et al. 1995); reviewed in (Ascherio and Munger 2007a; Christensen 2007; Fotheringham and Jacobson 2005; Sospedra and Martin 2005)]. The detection of HHV-6 genome in the absence of viral antigen is suggestive of a latent infection in controls, while the presence of HHV-6 genome and viral antigen in MS brains suggests an active infection [reviewed in (Fotheringham and Jacobson 2005)].

In studies examining blood and CSF, HHV-6 viral DNA and HHV-6 specific antibodies have been detected in the serum and CSF of MS patients [reviewed in (Ascherio and Munger 2007a; Christensen 2007; Fotheringham and Jacobson 2005)]. Recently, a correlation has been made between exacerbations of RR–MS and increased levels of HHV-6 DNA in the serum and HHV-6A RNA and DNA in the PBMC [(Álvarez-Lafuente et al. 2004); reviewed in (Christensen 2007; Fotheringham and Jacobson 2005)]. In comparing RR–MS to SP–MS, it has been reported that HHV-6 DNA (in DNA extracted from PBMC and in DNA extracted from serum) and transcripts of HHV-6 genes (in mRNA extracted from PBMC) can be found significantly more often in (1) RR–MS patients than in SP–MS patients, and (2) in RR–MS patients experiencing a relapse than in SP–MS patients (Álvarez-Lafuente et al. 2007). No significant differences were found for viral load of HHV-6 in the positive samples or for the distribution of the HHV-6B subtype (in DNA extracted from PBMC), and only HHV-6A was found in the positive serum samples. These results suggest that HHV-6A may contribute to exacerbations of RR–MS, but not to the pathogenesis of SP–MS (Álvarez-Lafuente et al. 2007).

Other studies of serum and CSF have shown that levels of the soluble form of the cellular receptor for HHV-6 (both HHV-6A and HHV-6B), CD46, a complement regulatory protein, are increased in serum and CSF in MS subjects. Serum levels of soluble CD46 were significantly increased in MS patients who were shown to be positive for HHV-6 DNA in their serum [(Soldan et al. 2001); reviewed in (Christensen 2007; Fotheringham and Jacobson 2005)]. However, no correlation was found between the levels of soluble CD46 in the serum and the extent of clinical disability, disease duration or the number of gadolinium-enhancing lesions as measured by magnetic resonance imaging (Soldan et al. 2001).

Finally, oligoclonal Ig bands, common in MS CSF and indicative of intrathecal antibody production, specific for HHV-6 (subtype not specified) have been detected in 20% of MS patients [(Derfuss et al. 2005); reviewed in (Christensen 2007)]. This result has been confirmed and enhanced by the narrowing of the specificity of the intrathecal antibodies to HHV-6A (Virtanen et al. 2007).

HHV-6 could act as a trigger for MS in genetically susceptible individuals through either molecular mimicry, as demonstrated by the cross-reaction of HHV-6 reactive CD4[+] T cells, isolated from the PBMC of MS patients, with MBP [(Cirone et al. 2002; Tejada-Simon et al. 2003); reviewed in (Fotheringham and Jacobson 2005; Libbey et al. 2007)], or through excessive complement activation [reviewed in (Fotheringham and Jacobson 2005)]. CD46 is a constitutively express soluble and membrane-bound protein that normally functions to inactivate C3b and C4b complement products, thus protecting healthy cells from lysis by autologous complement. CD46 has several isoforms, and activation of the CD46 isoform dominant in brain causes an increase in CD8[+] T cell cytotoxicity, a decrease in CD4[+] T cell proliferation and a decrease in IL-10 and IFN-γ production by CD4[+] T cells [reviewed in (Fotheringham and Jacobson 2005)]. Increased levels of soluble CD46 in MS patients may be an indication of activation of the complement system (Soldan et al. 2001). Engagement of CD46 by HHV-6 may result in increased activation of the complement cascade on autologous cells through downregulation of the receptor. The resulting complement-mediated tissue destruction may play a role in triggering MS (Soldan et al. 2001).

Thus, as with EBV, an association between HHV-6 infection and MS has been suggested, however, infection with HHV-6 alone cannot explain the epidemiological aspects, including the migration data, of MS. With respect to HHV-6, and also TTV, although a virus that is ubiquitous cannot be intrinsically primarily pathogenic, it cannot be ruled out that certain subtypes/genotypes may be specifically associated with certain diseases (Simmonds 2002). Finally, there is some evidence for an increased frequency of reactivation of latent HHV-6 in subjects with MS compared to controls; however, this may just be a consequence, not a cause, of MS [reviewed in (Ascherio and Munger 2007a)].

6   Basic Animal Model: Viral Persistence Within the CNS

All three human viruses discussed above, TTV, EBV and HHV-6, persist either as an active infection (TTV) or as latent infections that can be reactivated (EBV and HHV-6). These viruses persist either within the CNS (EBV and HHV-6) or in the periphery (TTV and EBV). An animal model of MS that has the characteristic of viral persistence within the CNS is Theiler's murine encephalomyelitis virus (TMEV) infection of mice.

TMEV is a naturally occurring enteric pathogen of the family *Picornaviridae* [reviewed in (Libbey and Fujinami 2003; Oleszak et al. 2004)]. TMEV is a nonenveloped, positive sense, single-stranded RNA virus. The ability of TMEV to cause disease and the disease characteristics are dependent on the age, sex and strain of the mouse [for a review of the genetics of TMEV infection, please see (Brahic et al. 2005] and the dose and strain of the virus. TMEV has been divided into two subgroups depending on neurovirulence. The more neurovirulent of the two, the GDVII subgroup, containing the GDVII and FA strains, causes an acute fatal polioencephalomyelitis. The less neurovirulent of the two, the Theiler's Original (TO) subgroup, containing the DA, TO, WW and BeAn strains, causes acute encephalomyelitis followed by CNS demyelination associated with viral persistence in susceptible mouse strains (SJL/J). Acute encephalomyelitis is associated with viral replication within the gray matter of the CNS. Resistant mouse strains (C57BL/6) clear the virus following the acute disease. The persistence of the virus within the CNS of susceptible strains of mice results in demyelination of the white matter of the CNS with features similar to what is observed in MS in humans [reviewed in (Oleszak et al. 2004)]. Although in the past axons were thought to be preserved or damaged secondarily to demyelination, studies by our group have shown that axonal damage can be detected as early as 1 week after infection with the DA strain of TMEV, long before demyelination is apparent (at 4 weeks after infection), the extent of axonal damage increased with time and the distribution of injured axons corresponded to regions where demyelination developed during the chronic phase (Tsunoda et al. 2003; Tsunoda and Fujinami 2002). Further studies demonstrated that induction of axonal degeneration, through injection of a toxic lectin, could target inflammatory demyelinating lesions to sites of the CNS that are normally unaffected by TMEV-induced demyelination (posterior funiculus of the spinal cord), thus suggesting that axonal damage plays a role in recruiting inflammatory cells and targeting demyelinating lesions to specific sites of the CNS (Tsunoda et al. 2007b). Clinically, the TMEV-infected mice with CNS demyelination display symptoms of extremity weakness, spasticity, incontinence and spastic paralysis that are similar to symptoms observed in MS in humans [reviewed in (Libbey and Fujinami 2003; Oleszak et al. 2004)].

In the TMEV model of MS, molecular mimicry between foreign and self epitopes (Olson et al. 2001) and epitope spreading (the end result of bystander activation) from viral to myelin antigens (Miller et al. 1997) are important for the initiation and progression, respectively, of the CD4$^+$ T cell-mediated autoimmune demyelinating disease [reviewed in (Oleszak et al. 2004)]. Olson et al (Olson et al. 2001) were able to directly demonstrate that molecular mimicry could initiate virus-induced demyelination. This was accomplished through the insertion of native and mimic sequences of the encephalitogenic myelin proteolipid protein (PLP) epitope, PLP$_{139-151}$, into a nonpathogenic, nonpersistent TMEV variant. Intracerebral infection of SJL/J mice with recombinant virus variants carrying the native and mimic sequences resulted in an early-onset demyelinating disease course characterized by the presence of CD4[+] Th1 cells specific both for the immunodominant viral epitope, $VP2_{70-86}$, and for the immunodominant PLP epitope, $PLP_{139-151}$. This held true even for the mimic sequence, derived from the protease IV protein of *Haemophilus influenza*, which only shared 6 out of 13 amino acids with $PLP_{139-151}$. It was also demonstrated that the recombinant virus variants carrying the native and mimic sequences were able to persist in the CNS subsequent to intracerebral infection. In contrast, intravenous, subcutaneous and intraperitoneal infection of SJL/J mice with the virus variant carrying the native sequence also resulted in early-onset of the disease course, suggesting that sites distant from the CNS can be the initial sites of infection and that the mimic-encoding virus may not need to infect the target organ, the CNS in this case, in order to induce the autoimmune disease. One factor that may be important in determining whether or not a viral infection, by a virus carrying a self mimic epitope, triggers an autoimmune disease is the primary host cell type infected (Olson et al. 2001). Miller et al. (Miller et al. 1997) were able to demonstrate that epitope spreading could play a major role in the progression of virus-induced demyelination. Through kinetic and functional studies, they showed that demyelination in TMEV-infected mice is initiated by CD4[+] T cells with specificity for the virus followed by de novo priming, through the secondary release of sequestered autoantigens, of self-reactive T cells specific for multiple myelin epitopes [$PLP_{139-151}$, PLP_{56-70}, $PLP_{178-191}$ and myelin oligodendrocyte glycoprotein $(MOG)_{92-106}$] which arise in an ordered progression starting 3–4 weeks after the onset of disease. It is important to note that persistence of the virus within the CNS is required for the development of this CNS autoimmunity (Miller et al. 1997). These two mechanisms: molecular mimicry and epitope spreading resulting from bystander activation, most likely function in concert within a TMEV-infected mouse to initiate and perpetuate virus-induced demyelination.

The viral capsid of TMEV plays a major role in persistence and persistent virus can be found in oligodendrocytes, astrocytes and microglia/macrophages [reviewed in (Oleszak et al. 2004)]. Viral persistence, though maybe not sufficient for the development of the demyelinating disease as viral persistence can be demonstrated without the development of the demyelinating disease (Oleszak et al. 1988; Patick et al. 1990), is a prerequisite to developing the chronic demyelinating disease and the extent to which the virus persists is directly proportional to the extent of demyelination and clinical disease.

7 Prime/Challenge Model: Acute Peripheral Infection

Multiple virus infections may be required for the development of MS, as proposed by the hygiene hypothesis (above). However, more specifically, multiple virus infections may interact in such a way that the initial infection may act to prime the immune response, and a subsequent challenge infection with the same or related virus or some other nonspecific immunologic stimulus may trigger the disease. Also, this prime/challenge model allows for the possibility that the initial infection could protect the host from developing disease subsequent to a challenge infection.

Early studies, including those by our group, investigating prime/challenge animal models (though they were not necessarily described as such) for autoimmune demyelinating disease were performed in the context of EAE. Two groups, using animals other than mice, demonstrated that a viral infection (prime) could enhance/allow EAE induction by injection of neuroantigens (challenge) [(Liebert et al. 1990; Massanari et al. 1979); reviewed in (Barnett and Fujinami 1992)]. The first group showed that a persistent (heat-inactivated virus did not work) measles virus infection (prime) in hamsters potentiated the development and severity of EAE induced by injection of neuroantigen (guinea pig spinal cord) in complete Freund's adjuvant (CFA; challenge) (Massanari et al. 1979). The second group showed that a subclinical measles virus infection (prime) in Lewis rats allowed for the induction of EAE upon injection of the normally nonencephalitogenic MBP peptide, MBP_{69-81}, in CFA (challenge), and this induction of EAE was dependent on measles virus replication in the CNS (intraperitoneal injection of virus and heat-inactivated virus did not work) (Liebert et al. 1990). One group, using mice, demonstrated that subclinical, transient infection of the CNS with Semliki Forest virus (prime) allowed for the induction of EAE in resistant C57BL/6 mice upon injection of whole MBP in CFA (challenge), or upon adoptive transfer of MBP-sensitized lymphocytes (challenge) (Mokhtarian and Swoveland 1987). Together, these studies demonstrate that a viral infection of the CNS can prime the immune system for the development of demyelinating disease upon subsequent challenge with neuroantigens in model systems where neuroantigen challenge alone induces disease rarely (guinea pig spinal cord in CFA in hamsters) or not at all (MBP_{69-81} in CFA in Lewis rats; whole MBP in CFA in C57BL/6 mice) (Liebert et al. 1990; Massanari et al. 1979; Mokhtarian and Swoveland 1987).

The above studies demonstrated that persistent or transient viral infection of the CNS could prime for the later induction of EAE. This priming effect could occur through several mechanisms. Viral-induced damage to the CNS could facilitate the subsequent activation and/or clonal expansion of preexisting myelin-reactive lymphocytes (Liebert et al. 1990; Mokhtarian and Swoveland 1987). Cellular components together with viral envelop proteins could be exposed on the surface of infected cells due to viral-induced alterations of the surface of infected cells (Liebert et al. 1990). Finally, changes in the integrity of the blood–brain barrier due to viral infection could facilitate antigen-specific lymphocyte infiltration of the CNS (Liebert et al. 1990).

Our group has continued to examine and develop this prime/challenge hypothesis in mice over the last several years. Our initial study demonstrated that peripheral infection, via tail scarification, of SJL/J mice with a recombinant vaccinia virus carrying the entire coding region of rat PLP did not induce CNS disease by itself, but instead primed the mice for the development of an earlier and enhanced CNS disease (EAE) upon subcutaneous injection of the

$PLP_{139-151}$ peptide (identical between rat and mouse) in CFA plus intravenous injection of *Bordetella pertussis* (challenge) (Barnett et al. 1993). The small amount of vaccinia virus used for infection/vaccination in this study resulted in localized dermal replication of the virus without infection of the CNS. Thus, prior peripheral infection, with a virus capable of coding for a self protein (molecular mimicry), can prime the host immune system for an enhanced autoimmune response triggered by a later challenge (Barnett et al. 1993).

The above study was extended by the use of other encephalitogenic PLP peptides ($PLP_{178-191}$, $PLP_{104-117}$) as challenge antigens in this system and by the examination of the humoral immune response in these mice (Wang and Fujinami 1997). Acute EAE induced in susceptible strains of mice by direct administration of $PLP_{178-191}$ is very similar to the acute EAE induced by $PLP_{139-151}$; however, EAE induced using $PLP_{104-117}$ is associated with a delayed onset and slow chronic progressive disease. The enhanced CNS disease (EAE) observed in vaccinated mice upon injection of the $PLP_{139-151}$ peptide in adjuvant (challenge) was also observed upon injection of the $PLP_{104-117}$ peptide in adjuvant, but it was not observed upon injection of the $PLP_{178-191}$ peptide in adjuvant. With respect to the humoral immune response, antibodies specific for PLP and the various PLP peptides were present in the sera of primed mice prior to the challenge, and the respective PLP peptide-specific antibodies increased subsequent to challenge with the respective peptides. Following peptide challenge, the anti PLP IgG response shifted from an IgG1 to an IgG2a (IgG2c) and 2b phenotype. Thus, this study demonstrates that a humoral immune response to encephalitogenic PLP peptides can be generated by infection with a virus capable of coding for a self CNS protein (Wang and Fujinami 1997). An interesting addition to the above study, in which we prime with vaccinia virus encoding PLP and challenge with the $PLP_{139-151}$ peptide in CFA plus *B. pertussis*, is the demonstration that, when the *B. pertussis* supplementation is removed, there is still an initial early and enhanced acute attack of CNS disease (EAE) (Wang et al. 1999). However, following the initial earlier and enhanced CNS disease, the mice seldom experienced a clinical relapse, and if a relapse occurred, it was very mild and the mouse recovered very quickly. At this later time point (35–60 days post challenge), there was found to be significantly less demyelination and inflammation in the CNS of the vaccinated mice and lymphoproliferation in response to PLP peptide was significantly decreased, compared to controls. This reduced immune responsiveness suggests either unresponsiveness or regulated suppression, resulting in a permanent remission state (Wang et al. 1999).

In contrast to the above studies in which a viral infection primes the immune system for autoimmunity subsequent to a triggering challenge event, it was found that peripheral infection, via tail scarification or intraperitoneal injection, of PL/J mice with a recombinant vaccinia virus encoding the self peptide MBP_{1-23}, which did not induce CNS disease by itself, protected the mice from developing CNS disease (EAE) upon challenge with either purified guinea pig MBP or the acetylated MBP_{1-20} peptide in CFA (subcutaneous injection) plus intravenous injection of *B. pertussis* (Barnett et al. 1996). This result is allowed for in the prime/challenge model. The observed protection was found to be both mouse strain specific, as

SJL/J mice were not protected, and challenge antigen specific, as the protection did not extend to challenge with whole spinal cord homogenate. Also, this protection did not extend to the adoptive transfer of MBP_{1-11}-specific activated T cells. Examination of the delayed-type hypersensitivity response in this system suggested an immunological basis for the protection. The inability of T cells from these vaccinated/challenged mice to, in turn, adoptively transfer disease to naïve mice suggests anergy or deletion of effector T cells in these protected mice; although $CD8^+$ T cells do not participate in the protection as in vivo depletion of $CD8^+$ T cells did not remove the protective effect. Thus, prior peripheral infection with a virus capable of coding for a self epitope (altered peptide ligand, the encoded self epitope is not acetylated) can protect the host from an autoimmune response that would normally be triggered by a particular challenge (Barnett et al. 1996).

The above studies demonstrate that either enhancement of or protection from disease can occur after a subsequent challenge depending on the CNS protein encoded by the vaccinia virus vaccine. In turn, the ultimate response to a challenge infection may depend on what CNS epitope a priming virus has in common with self. To explore further the idea that different epitopes present at the priming step could determine the outcome after challenge, cDNAs encoding self CNS epitopes were constructed and replaced vaccinia virus vaccination in our prime/challenge model (Theil et al. 2001; Tsunoda et al. 1998). In the initial study, cDNAs that encoded the entire PLP protein (PLP_{all}) and the PLP peptides $PLP_{139-151}$ and $PLP_{178-191}$ were constructed. It was found that all three DNA constructs, injected intramuscularly, individually could induce a PLP-specific lymphoproliferative response, which was shown to be $CD4^+$ T cell-mediated, without inducing CNS disease. All three DNA constructs, when injected as the priming step, enhanced the CNS disease (more severe disease and more frequent relapses) that developed subsequent to challenge with $PLP_{139-151}$ peptide in CFA, while only priming with PLP_{all} enhanced disease following challenge with $PLP_{178-191}$ peptide in CFA. Also, the PLP-specific lymphoproliferative responses present in primed mice prior to challenge increased subsequent to challenge with the PLP peptides (Tsunoda et al. 1998). In a subsequent study, a cDNA was constructed that encoded ubiquitin fused to PLP (UPLP), leading to enhanced degradation of PLP through the proteasome pathway and the resulting PLP peptides were presented through the major histocompatibility (MHC) class I pathway (Theil et al. 2001). This method mimics infection with an intracellular virus that has molecular mimicry with CNS antigens without causing cytopathic effects in the host. Mice primed with UPLP and then challenged with phosphate-buffered saline (PBS) in CFA or with $PLP_{139-151}$ in CFA developed clinical and pathological signs of CNS disease. Mice primed with UPLP and then challenged with recombinant vaccinia virus encoding β-galactosidase developed pathological signs of CNS disease without any clinical signs of the disease. In contrast, mice primed with PLP_{all} and challenged with PBS in CFA did not develop pathological or clinical signs of the disease, thus indicating that the initial priming event may require efficient MHC class I presentation of the mimicking antigen which occurs with the ubiquitinated construct (Theil et al. 2001). An extension of this study more closely examined mice primed with UPLP and then challenged with $PLP_{139-151}$ in CFA (Theil et al. 2008). It was found that these mice experienced a milder relapse phase (1 month post challenge), compared to controls. During the relapse phase, these mice had decreased clinical and pathological signs of disease, decreased in vitro lymphoproliferation to $PLP_{139-151}$, reduced IFN-γ production and increased IL-4 production. In addition, $CD8^+$ T cells were found to be activated and expanded in these mice, confirming that degradation through the proteasome pathway and presentation by the MHC class I pathway was occurring, and these $CD8^+$ T cells could contribute to the modulation of the disease observed in these mice (Theil et al. 2008).

In addition to exploring the effect of the priming epitope, our group has explored various challenge events, in addition to the standard injection of PLP peptide in CFA, in our prime/challenge model (Theil et al. 2001; Tsunoda et al. 2007a). For our initial study, recombinant vaccinia viruses carrying the entire coding regions of PLP, glial fibrillary acidic protein (GFAP) and myelin associated glycoprotein (MAG) were constructed (Theil et al. 2001). Peripheral infection, via tail scarification, of SJL/J mice with any of the three vaccinia virus vaccines did not induce CNS disease. A subsequent nonspecific immunologic challenge with CFA, plus *B. pertussis*, resulted in the induction of pathological inflammatory CNS lesions without any clinical signs of disease. This shows that CNS proteins that are less encephalitogenic myelin antigens (MAG) and nonmyelin antigens (GFAP) can prime for disease, and that the challenge event can be a nonspecific immunologic stimulus. In comparison, priming of PL/J mice with the vaccinia viruses carrying PLP and GFAP followed by challenge with CFA, with or without *B. pertussis*, did not result in pathological or clinical signs of disease, thus reinforcing the importance of genetic background in the development of disease (Theil et al. 2001). In a subsequent study, the priming event was peripheral infection, via intraperitoneal injection, of SJL/J mice with a vaccinia virus vaccine encoding PLP (Tsunoda et al. 2007a). The challenge event was viral infection, via intraperitoneal injection, with several different viruses that either do not cause CNS disease on their own or cause a meningitis type of disease. Mice challenged with murine cytomegalovirus (MCMV; Smith strain) infection developed pathological signs of CNS disease, while mice challenged with lymphocytic choriomeningitis virus (LCMV; Armstrong strain) and the wild type strain of vaccinia virus (WR strain) did not develop the disease (Tsunoda et al. 2007a). One explanation of these results is that certain kinds of viral infections (MCMV) that promote IL-12 production, which contributes to the preferential development of Th1 responses (IFN-γ) over Th2 responses (IL-4), and possibly production of other members of the IL-12 family, such as IL-23 and IL-27, may be responsible for exacerbating or triggering MS through bystander activation of autoreactive T cells [(Tsunoda et al. 2007a); reviewed in (Fujinami 2001b; McCoy et al. 2006)]. Other kinds of viral infections (LCMV) that induce IFN-α/β production may suppress autoimmunity [(Tsunoda et al. 2007a); reviewed in (Fujinami 2001b; McCoy et al. 2006)]. The induction of IFN-α/β by infection may suppress autoimmunity through the downregulation of IL-12 [reviewed in (Fujinami 2001b)]. The priming with vaccinia virus encoding PLP followed by challenge with wild type vaccinia virus may not have resulted in disease due to rapid clearance of the virus (Tsunoda et al. 2007a).

8   Conclusion

This prime/challenge hypothesis easily explains why no one causative viral infection has ever been unequivocally proven to exist as the etiological agent of MS. If an initial encounter with any number of infectious agents can prime genetically susceptible individuals for CNS disease without causing overt disease, and overt disease only occurs after an encounter with any number of specific or nonspecific agents, then the likelihood of isolating one specific causative viral infection for MS is very low. The variability in the priming and challenging organisms may also contribute to the variability in the clinical course of MS observed in individual patients.

The explanation for how many different microbial infections can induce and exacerbate a single autoimmune disease has been described as the fertile-field hypothesis [(von Herrath et al. 2003); reviewed in (Fujinami et al. 2006)]. Autoreactive cells could be initially primed by both molecular mimicry and bystander activation, thus setting up a fertile-field, and then subsequent infection by a broad range of microorganisms could activate the preexisting autoreactive cells to autoagression through bystander activation. The fertile-field, induced by infection of the host and interactions between the immune response and the virus, would be transient and could change quantitatively and qualitatively over the course of infection depending on the type, anatomical location and duration of the infection. The initial priming event and the subsequent challenge event need not occur within the target organ, may be greatly separated in time, and more than one subsequent challenge event may be required for the development of overt disease [(von Herrath et al. 2003); reviewed in (Fujinami et al. 2006)]. The identification and characterization of viruses that are able to set up a fertile field within the host and then the identification and characterization of viruses that are able to trigger disease from the fertile field are challenges that remain open for future experimentation.

The identification of common inflammatory responses to otherwise distinct pathogens could lead us to the mechanism for how the prime/challenge model results in autoimmune disease (von Herrath et al. 2003). As described above, it has been found that a challenge virus (MCMV) that induces the production of IL-12 also induces the development of disease in an animal primed for autoimmune disease by exposure to a virus encoding a molecular mimic to a CNS protein (Tsunoda et al. 2007a). With this in mind, it is interesting to note that several viruses that are known to induce the production of IL-12, including herpes simplex virus, HHV-6, influenza virus and coronavirus, are viruses that have either been isolated from MS patients or associated with exacerabations of MS [reviewed in (Fujinami 2001b)]. The identification of immunomodulatory molecules that play a role in either the induction and/or exacerbation or the amelioration of CNS disease is a step towards being able to intelligently design vaccines that prevent disease through immune regulation.

Acknowledgements   We are grateful to Ms. Kathleen Borick for preparation of the manuscript. This work was supported by NIH AI581501.

References

Álvarez-Lafuente R, de Las Hervas V, Bartolome M, Picazo JJ, Arroyo R (2004) Relapsing-remitting multiple sclerosis and human herpesvirus 6 active infection. Arch Neurol 61:1523–1527

Álvarez-Lafuente R, de Las Hervas V, Garcia-Montojo M, Bartolome M, Arroyo R (2007) Human herpesvirus-6 and multiple sclerosis: relapsing-remitting versus secondary progressive. Mult Scler 13:578–583

Álvarez-Lafuente R, Garcia-Montojo M, de Las Hervas V, Bartolome M, Arroyo R (2006) Clinical parameters and HHV-6 active replication in relapsing-remitting multiple sclerosis patients. J Clin Virol 37(Suppl 1):S24–S26

Andersen O, Lygner PE, Bergstrom T, Andersson M, Vahlne A (1993) Viral infections trigger multiple sclerosis relapses: a prospective seroepidemiological study. J Neurol 240:417–422

Ascherio A, Munch M (2000) Epstein-Barr virus and multiple sclerosis. Epidemiology 11:220–224

Ascherio A, Munger KL (2007a) Environmental risk factors for multiple sclerosis. Part I: the role of infection. Ann Neurol 61:288–299

Ascherio A, Munger KL (2007b) Environmental risk factors for multiple sclerosis. Part II: noninfectious factors. Ann Neurol 61:504–513

Asher DM (1991) Slow viral infections of the human nervous system. In: Scheld WM, Whitley RJ, Durack DT (eds) Infections of the central nervous system. Raven, New York, pp 145–166

Babbe H, Roers A, Waisman A, Lassmann H, Goebels N, Hohlfeld R, Friese M, Schroder R, Deckert M, Schmidt S, Ravid R, Rajewsky K (2000) Clonal expansions of CD8+ T cells dominate the T cell infiltrate in active multiple sclerosis lesions as shown by micromanipulation and single cell polymerase chain reaction. J Exp Med 192:393–404

Barnett LA, Fujinami RS (1992) Molecular mimicry: a mechanism for autoimmune injury. FASEB J 6:840–844

Barnett LA, Whitton JL, Wada Y, Fujinami RS (1993) Enhancement of autoimmune disease using recombinant vaccinia virus encoding myelin proteolipid protein [published erratum appears in J. Neuroimmunol. 48:120, 1993]. J Neuroimmunol 44:15–25

Barnett LA, Whitton JL, Wang LY, Fujinami RS (1996) Virus encoding an encephalitogenic peptide protects mice from experimental allergic encephalomyelitis. J Neuroimmunol 64:163–173

Berti R, Brennan MB, Soldan SS, Ohayon JM, Casareto L, McFarland HF, Jacobson S (2002) Increased detection of serum HHV-6 DNA sequences during multiple sclerosis (MS) exacerbations and correlation with parameters of MS disease progression. J NeuroVirol 8:250–256

Bjartmar C, Wujek JR, Trapp BD (2003) Axonal loss in the pathology of MS: consequences for understanding the progressive phase of the disease. J Neurol Sci 206:165–171

Brahic M, Bureau JF, Michiels T (2005) The genetics of the persistent infection and demyelinating disease caused by Theiler's virus. Annu Rev Microbiol 59:279–298

Buljevac D, Flach HZ, Hop WCJ, Hijdra D, Laman JD, Savelkoul HFJ, van der Meché FGA, Van Doorn PA, Hintzen RQ (2002) Prospective study on the relationship between infections and multiple sclerosis exacerbations. Brain 125:952–960

Buljevac D, van Doornum GJ, Flach HZ, Groen J, Osterhaus AD, Hop W, Van Doorn PA, van der Meché FGA, Hintzen RQ (2005) Epstein-Barr virus and disease activity in multiple sclerosis. J Neurol Neurosurg Psychiatr 76:1377–1381

Cepok S, Zhou D, Srivastava R, Nessler S, Stei S, Bussow K, Sommer N, Hemmer B (2005) Identification of Epstein-Barr virus proteins as putative targets of the immune response in multiple sclerosis. J Clin Invest 115:1352–1360

Challoner PB, Smith KT, Parker JD, MacLeod DL, Coulter SN, Rose TM, Schultz ER, Bennett JL, Garber RL, Chang M, Schad PA, Stewart PM, Nowinski RC, Brown JP, Burmer GC (1995) Plaque-associated expression of human herpesvirus 6 in multiple sclerosis. Proc Natl Acad Sci U S A 92:7440–7444

Christensen T (2007) Human herpesviruses in MS. Int MS J 14:41–47

Cirone M, Cuomo L, Zompetta C, Ruggieri S, Frati L, Faggioni A, Ragona G (2002) Human herpesvirus 6 and multiple sclerosis: a study of T cell cross-reactivity to viral and myelin basic protein antigens. J Med Virol 68:268–272

Coo H, Aronson KJ (2004) A systematic review of several potential non-genetic risk factors for multiple sclerosis. Neuroepidemiology 23:1–12

Correale J, Fiol M, Gilmore W (2006) The risk of relapses in multiple sclerosis during systemic infections. Neurology 67:652–659

De Keyser J, Zwanikken C, Boon M (1998) Effects of influenza vaccination and influenza illness on exacerbations in multiple sclerosis. J Neurol Sci 159:51–53

Derfuss T, Hohlfeld R, Meinl E (2005) Intrathecal antibody (IgG) production against human herpesvirus type 6 occurs in about 20% of multiple sclerosis patients and might be linked to a polyspecific B-cell response. J Neurol 252:968–971

Dutta R, Trapp BD (2007) Pathogenesis of axonal and neuronal damage in multiple sclerosis. Neurology 68:S22-S31

Edwards S, Zvartau M, Clarke H, Irving W, Blumhardt LD (1998) Clinical relapses and disease activity on magnetic resonance imaging associated with viral upper respiratory tract infections in multiple sclerosis. J Neurol Neurosurg Psychiatr 64:736–741

Evans BK, Donely DK, Whitaker JN (1991) Neurological manifestations of infection with the human immunodeficiency viruses. In: Scheld WM, Whitley RJ, Durack DT (eds) Infections of the central nervous system. Raven, New York, pp 201–232

Fotheringham J, Jacobson S (2005) Human herpesvirus 6 and multiple sclerosis: potential mechanisms for virus-induced disease. Herpes 12:4–9

Fujinami RS (1996) Molecular mimicry. In: Shoenfeld Y, Peter JB (eds) Autoantibodies. Elsevier, Amsterdam, pp 507–519

Fujinami RS (1998) Molecular Mimicry. In: Rose NR, Mackay IR (eds) The autoimmune diseases. 3rd edn. Academic Press, San Diego, CA, pp 141–149

Fujinami RS (2000) Molecular mimicry and central nervous system autoimmune disease. In: Cunningham MW, Fujinami RS (eds) Molecular mimicry, microbes and autoimmunity, ASM, Washington, D C, pp 27–38

Fujinami RS (2001a) Can virus infections trigger autoimmune disease? J Autoimmun 16:229–234

Fujinami RS (2001b) Viruses and autoimmune disease–two sides of the same coin? Trends Microbiol 9:377–381

Fujinami RS, von Herrath MG, Christen U, Whitton JL (2006) Molecular mimicry, bystander activation, or viral persistence: infections and autoimmune disease. Clin Microbiol Rev 19:80–94

Gilden DH (2002) Multiple sclerosis exacerbations and infection. Lancet Neurol 1:145

Goverman J, Woods A, Larson L, Weiner LP, Hood L, Zaller DM (1993) Transgenic mice that express a myelin basic protein-specific T cell receptor develop spontaneous autoimmunity. Cell 72:551–560

Granieri E, Casetta I, Tola MR, Ferrante P (2001) Multiple sclerosis: infectious hypothesis. Neurol Sci 22:179–185

Hafler DA, Compston A, Sawcer S, Lander ES, Daly MJ, de Jager PL, de Bakker PI, Gabriel SB, Mirel DB, Ivinson AJ, Pericak-Vance MA, Gregory SG, Rioux JD, McCauley JL, Haines JL, Barcellos LF, Cree B, Oksenberg JR, Hauser SL (2007) Risk alleles for multiple sclerosis identified by a genomewide study. N Engl J Med 357:851–862

Holmoy T, Kvale EO, Vartdal F (2004) Cerebrospinal fluid CD4+ T cells from a multiple sclerosis patient cross-recognize Epstein-Barr virus and myelin basic protein. J NeuroVirol 10:278–283

Johnson RT (1998) Viral infections of the nervous system. Lippincott-Raven, Philadelphia

Kantarci OH (2008) Genetics and natural history of multiple sclerosis. Semin Neurol 28:7–16

Kriesel JD, Sibley WA (2005) Editorial: the case for rhinoviruses in the pathogenesis of multiple sclerosis. Mult Scler 11:1–4

Kriesel JD, White A, Hayden FG, Spruance SL, Petajan J (2004) Multiple sclerosis attacks are associated with picornavirus infections. Mult Scler 10:145–148

Kurtzke JF (1993) Epidemiologic evidence for multiple sclerosis as an infection. Clin Microbiol Rev 6:382–427

Kurtzke JF (1997) The epidemiology of multiple sclerosis. In: Raine CS, McFarland HF, Tourtellotte WW (eds) Multiple sclerosis: clinical and pathogenetic basis. Chapman & Hall, London, pp 91–140

Kurtzke JF, Heltberg A (2001) Multiple sclerosis in the Faroe Islands: an epitome. J Clin Epidemiol 54:1–22

Leibowitz U, Antonovsky A, Medalie JM, Smith HA, Halpern L, Alter M (1966) Epidemiological study of multiple sclerosis in Israel. II. Multiple sclerosis and level of sanitation. J Neurol Neurosurg Psychiatr 29:60–68

Levin LI, Munger KL, Rubertone MV, Peck CA, Lennette ET, Spiegelman D, Ascherio A (2005) Temporal relationship between elevation of epstein-barr virus antibody titers and initial onset of neurological symptoms in multiple sclerosis. JAMA 293:2496–2500

Libbey JE, Fujinami RS (2002) Are virus infections triggers for autoimmune disease? Clin Microbiol Newslett 24:73–76

Libbey JE, Fujinami RS (2003) Viral demyelinating disease in experimental animals. In: Herndon RM (ed) Multiple sclerosis: immunology, pathology and pathophysiology. Demos, New York, pp 125–133

Libbey JE, McCoy LL, Fujinami RS (2007) Molecular mimicry in multiple sclerosis. Int Rev Neurobiol 79:127–147

Liebert UG, Hashim GA, ter Meulen V (1990) Characterization of measles virus-induced cellular autoimmune reactions against myelin basic protein in Lewis rats. J Neuroimmunol 29:139–147

Lunemann JD, Edwards N, Muraro PA, Hayashi S, Cohen JI, Munz C, Martin R (2006) Increased frequency and broadened specificity of latent EBV nuclear antigen-1-specific T cells in multiple sclerosis. Brain 129:1493–1506

Lunemann JD, Kamradt T, Martin R, Munz C (2007) Epstein-barr virus: environmental trigger of multiple sclerosis? J Virol 81:6777–6784

Lunemann JD, Munz C (2007) Epstein-Barr virus and multiple sclerosis. Curr Neurol Neurosci Rep 7:253–258

Maggi F, Fornai C, Vatteroni ML, Siciliano G, Menichetti F, Tascini C, Specter S, Pistello M, Bendinelli M (2001) Low prevalence of TT virus in the cerebrospinal fluid of viremic patients with central nervous system disorders. J Med Virol 65:418–422

Marrie RA, Wolfson C, Sturkenboom MC, Gout O, Heinzlef O, Roullet E, Abenhaim L (2000) Multiple sclerosis and antecedent infections: a case-control study. Neurology 54:2307–2310

Massanari RM, Paterson PY, Lipton HL (1979) Potentiation of experimental allergic encephalomyelitis in hamsters with persistent encephalitis due to measles virus. J Infect Dis 139:297–303

McCoy L, Tsunoda I, Fujinami RS (2006) Multiple sclerosis and virus induced immune responses: autoimmunity can be primed by molecular mimicry and augmented by bystander activation. Autoimmunity 39:9–19

McElroy JP, Oksenberg JR (2008) Multiple sclerosis genetics. Curr Top Microbiol Immunol 318:45–72

Metz LM, McGuinness SD, Harris C (1998) Urinary tract infections may trigger relapse in multiple sclerosis. Axone 19:67–70

Miller SD, Vanderlugt CL, Begolka WS, Pao W, Yauch RL, Neville KL, Katz-Levy Y, Carrizosa A, Kim BS (1997) Persistent infection with Theiler's virus leads to CNS autoimmunity via epitope spreading. Nat Med 3:1133–1136

Mokhtarian F, Swoveland P (1987) Predisposition to EAE induction in resistant mice by prior infection with Semliki forest virus. J Immunol 138:3264–3268

Niino M, Fukazawa T, Kikuchi S, Sasaki H (2007) Recent advances in genetic analysis of multiple sclerosis: genetic associations and therapeutic implications. Expert Rev Neurother 7:1175–1188

Oleszak EL, Chang JR, Friedman H, Katsetos CD, Platsoucas CD (2004) Theiler's virus infection: a model for multiple sclerosis. Clin Microbiol Rev 17:174–207

Oleszak EL, Leibowitz JL, Rodriguez M (1988) Isolation and characterization of two plaque size variants of Theiler's murine encephalomyelitis virus (DA strain). J Gen Virol 69(Pt 9):2413–2418

Olson JK, Croxford JL, Calenoff MA, Dal Canto MC, Miller SD (2001) A virus-induced molecular mimicry model of multiple sclerosis. J Clin Invest 108:311–318

Panitch HS (1994) Influence of infection on exacerbations of multiple sclerosis. Ann Neurol 36(Suppl):S25–S28

Patick AK, Oleszak EL, Leibowitz JL, Rodriguez M (1990) Persistent infection of a glioma cell line generates a Theiler's virus variant which fails to induce demyelinating disease in SJL/J mice. J Gen Virol 71:2123–2132

Pender MP, Greer JM (2007) Immunology of multiple sclerosis. Curr Allergy Asthma Rep 7:285–292

Peterson LK, Fujinami RS (2006) Molecular mimicry. In: Shoenfeld Y, Gershwin ME, Meroni P-L (eds) Autoantibodies, 2nd edn. Elsevier, Philadelphia, pp 13–20

Pohl D, Krone B, Rostasy K, Kahler E, Brunner E, Lehnert M, Wagner HJ, Gartner J, Hanefeld F (2006) High seroprevalence of Epstein-Barr virus in children with multiple sclerosis. Neurology 67:2063–2065

Poskanzer DC, Schapira K, Miller H (1963a) Comparison of the epidemiology of multiple sclerosis and of poliomyelitis. Trans Am Neurol Assoc 88:253–255

Poskanzer DC, Schapira K, Miller H (1966) Multiple sclerosis and poliomyelitis. Acta Neurol Scand 42;Suppl 19:85–90

Poskanzer DC, Schapria K, Miller H (1963b) Multiple sclerosis and poliomyelitis. Lancet 2:917–921

Robertson NP, Clayton D, Fraser M, Deans J, Compston DA (1996) Clinical concordance in sibling pairs with multiple sclerosis. Neurology 47:347–352

Rutschmann OT, McCrory DC, Matchar DB (2002) Immunization and MS: a summary of published evidence and recommendations. Immunization Panel of the Multiple Sclerosis Council for Clinical Practice and Guidelines. Neurology 59:1837–1843

Serafini B, Rosicarelli B, Franciotta D, Magliozzi R, Reynolds R, Cinque P, Andreoni L, Trivedi P, Salvetti M, Faggioni A, Aloisi F (2007) Dysregulated Epstein-Barr virus infection in the multiple sclerosis brain. J Exp Med 204:2899–2912

Sibley WA, Bamford CR, Clark K (1985) Clinical viral infections and multiple sclerosis. Lancet 1:1313–1315

Simmonds P (2002) TT virus infection: a novel virus-host relationship. J Med Microbiol 51:455–458

Soldan SS, Fogdell-Hahn A, Brennan MB, Mittleman BB, Ballerini C, Massacesi L, Seya T, McFarland HF, Jacobson S (2001) Elevated serum and cerebrospinal fluid levels of soluble human herpesvirus type 6 cellular receptor, membrane cofactor protein, in patients with multiple sclerosis. Ann Neurol 50:486–493

Sospedra M, Martin R (2005) Immunology of multiple sclerosis. Annu Rev Immunol 23:683–747

Sospedra M, Martin R (2006) Molecular mimicry in multiple sclerosis. Autoimmunity 39:3–8

Sospedra M, Zhao Y, Zur HH, Muraro PA, Hamashin C, de Villiers E-M, Pinilla C, Martin R (2005) Recognition of conserved amino acid motifs of common viruses and its role in autoimmunity. PLoS Pathog 1:e41

Tejada-Simon MV, Zang YCQ, Hong J, Rivera VM, Zhang JZ (2003) Cross-reactivity with myelin basic protein and human herpesvirus-6 in multiple sclerosis. Ann Neurol 53:189–197

Thacker EL, Mirzaei F, Ascherio A (2006) Infectious mononucleosis and risk for multiple sclerosis: a meta-analysis. Ann Neurol 59:499–503

Theil DJ, Libbey JE, Rodriguez F, Whitton JL, Tsunoda I, Derfuss TJ, Fujinami RS. (2008) Targeting myelin proteolipid protein to the MHC class I pathway by ubiquitination modulates the course of experimental autoimmune encephalomyelitis. J Neuroimmunol 204:92-100

Theil DJ, Tsunoda I, Rodriguez F, Whitton JL, Fujinami RS (2001) Viruses can silently prime for and trigger central nervous system autoimmune disease. J NeuroVirol 7:220–227

Torkildsen O, Nyland H, Myrmel H, Myhr KM (2008) Epstein-Barr virus reactivation and multiple sclerosis. Eur J Neurol 15:106–108

Tsunoda I, Fujinami RS (1996) Two models for multiple sclerosis: experimental allergic encephalomyelitis and Theiler's murine encephalomyelitis virus. J Neuropathol Exp Neurol 55:673–686

Tsunoda I, Fujinami RS (2002) Inside-out versus Outside-in models for virus induced demyelination: axonal damage triggering demyelination. Springer Semin Immunopathol 24:105–125

Tsunoda I, Kuang L-Q, Libbey JE, Fujinami RS (2003) Axonal injury heralds virus-induced demyelination. Am J Pathol 162:1259–1269

Tsunoda I, Kuang L-Q, Tolley ND, Whitton JL, Fujinami RS (1998) Enhancement of experimental allergic encephalomyelitis (EAE) by DNA immunization with myelin proteolipid protein (PLP) plasmid DNA. J Neuropathol Exp Neurol 57:758–767

Tsunoda I, Libbey JE, Fujinami RS (2007a) Sequential polymicrobial infections lead to CNS inflammatory disease: possible involvement of bystander activation in heterologous immunity. J Neuroimmunol 188:22–33

Tsunoda I, Tanaka T, Saijoh Y, Fujinami RS (2007b) Targeting inflammatory demyelinating lesions to sites of Wallerian degeneration. Am J Pathol 171:1563–1575

Vandevelde M, Zurbriggen A (2005) Demyelination in canine distemper virus infection: a review. Acta Neuropathol 109:56–68

Virtanen JO, Farkkila M, Multanen J, Uotila L, Jaaskelainen AJ, Vaheri A, Koskiniemi M (2007) Evidence for human herpesvirus 6 variant A antibodies in multiple sclerosis: diagnostic and therapeutic implications. J Neurovirol 13:347–352

von Herrath MG, Fujinami RS, Whitton JL (2003) Microorganisms and autoimmunity: making the barren field fertile? Nat Rev Microbiol 1:151–157

Wandinger K, Jabs W, Siekhaus A, Bubel S, Trillenberg P, Wagner H, Wessel K, Kirchner H, Hennig H (2000) Association between clinical disease activity and Epstein-Barr virus reactivation in MS. Neurology 55:178–184

Wang L-Y, Fujinami RS (1997) Enhancement of EAE and induction of autoantibodies to T-cell epitopes in mice infected with a recombinant vaccinia virus encoding myelin proteolipid protein. J Neuroimmunol 75:75–83

Wang L-Y, Theil DJ, Whitton JL, Fujinami RS (1999) Infection with a recombinant vaccinia virus encoding myelin proteolipid protein causes suppression of chronic relapsing-remitting experimental allergic encephalomyelitis. J Neuroimmunol 96:148–157

Whitley RJ, Schlitt M (1991) Encephalitis caused by herpesviruses, including B virus. In: Scheld WM, Whitley RJ, Durack DT (eds) Infections of the central nervous system. Raven, New York, pp 41–86

Whitton JL, Fujinami RS (1999) Viruses as triggers of autoimmunity: facts and fantasies. Curr Opin Microbiol 2:392–397

Immunological Basis for the Development of Tissue Inflammation and Organ-Specific Autoimmunity in Animal Models of Multiple Sclerosis

Thomas Korn, Meike Mitsdoerffer, and Vijay K. Kuchroo

Abstract Experimental autoimmune encephalomyelitis (EAE) is an animal model for multiple sclerosis (MS) that has shaped our understanding of autoimmune tissue inflammation in the central nervous system (CNS). Major therapeutic approaches to MS have been first validated in EAE. Nevertheless, EAE in all its modifications is not able to recapitulate the full range of clinical and histopathogenic aspects of MS. Furthermore, autoimmune reactions in EAE-prone rodent strains and MS patients may differ in terms of the relative involvement of various subsets of immune cells. However, the role of specific molecules that play a role in skewing the immune response towards pathogenic autoreactivity is very similar in mice and humans. Thus, in this chapter, we will focus on the identification of a novel subset of inflammatory T cells, called Th17 cells, in EAE and their interplay with other immune cells including protective regulatory T cells (T-regs). It is likely that the discovery of Th17 cells and their relationship with T-regs will change our understanding of organ-specific autoimmune diseases in the years to come.

1 Introduction

Multiple sclerosis (MS) is a chronic inflammatory demyelinating disease of the central nervous system (CNS). There is a large body of evidence to suggest that MS is an autoimmune disease in that autoreactive cells of the adaptive immune system provide

T. Korn(✉)
Technische Universität München, Department of Neurology, Klinikum rechts der Isar, Ismaninger Str. 22, 81675, München, Germany
e-mail: korn@lrz.tum.de

M. Mitsdoerffer and V.K. Kuchroo(✉)
Center for Neurologic Diseases, Brigham and Women's Hospital, Harvard Medical School, 77 Avenue Louis Pasteur, Boston, MA 02115 USA
e-mail: vkuchroo@rics.bwh.harvard.edu

Results Probl Cell Differ, doi:10.1007/400_2008_17
© Springer-Verlag Berlin Heidelberg 2009

the initial trigger to orchestrate a pathogenic cascade resulting in inflammatory demyelination (Sospedra and Martin 2005). MS is diagnosed on the basis of the clinical presentation. The assignment to the diagnosis MS requires dissemination in space and time of clinical episodes suggestive of inflammatory demyelination. Thus, at least one episode and one relapse affecting different neurological systems have to be experienced by the patient in order to make the diagnosis of definite MS. Paraclinical read-out markers such as cerebrospinal fluid (CSF) analysis and magnetic resonance imaging (MRI) can substitute for a clinical relapse and thereby allow diagnosis of MS before the first relapse (McDonald et al. 2001). This is relevant since early anti-inflammatory treatment appears to improve long-term disability (Kappos et al. 2006). The requirement of a combination of signs and symptoms for the diagnosis of MS suggests that the pathological process in MS is incompletely understood. Moreover, human MS is heterogeneous both clinically and histopathologically and hence meets the criteria of a syndrome rather than a defined disease entity. Heterogeneity has been proposed because of the differences in the clinical course of the disease (Cottrell et al. 1999a, b; Ebers et al. 2000; Kremenchutzky et al. 1999, 2006; Pittock et al. 2004; Weinshenker et al. 1989a, b, 1991a, b) and variable histopathological patterns (Lucchinetti et al. 2000), both of which may suggest the involvement of different effector cells/molecules in the induction of tissue inflammation in MS. Most patients have a relapsing–remitting disease course, but about 10% exhibit a progressive disease course from the beginning (primary progressive). Half of the relapsing patients develop secondary progressive disease 10–15 years after the initial diagnosis. A histopathological classification has been proposed according to whether CNS lesions showed predominant T cell infiltrates (pattern I), B cell infiltrates with complement deposits (pattern II), loss of oligodendrocytes in the absence of prominent inflammatory infiltrates (pattern III), or altered myelination in the periplaque white matter suggestive of primary oligendrocyte disorders (pattern IV) (Lucchinetti et al. 2000).

2 Experimental Autoimmune Encephalomyelitis as a Model for MS

A variety of animal models have been introduced over the years in order to simulate clinical and histopathological patterns of chronic demyelinating CNS inflammation and thereby to learn more about the crucial steps in the pathogenesis of MS. Many aspects of our current concept of the pathogenic cascade in MS have been derived from studies on animal models of inflammatory demyelination in the CNS. The most widely used animal model is experimental autoimmune encephalomyelitis (EAE) (Gold et al. 2006; Zamvil and Steinman 1990). EAE can be induced by immunization of susceptible experimental animal strains including primates and rodents with myelin autoantigens emulsified in complete Freund's adjuvant (CFA). Within a week after immunization, autoreactive myelin antigen-specific CD4+ T cells are activated, which expand in the peripheral lymphoid tissue (Wekerle et al. 1994).

Ten days after immunization, activated T cells begin to cross the blood–brain barrier in large numbers; in contrast, unactivated T cells are not able to cross the blood–brain barrier (Flugel et al. 2001). Myelin antigen-specific CD4+ T cells become reactivated in the perivascular space in an MHC (major histocompatibility complex) class II restricted manner (Kawakami et al. 2005). Here, myelin antigens are most likely presented to infiltrating T cells by blood-derived perivascular antigen presenting cells (APCs) rather than local APCs of the CNS (Greter et al. 2005; Hickey and Kimura 1988). This results in further activation of T cells and production of cytokines such as interleukin (IL)-17, interferon (IFN)-γ, IL-6, and tumor necrosis factor (TNF) as well as chemokines in the CNS. Subsequently, immune cells such as granulocytes and macrophages are attracted into the CNS parenchyma, mediating tissue inflammation and demyelination (Dijkstra et al. 1992). Secretion of toxic species such as nitric oxide (NO) and TNF, and degradation of myelin are consequences of this cascade (Probert et al. 2000; Willenborg et al. 2007). Similar to MS, the inflammatory episode in EAE wanes spontaneously and the animals recover, suggesting that T cell mediated inflammation in the CNS is regulated even in the absence of a therapeutic intervention. Depending on the strain of mice and the autoantigen used to immunize, relapses can occur. For example, immunization of female SJL mice with an epitope of proteolipoprotein (PLP139–151) leads to a relapsing disease (McRae et al. 1992; Tuohy et al. 1989). This model of EAE is very well established and demyelinating inflammation in this model is driven by autoreactive myelin antigen-specific CD4+ T cells since in vitro activated myelin antigen-specific CD4+ T cells are necessary and sufficient to transfer the disease to unmanipulated host animals (Kuchroo et al. 1993; McRae et al. 1992).

Since EAE shares many clinical and histopathological features with MS, it is believed that MS is an autoimmune disease that is driven by myelin antigen-specific CD4+ MHC class II restricted T cells. This hypothesis is supported by the fact that the most important genetic susceptibility factor for MS is a certain HLA class II haplotype, i.e., DR15 and DQw6 (Martin and McFarland 1995; Sospedra and Martin 2005). Enhanced frequencies of myelin antigen-specific CD4+ MHC class II restricted T cells are found in the blood and CSF fluid of MS patients (Allegretta et al. 1990; Burns et al. 1999; Zhang et al. 1994). Moreover, studies with altered peptide ligands of myelin antigen-specific T cell receptors (TcRs) that were designed to downmodulate T cell activation inadvertently resulted in the activation of CD4+ HLA class II-restricted myelin basic protein (MBP)-specific T cells leading to clinical relapses (Bielekova et al. 2000). The clinical picture as well as the MRI images were typical of MS relapses and were correlated with an increased frequency and activation status of CD4+ MBP-specific T cells in the peripheral blood (Bielekova et al. 2000).

In general, the autoimmune nature of MS and the CD4+ T cell- and macrophage-mediated immunopathology are widely accepted pathogenic concepts. However, major criticisms of the classical EAE model are that it is an induced model of autoimmunity whereas MS develops spontaneously, and also that classical EAE fails to adequately capture antibody-mediated demyelination (Archelos et al. 2000; Piddlesden et al. 1993; Storch et al. 1998) and CD8+ T cell-mediated immunopathology (Babbe et al. 2000) that have been suggested to be important effectors in MS.

Attempts to study CD8+ T cell-mediated autoimmune pathology have been limited. First, virus-induced encephalitis models have been exploited for this purpose. However, many virus-induced encephalitis models show direct cytopathic effects on cells of the CNS, and activation and regulation of anti-virus CD8+ responses causing secondary immunopathology might be profoundly different from autoimmune reactions. In fact, virus-induced demyelination and immune-mediated demyelination can be difficult to differentiate. In Theiler's murine encephalitis virus (TMEV) infection, especially with the widely used strain BeAn8386, a first phase of virus-induced poliomyelitis is followed by a chronic demyelinating disease which is in part immune mediated. Nevertheless, immune-mediated demyelination occurs only when the virus persists (Miller et al. 1997). Virus-specific CD8+ T cells appear to be important for initial clearance of the virus. Paradoxically, CD8+ T cells are probably dispensable in the chronic demyelinating phase of TMEV infection leading to overt clinical disease which is mediated by CD4+ T cells with reactivity to myelin antigens (Miller et al. 1997). The complexity of the immunopathology of this model called for more direct approaches to investigate MHC class I restricted CD8+ T cell-mediated autoimmune pathology. A few studies were designed to directly address the possibility of CD8+ T cell-driven effector functions in CNS autoimmunity yet did not address the importance of CD4+ T cell help since they were pure adoptive transfer models where myelin antigen-specific CD8+ T cells were injected into host mice. Transfer of MHC class I (H-2^k) restricted MBP79–87-specific CD8+ T cells into immunocompromised host mice induced severe EAE characterized by inflammation and demyelination predominantly in the brain and less in the spinal cord. Anti-IFN-γ, but not neutralization of TNF, could attenuate disease in this model (Huseby et al. 2001). In contrast, in CD4+ T cell-mediated EAE, anti-IFN-γ treatment worsens disease, and blocking of TNF was shown to be beneficial suggesting that effector mechanisms might be different in CD8+ T cell- vs. CD4+ T cell-mediated EAE. As reported by Sun and coworkers, myelin oligodendrocyte glycoprotein (MOG)-specific CD8+ T cells also appear to be pathogenic upon adoptive transfer into host mice (Sun et al. 2001). Whether MOG35–55 can serve as an MHC class I restricted epitope has not been rigorously investigated. Later studies showed that MOG37–46 seems to be the correct MHC class I restricted epitope of MOG (Ford and Evavold 2005).

The role of B cells in MS is unclear. Immunoglobulin deposits have been found in MS lesions (pattern II), and oligoclonal IgG bands have been shown to be present in the CSF of MS patients for almost 30 years (Thompson et al. 1979). In MS lesions and the CSF of MS patients, B cells have been found to express somatically mutated immunoglobulins and therefore must be antigen experienced (Baranzini et al. 1999; Owens et al. 1998; Qin et al. 1998). Moreover, clonal expansion of B cells in the CSF of MS patients has been observed (Colombo et al. 2000). The specificity of oligoclonal IgG bands is most likely aimed at EBV components and not against myelin antigens (Cepok et al. 2005). In the serum of MS patients, antibodies (IgM) directed against MOG and MBP have been detected. It was suggested that these antibodies could serve as a predictor whether or not patients with an isolated clinical syndrome, i.e., with a solitary episode suggestive of an inflammatory demyelinating disorder of the CNS, would go on to develop definite MS (Berger et al. 2003). However, later studies could not confirm these results (Kuhle et al. 2007). Thus, it remains to be determined whether immunoglobulins play a pathogenic role in the structural development of MS lesions or are involved in functional impairment of nerve conduction in the CNS of MS patients. Besides antibody production, other functions of B cells, for example as APCs to T cells or as producers of cytokines, particularly IL-10, within the CNS, may be equally important for the pathogenesis of MS. Some of these questions have been addressed experimentally in animal models focusing on B cell dynamics and function in neuroinflammation.

Apart from modelling clinicopathological features of MS, EAE has been extremely useful for our understanding of the biology of autoreactive CD4+ T cells in vivo. Upon activation, naïve T cells differentiate into effector T cell subsets, producing different cytokines and mediating distinct effector functions. Initially, it was suggested that effector T helper (Th) cells can acquire either an IFN-γ (Th1)- or IL-4 (Th2)-producing phenotype (Mosmann and Coffman 1989). This paradigm has recently been updated following the discovery of a third subset of effector Th cells that produces IL-17 and exhibits distinct effector functions (Korn et al. 2007b). In the last 5 years there has been a wealth of information regarding this T cell subset: the cytokines for their differentiation have been identified, the key transcription factors that are involved in their generation have been recognized, and their functions in tissue inflammation and autoimmunity have been established. Although myelin antigen-specific Th1 cells were initially considered to be exclusively pathogenic in EAE, more recently, myelin antigen-specific Th17 cells have been shown to be even more potent in mediating tissue inflammation in EAE. Besides effector T cells, the peripheral CD4+ T cell compartment is composed of a subset of regulatory T cells (T-regs) that regulate effector T cell responses and have been shown to suppress the development of EAE. Since the identification of Foxp3 as master transcription factor for the generation of T-regs (Fontenot et al. 2005; Schubert et al. 2001), exact definition of T-reg populations has become possible. Tracking of Foxp3+ T-regs in vivo has allowed us to study their interplay with pathogenic effector T cells and their role in regulating autoimmune tissue inflammation.

In this chapter, we will outline advances in the understanding of the pathogenesis of autoimmune inflammation in the CNS that have been made on the basis of results obtained from studies of animal models combining the powerful tool of genetically modified mouse strains and classical EAE. Here, we will focus on CD4+ T cells in terms of triggering autoimmune inflammatory demyelination in the CNS since in the last 5 years, new subsets of T helper cells have been defined and characterized at a molecular level allowing us to finely dissect the role of various pathogenic effector T cells in mediating CNS autoimmunity. In this overview, we will discuss immunopathology in CNS autoimmunity, but not degenerative aspects of MS that have been discovered in recent years and have also been observed, at least in part, in EAE.

3 Experimental Autoimmune Encephalomyelitis: Historical Perspective

EAE was first, although of course unintentionally, induced in humans. Acute disseminated encephalomyelitis was a rare, but severe, side effect observed in individuals that had been administered rabies vaccine invented by Louis Pasteur in the 1880s (Stuart and Krikorian 1928). The vaccine consisted of fixed, and therefore inactivated, rabies virus that had been grown on rabbit CNS tissue. Acute disseminated encephalomyelitis does not have the symptoms of rabies and it was therefore hypothesized that this disabling CNS inflammation might be caused by rabbit CNS components contaminating the vaccine preparations. EAE as an animal model was introduced by Rivers in the 1930s (Rivers and Schwentker 1935) and was substantially improved as an animal model for demyelinating inflammation of the CNS by adding Freund's adjuvant to emulsify myelin antigens that were used to immunize susceptible laboratory animals and elicit EAE (Kabat et al. 1946). A conceptual breakthrough was achieved in the 1980s by the groups of Wekerle and Cohen, who showed that myelin antigen-specific CD4+ T cell lines propagated in vitro could, upon adoptive transfer, induce demyelinating inflammation in the CNS parenchyma (Ben-Nun et al. 1981). This proved that EAE is induced by an autoimmune response to myelin antigens initiated by CD4+ T cells. In the following years, EAE became an invaluable tool to study the activation of autoreactive T cells in the peripheral immune compartment, their migration into the CNS, and subsequent inflammation in the target tissue. In murine and primate EAE models, major immunogenic and encephalitogenic CNS autoantigens have been identified by isolation of antigen-specific MHC class II restricted T cell clones. Most of these autoantigens are myelin antigens such as myelin basic protein (MBP), proteolipoprotein (PLP), myelin oligodendrocyte glycoprotein (MOG), and myelin-associated oligodendrocyte basic protein (MOBP) (Gold et al. 2006). However, under certain circumstances T cell and B cell responses specific for neuronal or astrocytic antigens such as neurofascin and S100β are also able to induce EAE (Kojima et al. 1994; Mathey et al. 2007). Interestingly, T cell responses against non-myelin autoantigens such as αB crystallin, a small heat shock protein with anti-apoptotic effects which is expressed in the eye lens, but also astrocytes and oligodendrocytes of active MS lesions, have been implicated in the induction of EAE (Ousman et al. 2007). The dominance of a particular autoantigen in inducing EAE is strain dependent. Since in experimental settings bovine or guinea pig antigen preparations are often used to induce EAE in rodents, it is not only required that an epitope of a given myelin protein is presented in the context of a specific MHC class II restriction element, but also that the endogenous target myelin protein is homologous to the immunizing epitope. A number of encephalitogenic epitopes in different strains of mice and rats have been identified (Table 1).

In the course of an autoimmune reaction directed against myelin antigens, so-called epitope spreading has been shown to occur, i.e., T cells with specificity to myelin

Table 1   Examples of commonly used encephalitogenic epitopes

Epitope	Restriction element	Reference
MBP1–11	I–A^u	(Zamvil et al. 1986)
MBP35–47	I–E^u	(Zamvil et al. 1988)
MBP68–88	RT1B^l, RT1D^l (Lewis rat)	(Happ and Heber-Katz 1988)
MBP87–114	I–A^s	(Kono et al. 1988)
MOG35–55	I–A^b	(Mendel et al. 1995)
MOG91–108	RT1B^a, RT1D^a (Dark agouti rat)	(de Graaf et al. 2008)
PLP139–151	I–A^s	(Tuohy et al. 1989)

antigens different from the initiating antigen are expanded and drive the subsequent immune reaction (Lehmann et al. 1993; McRae et al. 1995; Sercarz et al. 1993). This is most likely due to release and presentation by local APCs of new myelin antigens released after the primary damage. It was initially thought that in order for epitope spreading to occur, APCs must process newly released antigens and carry them to the peripheral immune system to expand T cells. However, more recent data suggest that T cells can be primed within the CNS itself by infiltrating dendritic cells (DCs) (McMahon et al. 2005).

Response to an immunodominant antigen/MHC class II complex does not automatically predict encephalitogenicity. The initial hypothesis was that specific Vβ chain-usage of the TcR in the responding T cells might define encephalitogenicity. The TcR usage to some encephalitogenic epitopes appeared to be quite restricted: for example, Vβ8.2 for the I-A^u-restricted MBP1–11 epitope in mice and the RT1B^l-restricted MBP68–86 epitope in Lewis rats. In HLA DRB1*1501, DQA1*0102, DQB1*0602, and DPB1*0401 positive MS patients, T cells isolated from MS lesions had predominantly Vβ5.2 chain positive TcRs which had an identical sequence to a known MHC class II restricted T cell clone reactive against MBP89–106, suggesting that encephalitogenicity would be linked to a specific expression pattern of TcR Vβ chains in humans as well (Oksenberg et al. 1993). However, analysis of Vβ-usage to other encephalitogenic epitopes of other target proteins such as MOG (Mendel Kerlero de Rosbo and Ben-Nun 1996) and PLP quickly refuted this hypothesis; indeed, TcR Vβ-usage to these epitopes was shown to be quite diverse. Thus, the initial idea to design TcR vaccination strategies directed against particular Vβ chain fragments was partly successful in MBP peptide-induced EAE models (Howell et al. 1989; Jung et al. 1993) but failed in human MS where the driving antigen is unknown (Vandenbark et al. 1996).

4   Th1 Cells

Although myelin antigen specificity is crucial for the encephalitogenicity of a T cell clone, other factors including the ability to release a specific panel of cytokines and to express certain adhesion molecules for T cell trafficking are essential as well.

The usage of genetically manipulated mouse strains that were deficient for specific cytokines or adhesion molecules involved in the generation and trafficking of potentially encephalitogenic T cell populations was instrumental for discovering the conditions and properties that determine the encephalitogenicity of autoreactive T cells. Thus, EAE was an important tool to unravel the fate of T helper cells in vivo and to define subsets of T helper cells that would induce different types of immune responses. Three of the four approved treatments for MS have been first validated in EAE and were aimed at altering either the functional phenotype or the migratory capabilities of autoreactive T cells. Whereas glatiramer acetate and mitoxantrone skew and dampen the T cell response, respectively, natalizumab, a monoclonal antibody to VLA-4, was specifically designed to inhibit trafficking of activated T cells across the blood–brain barrier (Baron et al. 1993; Yednock et al. 1992).

Upon antigenic stimulation, naive CD4+ T cells proliferate and differentiate into different effector subsets characterized by the production of specific cytokines and effector functions. Since the pioneering work of Mosmann and Coffman, Th1 and Th2 cells were recognized and characterized by distinct cytokine profiles and effector functions (Mosmann and Coffman 1989). Th1 cells produce large quantities of IFN-γ, elicit delayed type hypersensitivity (DTH) responses, activate macrophages, and are highly effective in clearing intracellular pathogens. Th2 cells, on the other hand, produce IL-4, IL-5, IL-13, and IL-25. Th2 cells are especially important for IgE production, eosinophilic inflammation, and the clearance of helminth parasitic infections (Abbas et al. 1996). The cytokine milieu that often depends on the anatomical niche and the composition of the local innate immune cell populations that are present during the initial TcR engagement determines the commitment of a naive T cell and usually overrides other factors such as antigen structure and dose as well as level and type of co-stimulation.

Uncontrolled Th1 responses result in organ-specific autoimmunity. Th1 cells with specificity to self myelin antigens could adoptively transfer EAE (Ben-Nun et al. 1981; Kuchroo et al. 1993, 2002). In fact, initially all T cell clones reported to adoptively transfer EAE have been Th1 clones, whereas Th2 clones invariably failed to induce EAE unless the recipient mice were immunocompromised (Lafaille et al. 1997). Since IFN-γ was also detected in MS lesions (Traugott and Lebon 1988a, b) and in the CNS of EAE mice at the peak of disease and disappeared in the CNS lesions during recovery (Issazadeh et al. 1995; Olsson 1992), it was concluded that IFN-γ-producing autoreactive Th1 cells were responsible for inducing organ-specific autoimmunity. The differentiation of Th1 cells is dependent on IFN-γ, which is initially provided by cells of the innate immune system such as natural killer (NK) cells. The Th1 phenotype is then stabilized by IL-12, which is secreted by activated APCs. Upon engagement of its receptor, IFN-γ via STAT1 induces the transcription factor T-bet in naive CD4+ T cells. T-bet is a strong transactivator of the IFN-γ gene and also induces IL-12Rβ2 to equip T cells with a functional receptor for responding to IL-12 (Glimcher 2001; Szabo et al. 2000). IL-12 produced by cells of the innate immune system then stabilizes the Th1 phenotype of T cells by also inducing IFN-γ via STAT4. IL-12 is indispensable in the development of effector Th1 cells, and accordingly mice succumb to infections that require a strong Th1 response, such as *Listeria monocytogenes,* when IL-12 is neutralized (Tripp et al. 1994). IL-12 is a heterodimeric protein composed of p40 and p35 subunits (Gately et al. 1998). Since p40 knockout (KO) mice or wild-type mice treated with antibodies to p40 are entirely resistant to EAE (Becher et al. 2002; Leonard et al. 1995), this further supported the paradigm that Th1 cells are responsible for inducing the disease.

Th1 cells express a specific panel of chemokine receptors distinct from Th2 cells including CCR5 and CXCR3 (Bonecchi et al. 1998) that might direct them to specific anatomical sites depending on the chemokine production. At the site of inflammation Th1 cells have been shown to elicit a specific cascade of effector mechanisms. First, IFN-γ, the hallmark cytokine of Th1 cells, induces MHC class I and II molecules on epithelial cells and APCs. Second, a series of chemokines including MIP-1α, MIP-1β, and TCA-3 (Kuchroo et al. 1993) are produced by Th1 cells attracting mononuclear cells and shaping a cellular infiltrate dominated by macrophages. Macrophages are prominent cells in the cellular infiltrates of EAE and MS lesions (pattern I) and are important effector cells of demyelination; activated macrophages produce TNF and NO and actively phagocytose opsonized myelin constituents in a process called antibody-dependent cell-mediated cytotoxicity. Thus, Th1-mediated autoimmunity successfully explained many clinical and histopathological features of MS.

5   Th17 Cells

The concept of Th1 driven organ-specific autoimmunity was challenged when it was found that mice deficient for p35, the second subunit of IL-12, were not protected from EAE but developed more severe disease (Becher et al. 2002). Mice deficient in IL-18 or IL-12Rβ2 are also susceptible to EAE although both molecules promote the differentiation of Th1 cells (Gutcher et al. 2006; Zhang et al. 2003). Although IFN-γ itself is not only an effector cytokine of Th1 cells but has also a role in the contraction of activated T cell populations (Dalton et al. 2000), the fact that IFN-γ KO or IFN-γ receptor KO mice were susceptible to EAE (Ferber et al. 1996; Willenborg et al. 1996) seriously questioned whether EAE was a Th1 driven disease. Indeed, one interpretation of these data was that IFN-γ producing T cells might have a regulatory rather than a pro-inflammatory function in vivo.

The discovery that another cytokine apart from IL-12 used the p40 subunit but combined with a specific subunit p19 to form IL-23 (Oppmann et al. 2000) was instrumental in resolving some of the contradictions in the results obtained with IL-12. Whereas IL-12p35 KO mice are susceptible to EAE, IL-23p19 KO mice are absolutely protected from the disease (Cua et al. 2003). In the follow-up work, IL-23 was suggested to support the differentiation of a T helper cell subset that produced IL-17 but not IFN-γ (Langrish et al. 2005). Because of the production of IL-17, this subset was termed Th17 cells. In further studies, it became clear that the lack of any component of the IL-23/Th17 axis either protected or substantially attenuated the development of EAE. Moreover, IL-17 was found highly expressed in tissue sites of many autoimmune diseases including MS lesions (Lock et al. 2002). The cumulative data suggested an important role of IL-17 and Th17 cells in inducing tissue inflammation and autoimmunity.

The IL-17 family of cytokines has long been known and comprises six members (IL-17A-F). Except for IL-17E (IL-25) that appears to be associated with Th2 cells, members of the IL-17 family could not be categorized according to the Th1/Th2 paradigm (Kolls and Linden 2004). Only with the discovery of Th17 cells the unique role of these cytokines in immunity and autoimmunity is beginning to be understood. IL-17A and IL-17F have pro-inflammatory properties by inducing cytokines such as IL-6, IL-1β, and TNF, as well as chemokines and metalloproteinases from the responding tissues. IL-17A uses the IL-17RA, whereas IL-17F binds to the IL-17RC. Apparently, in humans IL-17RA and IL-17RC can form heterodimers that bind both IL-17A and IL-17F. Since the receptors are not only expressed on hematopoietic cells but also on epithelial and endothelial cells, IL-17 cytokines induce a broad tissue response.

Although IL-23 maintains or expands memory Th17 cells in vitro and in vivo, it is not the differentiation factor of these cells since the IL-23R is not expressed on naive T cells. Indeed, it was simultaneously found by several independent groups that a combination of TGF-β and IL-6 differentiated IL-17-producing T cells from naive T cells and also induced the expression of the IL-23R (Bettelli et al. 2006b; Mangan et al. 2006; Veldhoen et al. 2006a). It was a considerable surprise that TGF-β, which is regarded as an immunosuppressive cytokine (Li et al. 2006), was (together with IL-6) absolutely required for the induction of Th17 cells. In fact, TGF-β1 alone – the isoform of TGF-β that has a role in the immune system – is able to induce Foxp3+ T-regs and inhibits the proliferation of naive T cells as well as committed Th1 and Th2 cells. Paradoxically, TGF-β has now been recognized to be part of the differentiation cocktail of a T helper cell subset with highly inflammatory properties.

Th17 cells were firmly established as a unique T helper cell subset by the identification of the cytokines, i.e., the combination of IL-6 and TGF-β (Bettelli et al. 2006b; Mangan et al. 2006; Veldhoen et al. 2006a), and the transcription factors, i.e., ROR-γt (Ivanov et al. 2006), ROR-α (Yang et al. 2008b), and STAT3 (Laurence et al. 2007; Yang et al. 2007) required for their differentiation. Similarly to the expansion of Th1 responses by IFN-γ and Th2 responses by IL-4, Th17 responses appear to have an autoamplification loop; yet, IL-17 itself has no role in the differentiation of Th17 cells. However, IL-21, which is produced in large amounts by Th17 cells, is – together with TGF-β - able to induce ROR-γt and IL-17 in naive T cells and therefore constitutes an amplification loop for Th17 cells (Korn et al. 2007a; Nurieva et al. 2007; Zhou et al. 2007).

Th17 effector functions are distinct from Th1- and Th2-mediated immunity. Although Th17 cells are critical to enhance host protection against certain extracellular bacteria and fungi, which are not efficiently cleared by Th1 and Th2 responses, Th17 cells are also important inducers of autoimmunity and tissue inflammation.

5.1 Reciprocity between Th17 Cells and Induced T-Regs

TGF-β induces the T-reg-specific transcription factor Foxp3 and is required for the maintenance of induced T-regs (iT-regs) in the peripheral immune compartment. However, addition of IL-6 to TGF-β inhibits the generation of T-regs and induces Th17 cells. On the basis of these data, we first proposed (Bettelli et al. 2006b) that there is a reciprocal relationship between T-regs and Th17 cells, and that IL-6 plays a pivotal role in dictating the balance between these two cell types (Bettelli et al. 2006b; Korn et al. 2007a) regulating the initiation of immunity (autoimmunity) versus tolerance. Support for this concept and its relevance in vivo came from several recent studies: IL-2, which is a growth factor for T-regs, has also been shown to inhibit the generation of Th17 cells (Laurence et al. 2007). Consistent with these data, mice that lack IL-2 or in which IL-2 signalling is compromised (*Stat5*−/−), harbour reduced numbers of T-regs and an increased proportion of Th17 cells in the peripheral repertoire (Laurence et al. 2007). Moreover, these mice develop multi-organ inflammatory diseases, which can be prevented by the passive transfer of T-regs (Antov et al. 2003). Moreover, retinoic acid, a vitamin A metabolite, can drive the generation of T-regs (Coombes et al. 2007) by enhancing TGF-β signalling and enhancing Foxp3 promoter activity while abrogating the differentiation of Th17 cells, but not of Th1 cells, through the inhibition of IL-6 signalling (Mucida et al. 2007). Therefore, retinoic acid can regulate the balance between pro-inflammatory Th17 cells and anti-inflammatory T-regs. Finally, ROR-γt and ROR-α, the transcription factors for Th17 cells, and Foxp3, the transcription factor for T-regs, can physically bind to each other and antagonize each other's functions (Du et al. 2008; Zhou et al. 2008). In line with this concept, conditional deletion of Foxp3 protein in 'T-regs' in vivo resulted in an increase in ROR-γt, IL-17, and IL-21 expression (Gavin et al. 2007; Williams and Rudensky 2007), further corroborating the reciprocal relationship between Th17 cells and T-regs.

5.2 How are Th17 Cells Generated In Vivo?

Although it is clear that TGF-β together with IL-6 is required for the differentiation of Th17 cells from naive T cells, the in vivo sources of these differentiation factors are unclear. Whereas it is likely that IL-6 is produced by cells of the innate immune system upon encounter of microbial agents that trigger innate immune receptors such as Toll like receptors or dectin-1 like molecules, the source of TGF-β1 is still not known. Neutralization of TGF-β at the site of immunization inhibits the generation of Th17 cells and the development of EAE (Veldhoen et al. 2006b). Similarly, CD4+ T cells from mice with a dominant negative form of the TGF-β receptor II restricted to CD4+ cells cannot respond to TGF-β and are unable to differentiate into Th17 cells (Veldhoen et al. 2006b). Subsequently, the animals are protected from EAE. Naive T cells and T-regs, but also DCs, are able to produce TGF-β1.

Experiments with conditional KO animals that lack the expression of TGF-β in CD4+ T cells showed that T cells themselves are a relevant source of TGF-β for the generation of Th17 cells since mice lacking TGF-β in CD4+ T cells are unable to mount a Th17 response sufficient to induce EAE (Li et al. 2007). This suggests that paracrine (autocrine) secretion of TGF-β by T cells is essential for the differentiation of Th17 cells. Unless we are able to define the most relevant source of TGF-β for the generation of pathogenic Th17 cells, we should be cautious in applying immune therapies that rely on the transfer of ex vivo expanded T-regs to patients with autoimmune inflammatory diseases. Indeed, the concept that T-regs might be a source of TGF-β for the generation of Th17 cells (Veldhoen et al. 2006a) is supported by the observation that transfer of naive T cells together with T-regs into host animals promoted the induction of IL-17 but not IFN-γ in co-transferred conventional T cells (Lohr et al. 2006). TGF-β is produced in its latent form and needs to be activated. This is achieved by either proteolytic degradation or conformational changes of latency associated protein (LAP) that is associated with TGF-β in a heterotetrameric form and constrains the binding of the TGF-β homodimer to its receptor. For example, binding of LAP to thrombospondin on vascular endothelium or to integrins αVβ6 on epithelial cells or αVβ8 on DCs can lead to the procession of LAP and activate TGF-β in vivo. In fact, when αVβ8 is conditionally deleted in DCs, the mice suffer from a lymphoproliferative syndrome and tissue inflammation that is reminiscent of TGF-β deficiency in T cells (Travis et al. 2007). Thus, it is believed that DCs, although not primary producers of TGF-β, are essential to raise the levels of active TGF-β in the local environment.

TGF-β and IL-6 are the differentiation factors of Th17 cells, yet IL-23 appears to be indispensable for the initiation of severe tissue inflammation by Th17 cells. The mechanism of IL-23 action is elusive. TGF-β, together with IL-6 and IL-21, induces the expression of IL-23R on T cells, which requires the activation of STAT3 and probably also ROR-γt (Zhou et al. 2007). Current efforts are now focused on understanding the role of IL-23 in Th17-driven immunopathology. Compelling evidence is lacking to prove whether IL-23 is a maintenance or growth factor of Th17 cells. One report suggests that IL-23 might confer pathogenicity to Th17 cells by restraining the production of IL-10, which is produced by Th17 cells restimulated in the presence of TGF-β plus IL-6 (McGeachy et al. 2007). Furthermore, IL-23 appears to be essential for the induction of IL-22 (Zheng et al. 2007) and possibly other yet unidentified cytokines produced by Th17 cells. More recent data suggest that IL-23 might also be involved in enabling Th17 cells to traffic out of the local lymph nodes. Thus, in the absence of IL-23 signalling, Th17 cells can differentiate from naive T cells but are not equipped to leave the secondary lymphoid tissue in order to reach their target organ. IL-23 also acts on cells of the innate immune system to induce IL-17 production and drives chronic inflammation in the gut in the absence of T cells. IL-23 might also restrict the conversion of T-regs and thereby prepare the field for chronic inflammation at mucosal interfaces by inhibiting the de novo generation of T-regs (Izcue et al. 2008). In fact, the de novo generation of T-regs at mucosal interfaces driven by specialized DCs that raise the local concentration of TGF-β and produce retinoic acid appears to be an important means to maintain immunological homeostasis in the presence of the commensal flora (Coombes et al. 2007; Mucida et al. 2007).

5.3 What are the Effector Functions of Th17 Cells?

The mechanisms by which Th17 cells exert their pro-inflammatory potential are only beginning to be understood. Although there are several hypotheses as to why Th17 cells might be especially prone to drive tissue inflammation, there is only very little experimental evidence. First, Th17 might have a unique arsenal of adhesion molecules that readily allow them to infiltrate the parenchyma of solid organs. Second, by secreting effector cytokines with a broad receptor distribution they might communicate not only with other immune cells but also directly impact on tissue cells. Third, by inducing chemokines or by educating and conditioning local APCs in the target tissue they might attract and regulate the activation of other immune cells including Th1 cells that mediate tissue damage. Fourth, Th17 cells might not easily be regulated by natural T-regs.

Th17 cells secrete cytokines such as IL-21 to support a feedback amplification loop in terms of their own generation (Korn et al. 2007a; Nurieva et al. 2007; Zhou et al. 2007) but also to communicate with other immune cells such as B cells that express high levels of IL-21R (Chtanova et al. 2004). Other cytokines secreted by Th17 cells are the species-defining IL-17A, IL-17F, and IL-22 (Ouyang et al. 2008). This latter group of cytokines conveys effector functions of Th17 cells. It is not yet clear whether IL-17A, F and IL-22 are simultaneously produced by Th17 cells at any given time point. IL-17A KO mice (that were reported to have regular induction of IL-17F (Aujla et al. 2008)) are only partially protected from EAE (Komiyama et al. 2006) and IL-22 KO mice are not protected at all (Kreymborg et al. 2007), whereas IL-23p19 and IL-23R KO mice are almost completely resistant to disease suggesting that either a combination of effector cytokines produced by Th17 cells or another as yet unidentified Th17-derived factor induced by IL-23 confers pathogenicity to Th17 cells.

IL-17 and IL-17F induce a panel of chemokines to attract other immune cells (Hymowitz et al. 2001; Park et al. 2005; Yao et al. 1995). Prototypic effector cells chemoattracted by Th17 cells are neutrophils. However, indirectly other immune cells including Th1 cells may be attracted to the inflammatory lesion to orchestrate tissue inflammation. In an animal model of tuberculosis infection, Th17 cells were essential in attracting Th1 cells which then cleared the pathogen (Khader et al. 2007). The function of IL-22 is not yet understood. Like the IL-17R, the receptor for IL-22, which is a heterodimer of a specfic IL-22R and the IL-10R2, is widely expressed, including epithelial and endothelial cells. IL-22 has been shown to induce antimicrobial agents in keratinocytes (Liang et al. 2006) and is essential in the immune barrier function of epithelia as shown in bronchial infection models with *Klebsiella* (Aujla et al. 2008) and gut infection with *Citrobacter* (Zheng et al. 2008). IL-22 has been suggested to promote the breach of the blood–brain barrier

(Kebir et al. 2007) and to induce keratosis in mouse models of psoriasis and also in human psoriasis. However, IL-22 has also been shown to be protective in inflammation of the liver, the gut, and the myocard (Chang et al. 2006; Zenewicz et al. 2007; Zheng et al. 2008). For example, in experimental concanavalin A-induced hepatitis in mice, Th17 cells are a vehicle to deliver IL-22 into the site of inflammation to constrain liver damage (Zenewicz et al. 2007).

It is clear that Th17 cells are a distinct T helper cell subset, and strong evidence suggests that Th17 cells play an important role in the induction of EAE. However, adoptive transfer of Th1 cells is also able to induce EAE. During actively induced EAE, both antigen-specific Th17 and Th1 cells are generated. We have observed that Th17 cells infiltrate the CNS early during the disease course (Korn et al. 2007c), whereas double producers, i.e., CD4+ T cells that secrete both IFN-γ and IL-17, can be detected at frequencies of about 30% at the peak of disease. Later in the disease course and during remission, the frequency of IL-17+ CD4+ T cells in the CNS declines and the fraction of CD4+ T cells that exclusively produce IFN-γ increases. In late remission, both the IL-17+ and the IFN-γ + CD4+ T cell subsets decline, whereas the fraction of IL-10+ CD4+ T cells increases. At present, it is unclear whether Th17 cells and Th1 cells have to cooperate in order to induce autoimmune inflammation in the CNS. It is possible that Th17 cells – owing to their exquisite capacity to infiltrate tissues – represent the first wave of effector T helper cells and prepare the field for the infiltration of later waves of effector T cells. Indeed, many chemokines are target genes of IL-17. Another hypothesis is that those CD4+ T cells that produce both IFN-γ and IL-17 go on to become single producers of IFN-γ, shifting the frequencies of IL-17 producers and IFN-γ producers in the course of disease. Thus, the concept of a certain degree of plasticity of effector T cells in vivo cannot be ruled out. The generation of reporter mice in which cells that produce a specific cytokine can be tracked in vivo as a result of the simultaneous expression of a fluorescent protein will provide an essential tool to answer this question.

5.4   How Does the Discovery of Th17 Cells in Mice Translate to Humans?

IL-17 has been found highly expressed in human MS lesions, and IL-17+ T cells are enriched in active and re-activated chronic MS lesions (Lock et al. 2002; Tzartos et al. 2008). Molecules of the IL-23/Th17 axis appear to be associated with chronic tissue inflammation as assessed by snip analysis of patients with inflammatory bowel disease (Duerr et al. 2006), rheumatoid arthritis (Farago et al. 2008), psoriasis (Liu et al. 2008), and MS (Illes et al. 2008). Th17 cells have been identified in humans and have been defined by the expression of the chemokine receptors CCR4 and CCR6 (Acosta-Rodriguez et al. 2007b) or CCR2 and lack of CCR5 (Sato et al. 2007). It appeared that Th17 cells are a fraction of the memory cell population in peripheral blood that is directed against a particular type of antigens such as

Candida albicans (Acosta-Rodriguez et al. 2007b). Similarly as in mouse, these memory Th17 cells express high levels of IL-23R and can be expanded by IL-23. In contrast, the factors that are required for the de novo generation of Th17 cells from naive (CD45RA) cells in humans are still controversial. TGF-β plus IL-6 have been reported to fail in the induction of IL-17 in naive human T cells, and it has been suggested that either IL-1β plus IL-6 or IL-1β plus IL-23 are the differentiation factors for Th17 cells in humans (Acosta-Rodriguez et al. 2007a; Wilson et al. 2007); however, whether TGF-β is entirely dispensable for this process is questionable since the experimental settings used suffer from a number of setbacks. First, it is difficult to define naive T cells (as starting population for T cell differentiation assays) in the human system. The closest approximation to this would be the usage of CD4+ CD45RA+ T cells sorted from cord blood. Second, the necessity of TGF-β for the generation of human Th17 cells can be definitely excluded only when a serum-free medium or – in the presence of serum and platelets – an antibody to TGF-β that reliably neutralizes its activity is used. Interestingly, like in mouse, TGF-β plus IL-6 induce ROR-γt, the master transcription factor of Th17 cells, in human naive T cells (Chen et al. 2007). Three recent papers show that TGF-β is indeed important for the differentiation of Th17 cells in humans. IL-21 together with TGF-β, or IL-21 plus IL-6 together with TGF-β, can induce Th17 differentiation of naive human T cells (Manel et al. 2008; Volpe et al. 2008; Yang et al. 2008a). On the other hand, TGF-β suppresses the expansion of Th17 cells from the memory T cell compartment, which is driven by IL-1β plus IL-6 (Acosta-Rodriguez et al. 2007a).

6 Regulation of T Cell-Mediated Autoimmunity by T-Regs

Effector T cells that mediate inflammation are counterbalanced by regulatory T cells. Different types of T-regs have been defined in recent years. Whereas the role of T-regs in CD8+ T cell-driven disease models and infectious diseases is still unclear, it has been shown that T-regs are essential in maintaining peripheral tolerance and regulating self-reactive CD4+ T cells that have escaped deletion in the thymus. It is clear that this subset of T-regs is thymus derived, and therefore has been called "natural T-regs" (Sakaguchi 2005). Initially, the high expression level of the high-affinity IL-2 receptor CD25 on these T-regs has been used to define this subset of T-regs. Currently, there is a debate whether T-regs can be generated from conventional T cells outside the thymus. Depending on the prevailing cytokine milieu, the de novo generation of induced T-regs seems possible in defined niches of the immune system. However, under inflammatory conditions, T-regs are not induced and natural T-regs carry the burden of immunoregulation in the face of inflammation (Korn et al. 2007c). The last 5 years have seen major advances in the understanding of the dynamics and function of T-regs as a result of the identification of Foxp3 as major transcription factor of T-regs. The introduction of reporter mouse models where Foxp3 positive cells could be genetically marked allowing the tracking of these cells in vivo has substantially enhanced our understanding of the dynamics of T-reg responses during autoimmunity (Fontenot et al. 2005).

In the absence of thymus-derived (naturally occurring) Foxp3+ CD4+ CD25+ T-regs, multi-organ autoimmune disease develops. Genetic deficiency of Foxp3 leads to the lack of functional T-regs resulting in the scurfy phenotype in mice (Russell et al. 1959; Schubert et al. 2001) and the IPEX (immunodysregulation polyendocrinopathy enteropathy X-linked) syndrome in humans (Bennett et al. 2001; Wildin et al. 2001), which are fatal because of an uncontrollable lymphoproliferative syndrome. Depletion of CD4+ CD25+ T-regs prior to immunization increases the severity of EAE and makes EAE-resistant mouse strains susceptible to disease (Reddy et al. 2004). Conversely, adoptive transfer of CD4+ CD25+ T-regs before the onset of disease attenuates EAE (Hori et al. 2002). Foxp3 is a transcription factor that is restricted to CD4+ T-regs and stabilizes their functional phenotype (Gavin et al. 2007). Despite the progress in identifying molecules essential for the functional phenotype of T-regs, many questions remain regarding their mechanism and site of action. For example, in EAE it has been proposed that T-regs do not enter the CNS but act in the peripheral lymphoid tissue preventing effector cells from migrating into the CNS (Kohm et al. 2002). Using reporter mice in which the expression of Foxp3 is coupled to the expression of GFP, we have shown that naturally occurring myelin antigen-specific T-regs are expanded during an autoimmune reaction and migrate to the target tissue where they acquire an IL-10-producing phenotype (Korn et al. 2007c). Interestingly, whereas the ratio of antigen-specific effector T cells vs T-regs is strongly in favour of effector T cells at the peak of disease, this changes during remission in that there is massive attrition of effector T cells while the T-reg numbers in the CNS stay constant. Moreover, the capacity of T-regs to produce IL-10 massively increases as from the peak of disease suggesting that T-regs may be involved in the contraction of pathogenic T cell populations in the target tissue. However, direct proof for this is still missing. The ratio of effector T cells vs. T-regs alone might not be the only parameter controlling the outcome of the autoimmune reaction since factors like IL-6 and TNF present at the site of inflammation in the target tissue were illustrated to impair T-reg-mediated suppression of effector T cells by making them resistant to immunomodulation (Korn et al. 2007c). This might also explain why in spite of the presence of T-regs in inflammatory tissues immunopathology cannot be appropriately controlled as has been observed in MS and rheumatoid arthritis.

A multitude of mechanisms has been suggested for the mechanism of T-reg-mediated suppression in vivo. It is likely that combinations of different mechanisms are operational. The dominant mechanism might also depend on the compartment where the suppression takes place (secondary lymphoid tissue vs. target organ) and on the effector T cells that need to be regulated (uncommitted T cells vs. terminally differentiated effector T cells). Using two-photon microscopy in the secondary lymphoid tissue, a direct antigen-specific interaction has been visualized between T-regs and DCs, suggesting that the prevention of immune reactions by T-regs in lymph nodes might be mediated by inhibiting the activation of APCs by a mechanism that requires close physical contact between T-regs and DCs (Tang et al. 2006).

In contrast, in the target tissue, T-regs might secrete soluble mediators such as IL-10 and IL-35 or enhance the pericellular generation of adenosine (Deaglio et al. 2007), which leads to a broad down-modulation of inflammation (Collison et al. 2007; Rubtsov et al. 2008).

7 Regulatory Cytokines

Th1- and Th2-associated cytokines cross-inhibit each other's generation. Thus, Th2 cells have been regarded beneficial in Th1-mediated organ-specific autoimmunity. For example, IL-10, which is produced by Th2 cells but also by other T cell subsets, macrophages, and glial cells, is believed to be an important factor to reduce inflammation: IL-10 leads to the down-regulation of MHC class II molecules and inhibits TNF production in macrophages as well as other inflammatory cytokines (Moore et al. 2001). With the discovery of Th17 cells and their potential pathogenic role in tissue inflammation and autoimmunity, the concept that organ-specific autoimmunity develops because of dysregulation of Th1 or Th2 cells and their associated cytokines is no longer appropriate. The signature cytokines of Th1 and Th2 cells, IFN-γ and IL-4, inhibit the generation of Th17 cells, and IL-4 seems to have a stronger effect than IFN-γ (Park et al. 2005). However, other cytokines have been shown to inhibit Th17 immunity and thereby dampen tissue inflammation and immunopathology. First, IL-25, which is itself a member of the IL-17 family of cytokines (IL-17E) and which has been recognized as a Th2-associated cytokine, inhibits the differentiation of Th17 cells and the severity of EAE (Kleinschek et al. 2007). This effect seems to be indirect via enhancement of IL-13, which down-regulates the expression of Th17-promoting factors such as IL-6, IL-1β, and IL-23 in DCs. Second, IL-27, which is a member of the IL-12 family of heterodimeric cytokines and is composed of the p28 subunit and the EBI3 gene product, directly inhibits the generation of Th17 cells in a STAT1-dependent manner. Consequently, mice deficient for WSX-1, the receptor for IL-27, develop enhanced frequencies of Th17 cells and exhibit exacerbated EAE and chronic neurotoxoplasmosis (Batten et al. 2006; Stumhofer et al. 2006). Although direct proof in vivo is missing in vivo, it is likely that IL-27 is derived from cells of the innate immune system, such as DCs. Recently, it has been shown that TGF-β together with IL-27 initiates the differentiation of IL-10 producing Tr-1-like cells (Awasthi et al. 2007; Stumhofer et al. 2007). Thus, besides a direct inhibitory effect on pro-inflammatory Th17 cells, IL-27 might dampen inflammation indirectly by enhancing the differentiation of antigen-specific IL-10-producing Tr-1 cells.

8 B Cells

The role of B cells in MS has been studied less in animal models. In general, B cells could serve as APCs to activate T cells, they could differentiate into cells that produce antibodies of pathogenic relevance, or by secretion of cytokines they could modulate inflammation. The most interesting insights into possible pathogenic functions of B cells were obtained from animal models that combined pathogenic T cells and B cells.

Early studies in Lewis rats that had been depleted of B cells and immunoglobulins by injection of anti-rat Ig from birth suggested a role of B cells, as these animals showed resistance to EAE induction by MBP emulsified in CFA (Willenborg and Prowse 1983). However, injection of MBP-specific serum following immunization with MBP made them susceptible to disease (Willenborg et al. 1986). Myers et al. found similar results in B cell-depleted SJL and PL/1 mice, which showed a resistance to EAE induction by MBP (Myers et al. 1992). Furthermore, they demonstrated that MBP-specific antibodies increased the efficiency of adoptive transfer EAE in B cell-depleted mice and in wild-type mice when the number of adoptively transferred T cells was limited (Myers et al. 1992). These results were in conflict with subsequent studies in μMT mice that were rendered deficient in B cells by genetic deletion of their μ chain transmembrane region. Immunization of B10.PL and B10.PLμMT with the NH_2-terminal MBP peptide Ac1–11 led to similar disease in both strains. However, a higher variety in disease onset, severity, and recovery was observed in B cell-deficient mice compared to control mice (Wolf et al. 1996). Studies in C57Bl/6 μMT mice showed no differences compared to wild-type mice regarding clinical disease and histological features after sensitization with MOG35–55 peptide but indicated that B cell deficient animals were protected from EAE when immunized with the whole recombinant MOG protein (Hjelmstrom et al. 1998; Lyons et al. 1999). Indeed, passive transfer of B cells, or even more efficiently serum, from MOG-primed wild-type mice restored the susceptibility to clinical and histological EAE in B cell-deficient mice (Lyons et al. 2002). However, Oliver et al. later specified that only human MOG protein-induced EAE was B cell dependent (Oliver et al. 2003). Other studies in B cell-deficient mice on the C57Bl/10 and DBA/1 background found reduced severity and incidence of disease in response to immunization with recombinant MOG (Svensson et al. 2002). Histological examination revealed decreased demyelination in B cell-deficient mice rather than reduced numbers of infiltrating cells (Svensson et al. 2002). Differences in the animal models used impede the comparison of these results. First, earlier studies used infusions of anti-IgM antibodies in neonatal animals to deplete B cells whereas later studies used B cell-deficient mice generated by targeted disruption of the IgM heavy chain gene transmembrane region, which prevents the development of mature B cells. Moreover, analysis of anti-μ-treated mice, i.e. mice treated with antibodies to IgM to diminish surface IgM bearing B cells which are also the precursors of B cells bearing other Ig isotypes, showed dysfunction of T cells as well, which might account for some of the observed differences (Kim et al. 1984). Second, varying preparations of antigen from different species were used, e.g., whole protein, homogenized spinal cord, or peptides. Taken together, B cells appear to play an important role in antigen presentation of protein antigens, but less so in presentation of peptide antigens. However, whether B cells skew the immune response or interact with a specific population of T cells during an autoimmune response remains to be determined.

8.1 Role of Antibodies

First evidence for a pathogenic role of antibodies in EAE emerged from studies with serum factors of rabbits immunized for EAE with bovine brain white matter emulsified in CFA which showed an effect on demyelination (Grundke-Iqbal et al. 1981). Thereafter, numerous studies addressed the identification and role of antibodies in EAE. Antibodies directed against MOG seem to have a prominent role, probably due to its accessibility on the outer surface of the myelin sheath (Gardinier et al. 1992). Administration of anti-MOG antibodies led to increased disease severity and demyelination in acute MBP-induced EAE in Lewis rats (Lassmann et al. 1988; Linington et al. 1988; Schluesener et al. 1987). Similarly, injection of MOG-specific antibodies caused pronounced demyelination in MBP-immunized primates (common marmorset *C. jacchus*) whereas immunization with MBP alone resulted in clinically mild EAE and no demyelination (Genain et al. 1995). SJL mice with relapsing EAE showed fatal relapses when an anti-MOG antibody was given during the recovery phase (Schluesener et al. 1987). Generation of a transgenic mouse that produces high titers of autoantibodies against MOG brought further insights. Albeit these mice did not develop spontaneous disease, immunization with encephalitogenic antigens provoked accelerated disease onset as well as increased incidence and severity of EAE (Litzenburger et al. 1998). In addition, antibodies to a variety of other myelin antigens including myelin lipids have been described (Endoh et al. 1986; Moore and Raine 1988; Tabira and Endoh 1985; Van der Goes et al. 1999; van der Veen et al. 1986; Wang and Fujinami 1997). Although there is no doubt that autoantibodies are produced and are involved in the pathogenesis of EAE, their exact mode of action is still unclear. A variety of mechanisms such as complement activation by MOG-specific antibodies (Piddlesden et al. 1993) as well as antibody-dependent cell-mediated cytotoxicity (Piddlesden et al. 1991) might be involved. Furthermore, autoantibodies might influence T cell responses to certain epitopes through their effect on antigen processing (Dai et al. 1999; Simitsek et al. 1995).

8.2 B Cells as APCs

There is no doubt that DCs efficiently present peptide antigens to T cells, and in that way are capable of stimulating naïve T cells (Aloisi et al. 1999; Eynon and Parker 1992; Fuchs and Matzinger 1992; Lassila et al. 1988; Morris et al. 1994; Ronchese and Hausmann 1993). On the other hand, the capacity of DCs to present protein antigens might be limited (Constant et al. 1995a, b). Although it is controversial whether B cells stimulate unprimed proliferative responses in T cells, recent studies have shown that the form of antigen is decisive for the APC type involved in T cell priming. Antigen-specific B cells seem to be most potent when the amount of antigen is limited (Constant et al. 1995a, b), and antigen-specific B cells can present 10,000-fold lower antigen concentrations compared to non-antigen-specific B cells or other APCs (Lanzavecchia 1990; Pierce et al. 1988). Other studies suggest that auto-reactive B cells are even the first APCs to present antigen to T cells, and DCs subsequently activate these auto-reactive T cells (Yan et al. 2006). Several groups showed normal priming of T cells to peptides in B cell-deficient mice measured by proliferation to recall antigen, yet response to most proteins seems to be impaired (Constant et al. 1995a; Epstein et al. 1995; Lyons et al. 1999; Phillips et al. 1996). These results are in concordance with earlier studies in mice treated with anti-IgM antibodies (Hayglass et al. 1986; Janeway et al. 1987; Kurt-Jones et al. 1988; Ron et al. 1981).

Another important function of autoantigen-reactive B cells might be the provision of IL-10 for recovery from EAE (Fillatreau et al. 2002). These results are further corroborated by recent work from Mann et al., which showed a delay in the emergence of T-reg cells and IL-10 in the CNS of B cell-deficient mice during EAE (Mann et al. 2007). Reconstitution with wild-type B cells but not B7-deficient B cells restored the capacity of the animals to recover from EAE.

But the most interesting data came from transgenic mice in which both T cells and B cells recognize the same autoantigen MOG. About 60% of these mice show a severe and spontaneous form of EAE. MOG-specific B cells act as very efficient APCs to transgenic T cells and undergo class switching to IgG1. Histological examination of the CNS showed an interesting pattern with a clear predominance of inflammatory lesions in spinal cord and optic nerves similar to the lesion distribution seen in human Devic syndrome (Bettelli et al. 2006a; Krishnamoorthy et al. 2006). Furthermore, sick mice showed lymphoid follicle-like structures in the spinal cord, suggesting a productive interaction between T cells and B cells (Bettelli et al. 2006a).

9 Concluding Remarks

Animal models have been extremely useful to understand pathogenic processes in organ-specific autoimmunity. However, it is also clear that EAE is not human MS, and however sophisticated the modifications of this animal model will be, it will not be possible to mimic all the clinical and histopathological aspects of the human disease. In fact, we do not even know the driving autoantigen in MS, and it will not likely be a single antigen. Thus, it is evident that preclinical studies of new therapeutic agents that are conducted in EAE are important, but we have to be cautious in interpreting the results, and in this regard this model certainly has its weaknesses. On the other hand, we have collected a tremendous amount of information on the biology of immune cells including T cells and B cells in vivo by exploiting the EAE model. Especially, the importance of CD4+ T helper cells in the pathogenic process of CNS autoimmunity has clearly been demonstrated in this model, even though other effector mechanisms including CD8+ T cells might play a role in the generation of autoimmune inflammatory lesions in the CNS. Therefore, we propose that EAE with all its imperfections is one of the most powerful and extremely useful models to study the biology of immune cells during organ-specific autoimmunity in vivo.

Acknowledgement TK and MM are supported by the Deutsche Forschungsgemeinschaft (KO 2964/2–1 and MI 1221/1–1).

References

Abbas AK, Murphy KM, Sher A (1996) Functional diversity of helper T lymphocytes. Nature 383:787–793

Acosta-Rodriguez EV, Napolitani G, Lanzavecchia A, Sallusto F (2007a) Interleukins 1beta and 6 but not transforming growth factor-beta are essential for the differentiation of interleukin 17-producing human T helper cells. Nat Immunol 8:942–949

Acosta-Rodriguez EV, Rivino L, Geginat J, Jarrossay D, Gattorno M, Lanzavecchia A, Sallusto F, Napolitani G (2007b) Surface phenotype and antigenic specificity of human interleukin 17-producing T helper memory cells. Nat Immunol 8:639–646

Allegretta M, Nicklas JA, Sriram S, Albertini RJ (1990) T cells responsive to myelin basic protein in patients with multiple sclerosis. Science 247:718–721

Aloisi F, Ria F, Columba-Cabezas S, Hess H, Penna G, Adorini L (1999) Relative efficiency of microglia, astrocytes, dendritic cells and B cells in naive CD4+ T cell priming and Th1/Th2 cell restimulation. Eur J Immunol 29:2705–2714

Antov A, Yang L, Vig M, Baltimore D, Van Parijs L (2003) Essential role for STAT5 signaling in CD25+ CD4+ regulatory T cell homeostasis and the maintenance of self-tolerance. J Immunol 171:3435–3441

Archelos JJ, Storch MK, Hartung HP (2000) The role of B cells and autoantibodies in multiple sclerosis. Ann Neurol 47:694–706

Aujla SJ, Chan YR, Zheng M, Fei M, Askew DJ, Pociask DA, Reinhart TA, McAllister F, Edeal J, Gaus K, Husain S, Kreindler JL, Dubin PJ, Pilewski JM, Myerburg MM, Mason CA, Iwakura Y, Kolls JK (2008) IL-22 mediates mucosal host defense against Gram-negative bacterial pneumonia. Nat Med 14:275–281

Awasthi A, Carrier Y, Peron JP, Bettelli E, Kamanaka M, Flavell RA, Kuchroo VK, Oukka M, Weiner HL (2007) A dominant function for interleukin 27 in generating interleukin 10-producing anti-inflammatory T cells. Nat Immunol 8:1380–1389

Babbe H, Roers A, Waisman A, Lassmann H, Goebels N, Hohlfeld R, Friese M, Schroder R, Deckert M, Schmidt S, Ravid R, Rajewsky K (2000) Clonal expansions of CD8(+) T cells dominate the T cell infiltrate in active multiple sclerosis lesions as shown by micromanipulation and single cell polymerase chain reaction. J Exp Med 192:393–404

Baranzini SE, Jeong MC, Butunoi C, Murray RS, Bernard CC, Oksenberg JR (1999) B cell repertoire diversity and clonal expansion in multiple sclerosis brain lesions. J Immunol 163:5133–5144

Baron JL, Madri JA, Ruddle NH, Hashim G, Janeway CA, Jr. (1993) Surface expression of alpha 4 integrin by CD4 T cells is required for their entry into brain parenchyma. J Exp Med 177:57–68

Batten M, Li J, Yi S, Kljavin NM, Danilenko DM, Lucas S, Lee J, de Sauvage FJ, Ghilardi N (2006) Interleukin 27 limits autoimmune encephalomyelitis by suppressing the development of interleukin 17-producing T cells. Nat Immunol 7:929–936

Becher B, Durell BG, Noelle RJ (2002) Experimental autoimmune encephalitis and inflammation in the absence of interleukin-12. J Clin Invest 110:493–497

Bennett CL, Christie J, Ramsdell F, Brunkow ME, Ferguson PJ, Whitesell L, Kelly TE, Saulsbury FT, Chance PF, Ochs HD (2001) The immune dysregulation, polyendocrinopathy, enteropathy, X-linked syndrome (IPEX) is caused by mutations of FOXP3. Nat Genet 27:20–21

Ben-Nun A, Wekerle H, Cohen IR (1981) The rapid isolation of clonable antigen-specific T lymphocyte lines capable of mediating autoimmune encephalomyelitis. Eur J Immunol 11:195–199

Berger T, Rubner P, Schautzer F, Egg R, Ulmer H, Mayringer I, Dilitz E, Deisenhammer F, Reindl M (2003) Antimyelin antibodies as a predictor of clinically definite multiple sclerosis after a first demyelinating event. N Engl J Med 349:139–145

Bettelli E, Baeten D, Jager A, Sobel RA, Kuchroo VK (2006a) Myelin oligodendrocyte glyco-protein-specific T and B cells cooperate to induce a Devic-like disease in mice. J Clin Invest 116:2393–2402

Bettelli E, Carrier Y, Gao W, Korn T, Strom TB, Oukka M, Weiner HL, Kuchroo VK (2006b) Reciprocal developmental pathways for the generation of pathogenic effector TH17 and regulatory T cells. Nature 441:235–238

Bielekova B, Goodwin B, Richert N, Cortese I, Kondo T, Afshar G, Gran B, Eaton J, Antel J, Frank JA, McFarland HF, Martin R (2000) Encephalitogenic potential of the myelin basic protein peptide (amino acids 83–99) in multiple sclerosis: results of a phase II clinical trial with an altered peptide ligand. Nat Med 6:1167–1175

Bonecchi R, Bianchi G, Bordignon PP, D'Ambrosio D, Lang R, Borsatti A, Sozzani S, Allavena P, Gray PA, Mantovani A, Sinigaglia F (1998) Differential expression of chemokine receptors and chemotactic responsiveness of type 1 T helper cells (Th1s) and Th2s. J Exp Med 187:129–134

Burns J, Bartholomew B, Lobo S (1999) Isolation of myelin basic protein-specific T cells predomi-nantly from the memory T cell compartment in multiple sclerosis. Ann Neurol 45:33–39

Cepok S, Zhou D, Srivastava R, Nessler S, Stei S, Bussow K, Sommer N, Hemmer B (2005) Identification of Epstein-Barr virus proteins as putative targets of the immune response in multiple sclerosis. J Clin Invest 115:1352–1360

Chang H, Hanawa H, Liu H, Yoshida T, Hayashi M, Watanabe R, Abe S, Toba K, Yoshida K, Elnaggar R, Minagawa S, Okura Y, Kato K, Kodama M, Maruyama H, Miyazaki J, Aizawa Y (2006) Hydrodynamic-based delivery of an interleukin-22-Ig fusion gene ameliorates experimental autoimmune myocarditis in rats. J Immunol 177:3635–3643

Chen Z, Tato CM, Muul L, Laurence A, O'Shea JJ (2007) Distinct regulation of interleukin-17 in human T helper lymphocytes. Arthritis Rheum 56:2936–2946

Chtanova T, Tangye SG, Newton R, Frank N, Hodge MR, Rolph MS, Mackay CR (2004) T follicular helper cells express a distinctive transcriptional profile, reflecting their role as non-Th1/Th2 effector cells that provide help for B cells. J Immunol 173:68–78

Collison LW, Workman CJ, Kuo TT, Boyd K, Wang Y, Vignali KM, Cross R, Sehy D, Blumberg RS, Vignali DA (2007) The inhibitory cytokine IL-35 contributes to regulatory T cell function. Nature 450:566–569

Colombo M, Dono M, Gazzola P, Roncella S, Valetto A, Chiorazzi N, Mancardi GL, Ferrarini M (2000) Accumulation of clonally related B lymphocytes in the cerebrospinal fluid of multiple sclerosis patients. J Immunol 164:2782–2789

Constant S, Sant'Angelo D, Pasqualini T, Taylor T, Levin D, Flavell R, Bottomly K (1995a) Peptide and protein antigens require distinct antigen-presenting cell subsets for the priming of CD4+ T cells. J Immunol 154:4915–4923

Constant S, Schweitzer N, West J, Ranney P, Bottomly K (1995b) B lymphocytes can be competent antigen-presenting cells for priming CD4+ T cells to protein antigens in vivo. J Immunol 155:3734–3741

Coombes JL, Siddiqui KR, Arancibia-Carcamo CV, Hall J, Sun CM, Belkaid Y, Powrie F (2007) A functionally specialized population of mucosal CD103+ DCs induces Foxp3+ regulatory T cells via a TGF-beta and retinoic acid-dependent mechanism. J Exp Med 204:1757–1764

Cottrell DA, Kremenchutzky M, Rice GP, Hader W, Baskerville J, Ebers GC (1999a) The natural history of multiple sclerosis: a geographically based study. 6. Applications to planning and interpretation of clinical therapeutic trials in primary progressive multiple sclerosis. Brain 122(Pt 4):641–647

Cottrell DA, Kremenchutzky M, Rice GP, Koopman WJ, Hader W, Baskerville J, Ebers GC (1999b) The natural history of multiple sclerosis: a geographically based study. 5. The clinical features and natural history of primary progressive multiple sclerosis. Brain 122(Pt 4):625–639

Cua DJ, Sherlock J, Chen Y, Murphy CA, Joyce B, Seymour B, Lucian L, To W, Kwan S, Churakova T, Zurawski S, Wiekowski M, Lira SA, Gorman D, Kastelein RA, Sedgwick JD (2003) Interleukin-23 rather than interleukin-12 is the critical cytokine for autoimmune inflammation of the brain. Nature 421:744–748

Dai Y, Carayanniotis KA, Eliades P, Lymberi P, Shepherd P, Kong Y, Carayanniotis G (1999) Enhancing or suppressive effects of antibodies on processing of a pathogenic T cell epitope in thyroglobulin. J Immunol 162:6987–6992

Dalton DK, Haynes L, Chu CQ, Swain SL, Wittmer S (2000) Interferon gamma eliminates responding CD4 T cells during mycobacterial infection by inducing apoptosis of activated CD4 T cells. J Exp Med 192:117–122

de Graaf KL, Barth S, Herrmann MM, Storch MK, Wiesmuller KH, Weissert R (2008) Characterization of the encephalitogenic immune response in a model of multiple sclerosis. Eur J Immunol 38:299–308

Deaglio S, Dwyer KM, Gao W, Friedman D, Usheva A, Erat A, Chen JF, Enjyoji K, Linden J, Oukka M, Kuchroo VK, Strom TB, Robson SC (2007) Adenosine generation catalyzed by CD39 and CD73 expressed on regulatory T cells mediates immune suppression. J Exp Med 204:1257–1265

Dijkstra CD, De Groot CJ, Huitinga I (1992) The role of macrophages in demyelination. J Neuroimmunol 40:183–188

Du J, Huang C, Zhou B, Ziegler SF (2008) Isoform-Specific Inhibition of ROR{alpha}-Mediated Transcriptional Activation by Human FOXP3. J Immunol 180:4785–4792

Duerr RH, Taylor KD, Brant SR, Rioux JD, Silverberg MS, Daly MJ, Steinhart AH, Abraham C, Regueiro M, Griffiths A, Dassopoulos T, Bitton A, Yang H, Targan S, Datta LW, Kistner EO, Schumm LP, Lee AT, Gregersen PK, Barmada MM, Rotter JI, Nicolae DL, Cho JH (2006) A genome-wide association study identifies IL23R as an inflammatory bowel disease gene. Science 314:1461–1463

Ebers GC, Koopman WJ, Hader W, Sadovnick AD, Kremenchutzky M, Mandalfino P, Wingerchuk DM, Baskerville J, Rice GP (2000) The natural history of multiple sclerosis: a geographically based study: 8: familial multiple sclerosis. Brain 123(Pt 3):641–649

Endoh M, Tabira T, Kunishita T (1986) Antibodies to proteolipid apoprotein in chronic relapsing experimental allergic encephalomyelitis. J Neurol Sci 73:31–38

Epstein MM, Di Rosa F, Jankovic D, Sher A, Matzinger P (1995) Successful T cell priming in B cell-deficient mice. J Exp Med 182:915–922

Eynon EE, Parker DC (1992) Small B cells as antigen-presenting cells in the induction of tolerance to soluble protein antigens. J Exp Med 175:131–138

Farago B, Magyari L, Safrany E, Csongei V, Jaromi L, Horvatovich K, Sipeky C, Maasz A, Radics J, Gyetvai A, Szekanecz Z, Czirjak L, Melegh B (2008) Functional variants of interleukin-23 receptor gene confer risk for rheumatoid arthritis but not for systemic sclerosis. Ann Rheum Dis 67:248–250

Ferber IA, Brocke S, Taylor-Edwards C, Ridgway W, Dinisco C, Steinman L, Dalton D, Fathman CG (1996) Mice with a disrupted IFN-gamma gene are susceptible to the induction of experimental autoimmune encephalomyelitis (EAE). J Immunol 156:5–7

Fillatreau S, Sweenie CH, McGeachy MJ, Gray D, Anderton SM (2002) B cells regulate autoimmunity by provision of IL-10. Nat Immunol 3:944–950

Flugel A, Berkowicz T, Ritter T, Labeur M, Jenne DE, Li Z, Ellwart JW, Willem M, Lassmann H, Wekerle H (2001) Migratory activity and functional changes of green fluorescent effector cells before and during experimental autoimmune encephalomyelitis. Immunity 14:547–560

Fontenot JD, Rasmussen JP, Williams LM, Dooley JL, Farr AG, Rudensky AY (2005) Regulatory T cell lineage specification by the forkhead transcription factor foxp3. Immunity 22:329–341

Ford ML, Evavold BD (2005) Specificity, magnitude, and kinetics of MOG-specific CD8+ T cell responses during experimental autoimmune encephalomyelitis. Eur J Immunol 35:76–85

Fuchs EJ, Matzinger P (1992) B cells turn off virgin but not memory T cells. Science 258:1156–1159

Gardinier MV, Amiguet P, Linington C, Matthieu JM (1992) Myelin/oligodendrocyte glycoprotein is a unique member of the immunoglobulin superfamily. J Neurosci Res 33:177–187

Gately MK, Renzetti LM, Magram J, Stern AS, Adorini L, Gubler U, Presky DH (1998) The interleukin-12/interleukin-12-receptor system: role in normal and pathologic immune responses. Annu Rev Immunol 16:495–521

Gavin MA, Rasmussen JP, Fontenot JD, Vasta V, Manganiello VC, Beavo JA, Rudensky AY (2007) Foxp3-dependent programme of regulatory T cell differentiation. Nature 445:771–775

Genain CP, Nguyen MH, Letvin NL, Pearl R, Davis RL, Adelman M, Lees MB, Linington C, Hauser SL (1995) Antibody facilitation of multiple sclerosis-like lesions in a nonhuman primate. J Clin Invest 96:2966–2974

Glimcher LH (2001) Lineage commitment in lymphocytes: controlling the immune response. J Clin Invest 108:s25–s30

Gold R, Linington C, Lassmann H (2006) Understanding pathogenesis and therapy of multiple sclerosis via animal models: 70 years of merits and culprits in experimental autoimmune encephalomyelitis research. Brain 129:1953–1971

Greter M, Heppner FL, Lemos MP, Odermatt BM, Goebels N, Laufer T, Noelle RJ, Becher B (2005) Dendritic cells permit immune invasion of the CNS in an animal model of multiple sclerosis. Nat Med 11:328–334

Grundke-Iqbal I, Raine CS, Johnson AB, Brosnan CF, Bornstein MB (1981) Experimental allergic encephalomyelitis. Characterization of serum factors causing demyelination and swelling of myelin. J Neurol Sci 50:63–79

Gutcher I, Urich E, Wolter K, Prinz M, Becher B (2006) Interleukin 18-independent engagement of interleukin 18 receptor-alpha is required for autoimmune inflammation. Nat Immunol 7:946–953

Happ MP, Heber-Katz E (1988) Differences in the repertoire of the Lewis rat T cell response to self and non-self myelin basic proteins. J Exp Med 167:502–513

Hayglass KT, Naides SJ, Scott CF, Jr., Benacerraf B, Sy MS (1986) T cell development in B cell-deficient mice. IV. The role of B cells as antigen-presenting cells in vivo. J Immunol 136:823–829

Hickey WF, Kimura H (1988) Perivascular microglial cells of the CNS are bone marrow-derived and present antigen in vivo. Science 239:290–292

Hjelmstrom P, Juedes AE, Fjell J, Ruddle NH (1998) B cell-deficient mice develop experimental allergic encephalomyelitis with demyelination after myelin oligodendrocyte glycoprotein sensitization. J Immunol 161:4480–4483

Hori S, Haury M, Coutinho A, Demengeot J (2002) Specificity requirements for selection and effector functions of CD25+ 4+ regulatory T cells in anti-myelin basic protein T cell receptor transgenic mice. Proc Natl Acad Sci U S A 99:8213–8218

Howell MD, Winters ST, Olee T, Powell HC, Carlo DJ, Brostoff SW (1989) Vaccination against experimental allergic encephalomyelitis with T cell receptor peptides. Science 246:668–670

Huseby ES, Liggitt D, Brabb T, Schnabel B, Ohlen C, Goverman J (2001) A pathogenic role for myelin-specific CD8(+) T cells in a model for multiple sclerosis. J Exp Med 194:669–676

Hymowitz SG, Filvaroff EH, Yin JP, Lee J, Cai L, Risser P, Maruoka M, Mao W, Foster J, Kelley RF, Pan G, Gurney AL, de Vos AM, Starovasnik MA (2001) IL-17s adopt a cystine knot fold: structure and activity of a novel cytokine, IL-17F, and implications for receptor binding. EMBO J 20:5332–5341

Illes Z, Safrany E, Peterfalvi A, Magyari L, Farago B, Pozsonyi E, Rozsa C, Komoly S, Melegh B (2008) 3′YTR C2370A allele of the IL-23 receptor gene is associated with relapsing-remitting multiple sclerosis. Neurosci Lett 431:36–38

Issazadeh S, Mustafa M, Ljungdahl A, Hojeberg B, Dagerlind A, Elde R, Olsson T (1995) Interferon gamma, interleukin 4 and transforming growth factor beta in experimental autoimmune encephalomyelitis in Lewis rats: dynamics of cellular mRNA expression in the central nervous system and lymphoid cells. J Neurosci Res 40:579–590

Ivanov, II, McKenzie BS, Zhou L, Tadokoro CE, Lepelley A, Lafaille JJ, Cua DJ, Littman DR (2006) The Orphan Nuclear Receptor RORgammat Directs the Differentiation Program of Proinflammatory IL-17(+) T Helper Cells. Cell 126:1121–1133

Izcue A, Hue S, Buonocore S, Arancibia-Carcamo CV, Ahern PP, Iwakura Y, Maloy KJ, Powrie F (2008) Interleukin-23 restrains regulatory T cell activity to drive T cell-dependent colitis. Immunity 28:559–570

Janeway CA, Jr., Ron J, Katz ME (1987) The B cell is the initiating antigen-presenting cell in peripheral lymph nodes. J Immunol 138:1051–1055

Jung S, Schluesener HJ, Toyka KV, Hartung HP (1993) Modulation of EAE by vaccination with T cell receptor peptides: V beta 8 T cell receptor peptide-specific CD4+ lymphocytes lack direct immunoregulatory activity. J Neuroimmunol 45:15–22

Kabat EA, Wolf A, Bezer AE (1946) Rapid Production of Acute Disseminated Encephalomyelitis in Rhesus Monkeys by Injection of Brain Tissue With Adjuvants. Science 104:362–363

Kappos L, Traboulsee A, Constantinescu C, Eralinna JP, Forrestal F, Jongen P, Pollard J, Sandberg-Wollheim M, Sindic C, Stubinski B, Uitdehaag B, Li D (2006) Long-term subcutaneous interferon beta-1a therapy in patients with relapsing-remitting MS. Neurology 67:944–953

Kawakami N, Nagerl UV, Odoardi F, Bonhoeffer T, Wekerle H, Flugel A (2005) Live imaging of effector cell trafficking and autoantigen recognition within the unfolding autoimmune encephalomyelitis lesion. J Exp Med 201:1805–1814

Kebir H, Kreymborg K, Ifergan I, Dodelet-Devillers A, Cayrol R, Bernard M, Giuliani F, Arbour N, Becher B, Prat A (2007) Human T(H)17 lymphocytes promote blood-brain barrier disruption and central nervous system inflammation. Nat Med 13:1173–1175

Khader SA, Bell GK, Pearl JE, Fountain JJ, Rangel-Moreno J, Cilley GE, Shen F, Eaton SM, Gaffen SL, Swain SL, Locksley RM, Haynes L, Randall TD, Cooper AM (2007) IL-23 and IL-17 in the establishment of protective pulmonary CD4+ T cell responses after vaccination and during Mycobacterium tuberculosis challenge. Nat Immunol 8:369–377

Kim KJ, Rollwagen F, Asofsky R, Lefkovits I (1984) The abnormal function of T cells in chronically anti-mu-treated mice with no mature B lymphocytes. Eur J Immunol 14:476–482

Kleinschek MA, Owyang AM, Joyce-Shaikh B, Langrish CL, Chen Y, Gorman DM, Blumenschein WM, McClanahan T, Brombacher F, Hurst SD, Kastelein RA, Cua DJ (2007) IL-25 regulates Th17 function in autoimmune inflammation. J Exp Med 204:161–170

Kohm AP, Carpentier PA, Anger HA, Miller SD (2002) Cutting edge: CD4+ CD25+ regulatory T cells suppress antigen-specific autoreactive immune responses and central nervous system inflammation during active experimental autoimmune encephalomyelitis. J Immunol 169:4712–4716

Kojima K, Berger T, Lassmann H, Hinze-Selch D, Zhang Y, Gehrmann J, Reske K, Wekerle H, Linington C (1994) Experimental autoimmune panencephalitis and uveoretinitis transferred to the Lewis rat by T lymphocytes specific for the S100 beta molecule, a calcium binding protein of astroglia. J Exp Med 180:817–829

Kolls JK, Linden A (2004) Interleukin-17 family members and inflammation. Immunity 21:467–476

Komiyama Y, Nakae S, Matsuki T, Nambu A, Ishigame H, Kakuta S, Sudo K, Iwakura Y (2006) IL-17 plays an important role in the development of experimental autoimmune encephalomy-elitis. J Immunol 177:566–573

Kono DH, Urban JL, Horvath SJ, Ando DG, Saavedra RA, Hood L (1988) Two minor determinants of myelin basic protein induce experimental allergic encephalomyelitis in SJL/J mice. J Exp Med 168:213–227

Korn T, Bettelli E, Gao W, Awasthi A, Jager A, Strom TB, Oukka M, Kuchroo VK (2007a) IL-21 initiates an alternative pathway to induce proinflammatory T(H)17 cells. Nature 448:484–487

Korn T, Oukka M, Kuchroo V, Bettelli E (2007b) Th17 cells: Effector T cells with inflammatory properties. Semin Immunol 19:362–371

Korn T, Reddy J, Gao W, Bettelli E, Awasthi A, Petersen TR, Backstrom BT, Sobel RA, Wucherpfennig KW, Strom TB, Oukka M, Kuchroo VK (2007c) Myelin-specific regulatory T cells accumulate in the CNS but fail to control autoimmune inflammation. Nat Med 13:423–431

Kremenchutzky M, Cottrell D, Rice G, Hader W, Baskerville J, Koopman W, Ebers GC (1999) The natural history of multiple sclerosis: a geographically based study. 7. Progressive-relapsing and relapsing-progressive multiple sclerosis: a re-evaluation. Brain 122 (Pt 10):1941–1950

Kremenchutzky M, Rice GP, Baskerville J, Wingerchuk DM, Ebers GC (2006) The natural history of multiple sclerosis: a geographically based study 9: observations on the progressive phase of the disease. Brain 129:584–594

Kreymborg K, Etzensperger R, Dumoutier L, Haak S, Rebollo A, Buch T, Heppner FL, Renauld JC, Becher B (2007) IL-22 is expressed by Th17 cells in an IL-23-dependent fashion, but not required for the development of autoimmune encephalomyelitis. J Immunol 179:8098–8104

Krishnamoorthy G, Lassmann H, Wekerle H, Holz A (2006) Spontaneous opticospinal encephalomyelitis in a double-transgenic mouse model of autoimmune T cell/B cell cooperation. J Clin Invest 116:2385–2392

Kuchroo VK, Martin CA, Greer JM, Ju ST, Sobel RA, Dorf ME (1993) Cytokines and adhesion molecules contribute to the ability of myelin proteolipid protein-specific T cell clones to mediate experimental allergic encephalomyelitis. J Immunol 151:4371–4382

Kuchroo VK, Anderson AC, Waldner H, Munder M, Bettelli E, Nicholson LB (2002) T cell response in experimental autoimmune encephalomyelitis (EAE): role of self and cross-reactive antigens in shaping, tuning, and regulating the autopathogenic T cell repertoire. Annu Rev Immunol 20:101–123

Kuhle J, Pohl C, Mehling M, Edan G, Freedman MS, Hartung HP, Polman CH, Miller DH, Montalban X, Barkhof F, Bauer L, Dahms S, Lindberg R, Kappos L, Sandbrink R (2007) Lack of association between antimyelin antibodies and progression to multiple sclerosis. N Engl J Med 356:371–378

Kurt-Jones EA, Liano D, HayGlass KA, Benacerraf B, Sy MS, Abbas AK (1988) The role of antigen-presenting B cells in T cell priming in vivo. Studies of B cell-deficient mice. J Immunol 140:3773–3778

Lafaille JJ, Keere FV, Hsu AL, Baron JL, Haas W, Raine CS, Tonegawa S (1997) Myelin basic protein-specific T helper 2 (Th2) cells cause experimental autoimmune encephalomyelitis in immunodeficient hosts rather than protect them from the disease. J Exp Med 186:307–312

Langrish CL, Chen Y, Blumenschein WM, Mattson J, Basham B, Sedgwick JD, McClanahan T, Kastelein RA, Cua DJ (2005) IL-23 drives a pathogenic T cell population that induces autoimmune inflammation. J Exp Med 201:233–240

Lanzavecchia A (1990) Receptor-mediated antigen uptake and its effect on antigen presentation to class II-restricted T lymphocytes. Annu Rev Immunol 8:773–793

Lassila O, Vainio O, Matzinger P (1988) Can B cells turn on virgin T cells? Nature 334:253–255

Lassmann H, Brunner C, Bradl M, Linington C (1988) Experimental allergic encephalomyelitis: the balance between encephalitogenic T lymphocytes and demyelinating antibodies determines size and structure of demyelinated lesions. Acta Neuropathol 75:566–576

Laurence A, Tato CM, Davidson TS, Kanno Y, Chen Z, Yao Z, Blank RB, Meylan F, Siegel R, Hennighausen L, Shevach EM, O'Shea J J (2007) Interleukin-2 signaling via STAT5 constrains T helper 17 cell generation. Immunity 26:371–381

Lehmann PV, Sercarz EE, Forsthuber T, Dayan CM, Gammon G (1993) Determinant spreading and the dynamics of the autoimmune T cell repertoire. Immunol Today 14:203–208

Leonard JP, Waldburger KE, Goldman SJ (1995) Prevention of experimental autoimmune encephalomyelitis by antibodies against interleukin 12. J Exp Med 181:381–386

Li MO, Wan YY, Sanjabi S, Robertson AK, Flavell RA (2006) Transforming growth factor-beta regulation of immune responses. Annu Rev Immunol 24:99–146

Li MO, Wan YY, Flavell RA (2007) T Cell-Produced Transforming Growth Factor-beta1 Controls T Cell Tolerance and Regulates Th1- and Th17-Cell Differentiation. Immunity 26:579–591

Liang SC, Tan XY, Luxenberg DP, Karim R, Dunussi-Joannopoulos K, Collins M, Fouser LA (2006) Interleukin (IL)-22 and IL-17 are coexpressed by Th17 cells and cooperatively enhance expression of antimicrobial peptides. J Exp Med 203:2271–2279

Linington C, Bradl M, Lassmann H, Brunner C, Vass K (1988) Augmentation of demyelination in rat acute allergic encephalomyelitis by circulating mouse monoclonal antibodies directed against a myelin/oligodendrocyte glycoprotein. Am J Pathol 130:443–454

Litzenburger T, Fassler R, Bauer J, Lassmann H, Linington C, Wekerle H, Iglesias A (1998) B lymphocytes producing demyelinating autoantibodies: development and function in gene-targeted transgenic mice. J Exp Med 188:169–180

Liu Y, Helms C, Liao W, Zaba LC, Duan S, Gardner J, Wise C, Miner A, Malloy MJ, Pullinger CR, Kane JP, Saccone S, Worthington J, Bruce I, Kwok PY, Menter A, Krueger J, Barton A, Saccone NL, Bowcock AM (2008) A genome-wide association study of psoriasis and psoriatic arthritis identifies new disease Loci. PLoS Genet 4:e1000041

Lock C, Hermans G, Pedotti R, Brendolan A, Schadt E, Garren H, Langer-Gould A, Strober S, Cannella B, Allard J, Klonowski P, Austin A, Lad N, Kaminski N, Galli SJ, Oksenberg JR, Raine CS, Heller R, Steinman L (2002) Gene-microarray analysis of multiple sclerosis lesions yields new targets validated in autoimmune encephalomyelitis. Nat Med 8:500–508

Lohr J, Knoechel B, Wang JJ, Villarino AV, Abbas AK (2006) Role of IL-17 and regulatory T lymphocytes in a systemic autoimmune disease. J Exp Med 203:2785–2791

Lucchinetti C, Bruck W, Parisi J, Scheithauer B, Rodriguez M, Lassmann H (2000) Heterogeneity of multiple sclerosis lesions: implications for the pathogenesis of demyelination. Ann Neurol 47:707–717

Lyons JA, San M, Happ MP, Cross AH (1999) B cells are critical to induction of experimental allergic encephalomyelitis by protein but not by a short encephalitogenic peptide. Eur J Immunol 29:3432–3439

Lyons JA, Ramsbottom MJ, Cross AH (2002) Critical role of antigen-specific antibody in experimental autoimmune encephalomyelitis induced by recombinant myelin oligodendrocyte glycoprotein. Eur J Immunol 32:1905–1913

Manel N, Unutmaz D, Littman DR (2008) The differentiation of human T(H)-17 cells requires transforming growth factor-beta and induction of the nuclear receptor RORgammat. Nat Immunol 9:641–649

Mangan PR, Harrington LE, O'Quinn DB, Helms WS, Bullard DC, Elson CO, Hatton RD, Wahl SM, Schoeb TR, Weaver CT (2006) Transforming growth factor-beta induces development of the T(H)17 lineage. Nature 441:231–234

Mann MK, Maresz K, Shriver LP, Tan Y, Dittel BN (2007) B cell regulation of CD4+ CD25+ T regulatory cells and IL-10 via B7 is essential for recovery from experimental autoimmune encephalomyelitis. J Immunol 178:3447–3456

Martin R, McFarland HF (1995) Immunological aspects of experimental allergic encephalomyelitis and multiple sclerosis. Crit Rev Clin Lab Sci 32:121–182

Mathey EK, Derfuss T, Storch MK, Williams KR, Hales K, Woolley DR, Al-Hayani A, Davies SN, Rasband MN, Olsson T, Moldenhauer A, Velhin S, Hohlfeld R, Meinl E, Linington C (2007) Neurofascin as a novel target for autoantibody-mediated axonal injury. J Exp Med 204:2363–2372

McDonald WI, Compston A, Edan G, Goodkin D, Hartung HP, Lublin FD, McFarland HF, Paty DW, Polman CH, Reingold SC, Sandberg-Wollheim M, Sibley W, Thompson A, van den Noort S, Weinshenker BY, Wolinsky JS (2001) Recommended diagnostic criteria for multiple sclerosis: guidelines from the International Panel on the diagnosis of multiple sclerosis. Ann Neurol 50:121–127

McGeachy MJ, Bak-Jensen KS, Chen Y, Tato CM, Blumenschein W, McClanahan T, Cua DJ (2007) TGF-beta and IL-6 drive the production of IL-17 and IL-10 by T cells and restrain T(H)-17 cell-mediated pathology. Nat Immunol 8:1390–1397

McMahon EJ, Bailey SL, Castenada CV, Waldner H, Miller SD (2005) Epitope spreading initiates in the CNS in two mouse models of multiple sclerosis. Nat Med 11:335–339

McRae BL, Kennedy MK, Tan LJ, Dal Canto MC, Picha KS, Miller SD (1992) Induction of active and adoptive relapsing experimental autoimmune encephalomyelitis (EAE) using an encephalitogenic epitope of proteolipid protein. J Neuroimmunol 38:229–240

McRae BL, Vanderlugt CL, Dal Canto MC, Miller SD (1995) Functional evidence for epitope spreading in the relapsing pathology of experimental autoimmune encephalomyelitis. J Exp Med 182:75–85

Mendel I, Kerlero de Rosbo N, Ben-Nun A (1995) A myelin oligodendrocyte glycoprotein peptide induces typical chronic experimental autoimmune encephalomyelitis in H-2b mice: fine specificity and T cell receptor V beta expression of encephalitogenic T cells. Eur J Immunol 25:1951–1959

Mendel Kerlero de Rosbo N, Ben-Nun A (1996) Delineation of the minimal encephalitogenic epitope within the immunodominant region of myelin oligodendrocyte glycoprotein: diverse V beta gene usage by T cells recognizing the core epitope encephalitogenic for T cell receptor V beta b and T cell receptor V beta a H-2b mice. Eur J Immunol 26:2470–2479

Miller SD, Vanderlugt CL, Begolka WS, Pao W, Yauch RL, Neville KL, Katz-Levy Y, Carrizosa A, Kim BS (1997) Persistent infection with Theiler's virus leads to CNS autoimmunity via epitope spreading. Nat Med 3:1133–1136

Moore GR, Raine CS (1988) Immunogold localization and analysis of IgG during immune-mediated demyelination. Lab Invest 59:641–648

Moore KW, de Waal Malefyt R, Coffman RL, O'Garra A (2001) Interleukin-10 and the interleukin-10 receptor. Annu Rev Immunol 19:683–765

Morris SC, Lees A, Finkelman FD (1994) In vivo activation of naive T cells by antigen-presenting B cells. J Immunol 152:3777–3785

Mosmann TR, Coffman RL (1989) TH1 and TH2 cells: different patterns of lymphokine secretion lead to different functional properties. Annu Rev Immunol 7:145–173

Mucida D, Park Y, Kim G, Turovskaya O, Scott I, Kronenberg M, Cheroutre H (2007) Reciprocal TH17 and regulatory T cell differentiation mediated by retinoic acid. Science 317:256–260

Myers KJ, Sprent J, Dougherty JP, Ron Y (1992) Synergy between encephalitogenic T cells and myelin basic protein-specific antibodies in the induction of experimental autoimmune encephalomyelitis. J Neuroimmunol 41:1–8

Nurieva R, Yang XO, Martinez G, Zhang Y, Panopoulos AD, Ma L, Schluns K, Tian Q, Watowich SS, Jetten AM, Dong C (2007) Essential autocrine regulation by IL-21 in the generation of inflammatory T cells. Nature 448:480–483

Oksenberg JR, Panzara MA, Begovich AB, Mitchell D, Erlich HA, Murray RS, Shimonkevitz R, Sherritt M, Rothbard J, Bernard CC et-al. (1993) Selection for T cell receptor V beta-D beta-J beta gene rearrangements with specificity for a myelin basic protein peptide in brain lesions of multiple sclerosis. Nature 362:68–70

Oliver AR, Lyon GM, Ruddle NH (2003) Rat and human myelin oligodendrocyte glycoproteins induce experimental autoimmune encephalomyelitis by different mechanisms in C57BL/6 mice. J Immunol 171:462–468

Olsson T (1992) Cytokines in neuroinflammatory disease: role of myelin autoreactive T cell production of interferon-gamma. J Neuroimmunol 40:211–218

Oppmann B, Lesley R, Blom B, Timans JC, Xu Y, Hunte B, Vega F, Yu N, Wang J, Singh K, Zonin F, Vaisberg E, Churakova T, Liu M, Gorman D, Wagner J, Zurawski S, Liu Y, Abrams JS, Moore KW, Rennick D, de Waal-Malefyt R, Hannum C, Bazan JF, Kastelein RA (2000) Novel p19 protein engages IL-12p40 to form a cytokine, IL-23, with biological activities similar as well as distinct from IL-12. Immunity 13:715–725

Ousman SS, Tomooka BH, van Noort JM, Wawrousek EF, O'Connor KC, Hafler DA, Sobel RA, Robinson WH, Steinman L (2007) Protective and therapeutic role for alphaB-crystallin in autoimmune demyelination. Nature 448:474–479

Ouyang W, Kolls JK, Zheng Y (2008) The biological functions of T helper 17 cell effector cytokines in inflammation. Immunity 28:454–467

Owens GP, Kraus H, Burgoon MP, Smith-Jensen T, Devlin ME, Gilden DH (1998) Restricted use of VH4 germline segments in an acute multiple sclerosis brain. Ann Neurol 43:236–243

Park H, Li Z, Yang XO, Chang SH, Nurieva R, Wang YH, Wang Y, Hood L, Zhu Z, Tian Q, Dong C (2005) A distinct lineage of CD4 T cells regulates tissue inflammation by producing interleukin 17. Nat Immunol 6:1133–1141

Phillips JA, Romball CG, Hobbs MV, Ernst DN, Shultz L, Weigle WO (1996) CD4+ T cell activation and tolerance induction in B cell knockout mice. J Exp Med 183:1339–1344

Piddlesden S, Lassmann H, Laffafian I, Morgan BP, Linington C (1991) Antibody-mediated demyelination in experimental allergic encephalomyelitis is independent of complement membrane attack complex formation. Clin Exp Immunol 83:245–250

Piddlesden SJ, Lassmann H, Zimprich F, Morgan BP, Linington C (1993) The demyelinating potential of antibodies to myelin oligodendrocyte glycoprotein is related to their ability to fix complement. Am J Pathol 143:555–564

Pierce SK, Morris JF, Grusby MJ, Kaumaya P, van Buskirk A, Srinivasan M, Crump B, Smolenski LA (1988) Antigen-presenting function of B lymphocytes. Immunol Rev 106:149–180

Pittock SJ, McClelland RL, Mayr WT, Jorgensen NW, Weinshenker BG, Noseworthy J, Rodriguez M (2004) Clinical implications of benign multiple sclerosis: a 20-year population-based follow-up study. Ann Neurol 56:303–306

Probert L, Eugster HP, Akassoglou K, Bauer J, Frei K, Lassmann H, Fontana A (2000) TNFR1 signalling is critical for the development of demyelination and the limitation of T cell responses during immune-mediated CNS disease. Brain 123(Pt 10):2005–2019

Qin Y, Duquette P, Zhang Y, Talbot P, Poole R, Antel J (1998) Clonal expansion and somatic hypermutation of V(H) genes of B cells from cerebrospinal fluid in multiple sclerosis. J Clin Invest 102:1045–1050

Reddy J, Illes Z, Zhang X, Encinas J, Pyrdol J, Nicholson L, Sobel RA, Wucherpfennig KW, Kuchroo VK (2004) Myelin proteolipid protein-specific CD4+ CD25+ regulatory cells mediate genetic resistance to experimental autoimmune encephalomyelitis. Proc Natl Acad Sci U S A 101:15434–15439

Rivers TM, Schwentker FF (1935) Encephalomyelitis accompanied by myelin destruction experimentally produced in monkeys. J Exp Med 61:689–702

Ron Y, De Baetselier P, Gordon J, Feldman M, Segal S (1981) Defective induction of antigen-reactive proliferating T cells in B cell-deprived mice. Eur J Immunol 11:964–968

Ronchese F, Hausmann B (1993) B lymphocytes in vivo fail to prime naive T cells but can stimulate antigen-experienced T lymphocytes. J Exp Med 177:679–690

Rubtsov YP, Rasmussen JP, Chi EY, Fontenot J, Castelli L, Ye X, Treuting P, Siewe L, Roers A, Henderson WR, Jr., Muller W, Rudensky AY (2008) Regulatory T cell-derived interleukin-10 limits inflammation at environmental interfaces. Immunity 28:546–558

Russell WL, Russell LB, Gower JS (1959) Exceptional Inheritance of a Sex-Linked Gene in the Mouse Explained on the Basis That the X/O Sex-Chromosome Constitution Is Female. Proc Natl Acad Sci U S A 45:554–560

Sakaguchi S (2005) Naturally arising Foxp3-expressing CD25+ CD4+ regulatory T cells in immunological tolerance to self and non-self. Nat Immunol 6:345–352

Sato W, Aranami T, Yamamura T (2007) Cutting edge: Human Th17 cells are identified as bearing CCR2+ CCR5- phenotype. J Immunol 178:7525–7529

Schluesener HJ, Sobel RA, Linington C, Weiner HL (1987) A monoclonal antibody against a myelin oligodendrocyte glycoprotein induces relapses and demyelination in central nervous system autoimmune disease. J Immunol 139:4016–4021

Schubert LA, Jeffery E, Zhang Y, Ramsdell F, Ziegler SF (2001) Scurfin (FOXP3) acts as a repressor of transcription and regulates T cell activation. J Biol Chem 276:37672–37679

Sercarz EE, Lehmann PV, Ametani A, Benichou G, Miller A, Moudgil K (1993) Dominance and crypticity of T cell antigenic determinants. Annu Rev Immunol 11:729–766

Simitsek PD, Campbell DG, Lanzavecchia A, Fairweather N, Watts C (1995) Modulation of antigen processing by bound antibodies can boost or suppress class II major histocompatibility complex presentation of different T cell determinants. J Exp Med 181:1957–1963

Sospedra M, Martin R (2005) Immunology of multiple sclerosis. Annu Rev Immunol 23:683–747

Storch MK, Piddlesden S, Haltia M, Iivanainen M, Morgan P, Lassmann H (1998) Multiple sclerosis: in situ evidence for antibody- and complement-mediated demyelination. Ann Neurol 43:465–471

Stuart G, Krikorian KS (1928) The neuro-paralytic accidents of anti-rabies treatment. Ann Trop Med Parasitol 22:327–377

Stumhofer JS, Laurence A, Wilson EH, Huang E, Tato CM, Johnson LM, Villarino AV, Huang Q, Yoshimura A, Sehy D, Saris CJ, O'Shea JJ, Hennighausen L, Ernst M, Hunter CA (2006) Interleukin 27 negatively regulates the development of interleukin 17-producing T helper cells during chronic inflammation of the central nervous system. Nat Immunol 7:937–945

Stumhofer JS, Silver JS, Laurence A, Porrett PM, Harris TH, Turka LA, Ernst M, Saris CJ, O'Shea JJ, Hunter CA (2007) Interleukins 27 and 6 induce STAT3-mediated T cell production of interleukin 10. Nat Immunol 8:1363–1371

Sun D, Whitaker JN, Huang Z, Liu D, Coleclough C, Wekerle H, Raine CS (2001) Myelin antigen-specific CD8+ T cells are encephalitogenic and produce severe disease in C57BL/6 mice. J Immunol 166:7579–7587

Svensson L, Abdul-Majid KB, Bauer J, Lassmann H, Harris RA, Holmdahl R (2002) A comparative analysis of B cell-mediated myelin oligodendrocyte glycoprotein-experimental autoimmune encephalomyelitis pathogenesis in B cell-deficient mice reveals an effect on demyelination. Eur J Immunol 32:1939–1946

Szabo SJ, Kim ST, Costa GL, Zhang X, Fathman CG, Glimcher LH (2000) A novel transcription factor, T-bet, directs Th1 lineage commitment. Cell 100:655–669

Tabira T, Endoh M (1985) Humoral immune responses to myelin basic protein, cerebroside and ganglioside in chronic relapsing experimental allergic encephalomyelitis of the guinea pig. J Neurol Sci 67:201–212

Tang Q, Adams JY, Tooley AJ, Bi M, Fife BT, Serra P, Santamaria P, Locksley RM, Krummel MF, Bluestone JA (2006) Visualizing regulatory T cell control of autoimmune responses in nonobese diabetic mice. Nat Immunol 7:83–92

Thompson EJ, Kaufmann P, Shortman RC, Rudge P, McDonald WI (1979) Oligoclonal immunoglobulins and plasma cells in spinal fluid of patients with multiple sclerosis. Br Med J 1:16–17

Traugott U, Lebon P (1988a) Multiple sclerosis: involvement of interferons in lesion pathogenesis. Ann Neurol 24:243–251

Traugott U, Lebon P (1988b) Interferon-gamma and Ia antigen are present on astrocytes in active chronic multiple sclerosis lesions. J Neurol Sci 84:257–264

Travis MA, Reizis B, Melton AC, Masteller E, Tang Q, Proctor JM, Wang Y, Bernstein X, Huang X, Reichardt LF, Bluestone JA, Sheppard D (2007) Loss of integrin alpha(v)beta8 on dendritic cells causes autoimmunity and colitis in mice. Nature 449:361–365

Tripp CS, Gately MK, Hakimi J, Ling P, Unanue ER (1994) Neutralization of IL-12 decreases resistance to Listeria in SCID and C.B-17 mice. Reversal by IFN-gamma. J Immunol 152:1883–1887

Tuohy VK, Lu Z, Sobel RA, Laursen RA, Lees MB (1989) Identification of an encephalitogenic determinant of myelin proteolipid protein for SJL mice. J Immunol 142:1523–1527

Tzartos JS, Friese MA, Craner MJ, Palace J, Newcombe J, Esiri MM, Fugger L (2008) Interleukin-17 production in central nervous system-infiltrating T cells and glial cells is associated with active disease in multiple sclerosis. Am J Pathol 172:146–155

Van der Goes A, Kortekaas M, Hoekstra K, Dijkstra CD, Amor S (1999) The role of anti-myelin (auto)-antibodies in the phagocytosis of myelin by macrophages. J Neuroimmunol 101:61–67

Van der Veen RC, Sobel RA, Lees MB (1986) Chronic experimental allergic encephalomyelitis and antibody responses in rabbits immunized with bovine proteolipid apoprotein. J Neuroimmunol 11:321–333

Vandenbark AA, Chou YK, Whitham R, Mass M, Buenafe A, Liefeld D, Kavanagh D, Cooper S, Hashim GA, Offner H (1996) Treatment of multiple sclerosis with T cell receptor peptides: results of a double-blind pilot trial. Nat Med 2:1109–1115

Veldhoen M, Hocking RJ, Atkins CJ, Locksley RM, Stockinger B (2006a) TGFbeta in the context of an inflammatory cytokine milieu supports de novo differentiation of IL-17-producing T cells. Immunity 24:179–189

Veldhoen M, Hocking RJ, Flavell RA, Stockinger B (2006b) Signals mediated by transforming growth factor-beta initiate autoimmune encephalomyelitis, but chronic inflammation is needed to sustain disease. Nat Immunol 7:1151–1156

Volpe E, Servant N, Zollinger R, Bogiatzi SI, Hupe P, Barillot E, Soumelis V (2008) A critical function for transforming growth factor-beta, interleukin 23 and proinflammatory cytokines in driving and modulating human T(H)-17 responses. Nat Immunol 9:650–657

Wang LY, Fujinami RS (1997) Enhancement of EAE and induction of autoantibodies to T cell epitopes in mice infected with a recombinant vaccinia virus encoding myelin proteolipid protein. J Neuroimmunol 75:75–83

Weinshenker BG, Bass B, Rice GP, Noseworthy J, Carriere W, Baskerville J, Ebers GC (1989a) The natural history of multiple sclerosis: a geographically based study. I. Clinical course and disability. Brain 112(Pt 1):133–146

Weinshenker BG, Bass B, Rice GP, Noseworthy J, Carriere W, Baskerville J, Ebers GC (1989b) The natural history of multiple sclerosis: a geographically based study. 2. Predictive value of the early clinical course. Brain 112(Pt 6):1419–1428

Weinshenker BG, Rice GP, Noseworthy JH, Carriere W, Baskerville J, Ebers GC (1991a) The natural history of multiple sclerosis: a geographically based study. 3. Multivariate analysis of predictive factors and models of outcome. Brain 114(Pt 2):1045–1056

Weinshenker BG, Rice GP, Noseworthy JH, Carriere W, Baskerville J, Ebers GC (1991b) The natural history of multiple sclerosis: a geographically based study. 4. Applications to planning and interpretation of clinical therapeutic trials. Brain 114(Pt 2):1057–1067

Wekerle H, Kojima K, Lannes-Vieira J, Lassmann H, Linington C (1994) Animal models. Ann Neurol 36(Suppl):S47–S53

Wildin RS, Ramsdell F, Peake J, Faravelli F, Casanova JL, Buist N, Levy-Lahad E, Mazzella M, Goulet O, Perroni L, Bricarelli FD, Byrne G, McEuen M, Proll S, Appleby M, Brunkow ME (2001) X-linked neonatal diabetes mellitus, enteropathy and endocrinopathy syndrome is the human equivalent of mouse scurfy. Nat Genet 27:18–20

Willenborg DO, Prowse SJ (1983) Immunoglobulin-deficient rats fail to develop experimental allergic encephalomyelitis. J Neuroimmunol 5:99–109

Willenborg DO, Sjollema P, Danta G (1986) Immunoglobulin deficient rats as donors and recipients of effector cells of allergic encephalomyelitis. J Neuroimmunol 11:93–103

Willenborg DO, Fordham S, Bernard CC, Cowden WB, Ramshaw IA (1996) IFN-gamma plays a critical down-regulatory role in the induction and effector phase of myelin oligodendrocyte glycoprotein-induced autoimmune encephalomyelitis. J Immunol 157:3223–3227

Willenborg DO, Staykova M, Fordham S, O'Brien N, Linares D (2007) The contribution of nitric oxide and interferon gamma to the regulation of the neuro-inflammation in experimental autoimmune encephalomyelitis. J Neuroimmunol 191:16–25

Williams LM, Rudensky AY (2007) Maintenance of the Foxp3-dependent developmental program in mature regulatory T cells requires continued expression of Foxp3. Nat Immunol 8:277–284

Wilson NJ, Boniface K, Chan JR, McKenzie BS, Blumenschein WM, Mattson JD, Basham B, Smith K, Chen T, Morel F, Lecron JC, Kastelein RA, Cua DJ, McClanahan TK, Bowman EP, de Waal Malefyt R (2007) Development, cytokine profile and function of human interleukin 17-producing helper T cells. Nat Immunol 8:950–957

Wolf SD, Dittel BN, Hardardottir F, Janeway CA, Jr. (1996) Experimental autoimmune encephalomyelitis induction in genetically B cell-deficient mice. J Exp Med 184:2271–2278

Yan J, Harvey BP, Gee RJ, Shlomchik MJ, Mamula MJ (2006) B cells drive early T cell autoimmunity in vivo prior to dendritic cell-mediated autoantigen presentation. J Immunol 177:4481–4487

Yang XO, Panopoulos AD, Nurieva R, Chang SH, Wang D, Watowich SS, Dong C (2007) STAT3 regulates cytokine-mediated generation of inflammatory helper T cells. J Biol Chem 282:9358–9363

Yang L, Anderson DE, Baecher-Allan C, Hastings WD, Bettelli E, Oukka M, Kuchroo VK, Hafler DA (2008a) IL-21 and TGF-beta are required for differentiation of human T(H)17 cells. Nature 454:350–352

Yang XO, Pappu BP, Nurieva R, Akimzhanov A, Kang HS, Chung Y, Ma L, Shah B, Panopoulos AD, Schluns KS, Watowich SS, Tian Q, Jetten AM, Dong C (2008b) T Helper 17 Lineage Differentiation Is Programmed by Orphan Nuclear Receptors RORalpha RORgamma. Immunity 28:29–39

Yao Z, Fanslow WC, Seldin MF, Rousseau AM, Painter SL, Comeau MR, Cohen JI, Spriggs MK (1995) Herpesvirus Saimiri encodes a new cytokine, IL-17, which binds to a novel cytokine receptor. Immunity 3:811–821

Yednock TA, Cannon C, Fritz LC, Sanchez-Madrid F, Steinman L, Karin N (1992) Prevention of experimental autoimmune encephalomyelitis by antibodies against alpha 4 beta 1 integrin. Nature 356:63–66

Zamvil SS, Mitchell DJ, Moore AC, Kitamura K, Steinman L, Rothbard JB (1986) T cell epitope of the autoantigen myelin basic protein that induces encephalomyelitis. Nature 324:258–260

Zamvil SS, Mitchell DJ, Powell MB, Sakai K, Rothbard JB, Steinman L (1988) Multiple discrete encephalitogenic epitopes of the autoantigen myelin basic protein include a determinant for I-E class II-restricted T cells. J Exp Med 168:1181–1186

Zamvil SS, Steinman L (1990) The T lymphocyte in experimental allergic encephalomyelitis. Annu Rev Immunol 8:579–621

Zenewicz LA, Yancopoulos GD, Valenzuela DM, Murphy AJ, Karow M, Flavell RA (2007) Interleukin-22 but not interleukin-17 provides protection to hepatocytes during acute liver inflammation. Immunity 27:647–659

Zhang J, Markovic-Plese S, Lacet B, Raus J, Weiner HL, Hafler DA (1994) Increased frequency of interleukin 2-responsive T cells specific for myelin basic protein and proteolipid protein in peripheral blood and cerebrospinal fluid of patients with multiple sclerosis. J Exp Med 179:973–984

Zhang GX, Gran B, Yu S, Li J, Siglienti I, Chen X, Kamoun M, Rostami A (2003) Induction of experimental autoimmune encephalomyelitis in IL-12 receptor-beta 2-deficient mice: IL-12 responsiveness is not required in the pathogenesis of inflammatory demyelination in the central nervous system. J Immunol 170:2153–2160

Zheng Y, Danilenko DM, Valdez P, Kasman I, Eastham-Anderson J, Wu J, Ouyang W (2007) Interleukin-22, a T(H)17 cytokine, mediates IL-23-induced dermal inflammation and acanthosis. Nature 445:648–651

Zheng Y, Valdez PA, Danilenko DM, Hu Y, Sa SM, Gong Q, Abbas AR, Modrusan Z, Ghilardi N, de Sauvage FJ, Ouyang W (2008) Interleukin-22 mediates early host defense against attaching and effacing bacterial pathogens. Nat Med 14:282–289

Zhou L, Ivanov, II, Spolski R, Min R, Shenderov K, Egawa T, Levy DE, Leonard WJ, Littman DR (2007) IL-6 programs T(H)-17 cell differentiation by promoting sequential engagement of the IL-21 and IL-23 pathways. Nat Immunol 8:967–974

Zhou L, Lopes JE, Chong MM, Ivanov, II, Min R, Victora GD, Shen Y, Du J, Rubtsov YP, Rudensky AY, Ziegler SF, Littman DR (2008) TGF-beta-induced Foxp3 inhibits T(H)17 cell differentiation by antagonizing RORgammat function. Nature 453:236–240

T-Cells in Multiple Sclerosis

Christopher Severson and David A. Hafler

Abstract Multiple sclerosis (MS) is a multifocal demyelinating disease of the central nervous system pathologically characterized by lesions of infiltrating macrophages and T cells. Multiple lines of evidence implicate that T cells play a central role in both mediating and regulating MS pathophysiology, and efforts to develop rational therapeutic strategies for MS have focused on understanding factors which control T cell function. T cells are a highly heterogeneous population comprised of multiple cell subtypes which mediate both adaptive immunity and specific tolerance. Much has been learned about the molecular signals that induce T cell activation and differentiation, and several effective treatments for MS act by altering these activation and differentiation pathways. In recent years, increasing recognition has been given to T cell subsets which serve immunosuppressive or regulatory functions, and it has been discovered that patients with MS have a functional defect in these cells. Current work is beginning to shed light on interactions of pathogenic and regulatory T cells with the intrinsic cells of the CNS to provide a more comprehensive picture of MS pathogenesis.

1 Background

Multiple Sclerosis (MS) is a multifocal demyelinating disease of the central nervous system (CNS) characterized by perivascular inflammatory cell infiltrates and demyelination with accompanying oligodendrocyte loss and more limited axonal transactions (Charcot 1868; Lassmann et al. 1998). The typical clinical course of MS consists of episodic focal neurologic deficits disseminated in time and space later evolving into a condition of progressive generalized neurologic decline (Hafler 2004). Evidence for the inflammatory nature of MS comes from both the inflammatory histopathology and the presence of oligoclonal immunoglobulin production in the cerebrospinal fluid (CSF) of MS patients (Kabat et al. 1948). Extensive attempts to

C. Severson and D. Hafler (✉)

Division of Molecular Immunology, Center for Neurologic Diseases, Brigham and Women's Hospital, Harvard Medical School, Boston, MA, 02115, USA

e-mail: dhafler@rics.bwh.harvard.edu

Results Probl Cell Differ, DOI 10.1007/400_2009_12

identify an infectious pathogen in MS lesions have yielded in essence negative results (Giovannoni et al. 2006). The possibility that CNS inflammation and tissue destruction can result from an autoimmune mechanism was established through the development of the model of experimental autoimmune encephalitis (EAE), in which animals immunized against myelin components develop an inflammatory, at times demyelinating, CNS disease (Rivers 1935). The causal role of the immune system in the development of MS is supported by whole genome association studies, which have uncovered numerous MS risk alleles in immune related loci (International Multiple Sclerosis Genetics Consortium et al. 2007). The clinical relevance of immune physiology in MS is highlighted by the efficacy of several immunomodulatory therapies to impact the natural history of the disease (De Jager and Hafler 2007). However, our ability to treat MS remains quite limited and our understanding of the specific nature of the immunologic defects in MS is just beginning to be developed.

The importance of T-cells in the pathophysiology of MS is suggested by several lines of evidence. T-cells are located at the active edge of MS lesions, and the presence of perivascular infiltrates of T-cells throughout the CNS is a consistent feature in all stages of the disease (Prineas 1975). Patients with MS have a higher frequency of activated T-cells in their peripheral blood, specifically a higher percentage of activated myelin-reactive T-cells (Hafler et al. 1985; Zhang et al. 1994). The strongest genetic risk allele for the development of MS is in the class II HLA-DR locus, and recently another risk allele has been identified in the class I HLA locus (Yeo et al. 2007), suggesting that antigen presentation to CD4$^+$ and CD8$^+$ T-cells plays a causal role in the development of disease. In the EAE model, adoptive transfer of myelin-reactive T-cells into the peripheral blood of a previously healthy animal is sufficient to transfer disease (Zamvil et al. 1985). Together, these findings have led to the hypothesis that MS is a T-cell-mediated autoimmune disease.

Advances in our understanding of T-cell physiology have expanded our understanding of their role in MS pathophysiology. It is becoming increasingly clear that T-cells are not a monomorphic population differing only in their T-cell receptor (TCR) specificities, but rather they are a diverse mix of proinflammatory and antiinflammatory subtypes, whose reciprocal interactions we are just beginning to understand. In this chapter, we review what is known about the specificity and functional state of T-cells in MS patients. To understand more fully the physiology of these cells, we examine the factors that enhance or inhibit T-cell activation and the effector profiles T-cells assume after they are activated. We explore in detail the regulatory role certain T-cells play in inhibiting autoimmune and other inflammatory processes and the functional deficiency of these cells in MS. Finally, we discuss how T-cells specifically affect, and are affected by, the local environment of the CNS.

2   Immunopathophysiology of MS

Gross examination of MS brain and spinal tissue reveals multiple sharply demarcated plaques with a predilection for the optic nerves and the white mater tracts of the periventricular region, brainstem, and spinal cord. Histologic evaluation of the plaques reveals focal demyelination, oligodendrocyte loss, with more selective axonal transection, and an accompanying inflammatory cell infiltrate. The inflammatory cell profile of active lesions is characterized by oligoclonal T-cells (Wucherpfennig et al. 1992b), consisting of CD4[+] and CD8[+] α/β (Traugott et al. 1983; Hauser et al. 1986) and γ/δ T-cells (Wucherpfennig et al. 1992a), macrophages, and occasional B-cells and infrequent plasma cells (Prineas and Wright 1978). Macrophages predominate in the center of active plaques and are seen to contain myelin debris, while T-cells predominate at the plaque margin. In chronic active lesions, the inflammatory cell infiltrate is less prominent, but still persists around the lesion edge. In addition to their presence in demyelinated plaques, T-cell infiltrates cluster in perivascular spaces in both affected and unaffected regions throughout the brain and spinal cord, and individual T-cells can also be found diffusely scattered in normal appearing white matter (Babinski 1885; Traugott et al. 1983).

Perivascular infiltration of T-cells and other inflammatory cells in the CNS requires adhesion and transmigration of these cells across the blood-brain barrier (Raine 1994; Springer 1994). Extravasation of cells involves a highly regulated sequence of molecular interactions between inducible ligand-receptor pairs on the surface of the migrating cell and the endothelial barrier (Springer 1994; Engelhardt and Ransohoff 2005). Selective expression of adhesion molecules, chemokines and chemokine receptors, and matrix metalloproteinases (MMPs) are important mediators of this transmigration. In MS lesions, altered levels of the adhesion molecules, ICAM-1 and VCAM-1, have been identified on endothelial cells (Sobel et al. 1990; Washington et al. 1994; Zhang et al. 2000), and their respective ligands, LFA-1 and VLA-4, have been identified on the perivascular inflammatory cells (Bo et al. 1996). LFA-1 and VLA-4 are also expressed at higher levels in ex vivo T-cells from MS patients' blood and CSF (Elovaara et al. 2000). Chemokines enhance immune cell migration through direct chemoattraction and by activating leukocyte integrins to bind their adhesion molecule receptors on endothelial cells. An increase in proinflammatory chemokines is associated with demyelination in MS, and chemokine receptors have also been identified on macrophages, microglia, astrocytes, neurons, and endothelial cells in MS lesions. MMPs comprise a family of tightly regulated proteolytic enzymes that are secreted into the extracellular matrix (ECM). MMPs are expressed by activated T-cells, monocytes, astrocytes, and microglial cells (Leppert et al. 1995; Cuzner et al. 1996; Maeda and Sobel 1996). Potential mechanisms of MMP contribution to MS pathophysiology include disruption of the basement membrane of the BBB (Yong et al. 1998; Kieseier et al. 1999), thereby facilitating transmigration of inflammatory cells, and breakdown of the ECM, enabling infiltration into the neutrophil, and release of membrane-bound proinflammatory cytokines (e.g., TNF), which may also contribute to tissue damage (Leppert et al. 1995; Anthony et al. 1998)

After crossing the blood-brain barrier, invading T-cells would need to encounter cognate antigen in the CNS in order to be locally reactivated. T-cell expression of activation markers with evidence of cytokine production and clonal expansion suggest that such antigenic encounter has occurred (Raine 1994; Babbe et al. 2000). The effector profile of such cells is likely to depend on several factors in the microenvironment, including the cytokine milieu and the costimulatory profile of local and infiltrating antigen presenting cells (APCs). Mechanisms of myelin destruction and axonal damage are likely to be multiple and include direct effects of proinflammatory cytokines; digestion of surface myelin antigens by macrophages, which may include the binding of antibodies, complement-mediated injury, direct injury by CD4[+] and CD8[+] T-cells, oxygen radicals, nonspecific cytotoxicity, and apoptosis. Activation of resident CNS glial cells, such as microglia, may provide the basis for the generation or maintenance of pathologic responses, even in the absence of further infiltration of exogenous inflammatory cells.

3   Autoreactive T-Cells in MS

The presence of activated T-cells in MS lesions and the genetic effect of the HLA locus on disease susceptibility implicate TCR recognition of cognate antigen in the pathogenesis of disease. The analogy with EAE raises the question of whether T-cells in MS could be targeting myelin-derived self-antigen. Work in the early 1990s revealed that, rather than being completely deleted by negative thymic selection, myelin reactive T-cells are present in the peripheral blood of both MS patients and healthy controls (Pette et al. 1990), though these myelin reactive T-cells have lower affinities for self antigen/MHC complexes as compared to T-cells recognizing nonself microbial antigens. Thus, the mere presence of autoreactive cells in the periphery is an insufficient explanation for the development of autoimmune disease. If myelin-reactive T-cells are nonetheless involved in MS immunopathogenesis, the prediction would be that these cells must differ in some way from those found in healthy individuals. Indeed, Zhang et al. found that the IL-2 receptor, a hallmark of activated T-cells, was expressed on MBP-reactive T-cells from MS patients but not on those obtained from normal individuals (Zhang et al. 1994). Similar frequencies of MBP- and PLP-reactive CD4 T-cells were obtained from PBMCs of RRMS patients and controls after primary stimulation with antigen. However, if the PBMCs were first expanded with IL-2 and then further expanded with antigen, much higher frequencies of MBP- and PLP-reactive CD4 T-cells were obtained from the PBMCs of MS patients. Furthermore, after initial expansion in IL-2, MBP-reactive T-cells were obtainable from the CSF of MS patients but not from the CSF of control individuals. Thus, MBP-reactive T-cells obtained from MS patients are in an enhanced state of activation compared to those isolated from normal individuals.

4   Costimulation

In order for myelin-reactive T-cells to be activated in MS patients, they need to encounter antigen in the context of appropriate costimulatory signals. The concept of T-cell costimulation was developed from the work of Lafferty in the 1970s. While studying organ transplant rejection, he noticed that fixed APCs could tolerize

T-cells that would otherwise be activated. He postulated that naïve T-cells must require a second signal in order to be activated by antigen, and absent that second signal, encounter with antigen results in prolonged T-cell hyporesponsiveness and anergy (Lafferty and Woolnough 1977). The second signal he predicted was later shown to be mediated by costimulatory molecules up-regulated on functional APCs in response to inflammatory stimuli (Jenkins et al. 1991; Harding et al. 1992). This original concept of costimulation, as a second signal required by naïve cells, has evolved, as it has been recognized that costimulatory molecules augment activation of antigen-experienced as well as naïve cells, and T-cells also receive inhibitory costimulatory signals that attenuate or terminate their activation. The balance of activating and inhibiting costimulatory signals helps determine whether a T-cell-mediated response is one of tolerance or immunity.

4.1 The B7/CD28/CTLA-4 Axis

The most well described costimulatory pathway is the B7/CD28/CTLA-4 axis. The costimulation molecule CD28 is constitutively present on ~95% of T-cells, and its ligands B7–1(CD80) and B7–2(CD86) are variably expressed on APCs. B7–2 is expressed at low levels on resting APCs, but it is quickly up-regulated in response to inflammatory stimuli, while B7–1 is only induced after prolonged stimulation (Anderson et al. 1999). When CD28 binds with B7 ligand in the context of TCR engagement, it enhances T-cell proliferation and differentiation through up-regulation of the antiapoptotic factor bcl-x_L, increased production of the growth factor IL-2, and facilitation of cell cycle progression (Keir and Sharpe 2005). After activation, T-cells are induced to express the inhibitory costimulation molecule CTLA-4. CTLA-4 also binds with the ligands B7–1 and B7–2, but in contrast to CD28, CTLA-4 ligation results in an attenuation of TCR signaling and a proliferation block (Walunas et al. 1994; Krummel and Allison 1996). The activational history of a T-cell affects the relative expression of CD28 and CTLA-4 and the strength of signal they each transduce in response to B7 ligation.

The brains of MS patients have been shown to express significant levels of both B7–1 and B7–2, in comparison to inflammatory infarcts in which only B7–2 could be detected (Windhagen et al. 1995). In the EAE model, B7–1 expression has been shown to be necessary for the phenomenon of epitope spreading, in which an animal immunized against a single myelin peptide will develop activated T-cells reacting with multiple other myelin antigens (Miller et al. 1995). The expression of both B7–1 and B7–2 molecules in MS patients' brains, therefore, would be expected to create the environment of an adjuvant, leading to activation of multiple myelin reactive T-cells, which could explain the diversity of myelin-reactive T-cells found in MS patients' blood. When the T-cell requirement for B7 costimulation was examined, it was found that MBP reactive T-cells from MS patients are less dependent on B7 stimulation to induce clonal expansion (Scholz et al. 1998), and they do not seem to be as restrained by CTLA-4 signaling as MBP reactive T-cells from healthy controls (Oliveira et al. 2003). While these findings may be the consequence of chronic activation of autoreactive cells, the finding of an increased percentage of CD28 negative T-cells in MS patients has raised the possibility that costimulatory dysregulation plays a role in the pathogenesis of the disease (Markovic-Plese et al. 2001).

4.2   Other Costimulatory Pathways

In recent years, it has been recognized that the B7/CD28/CTLA-4 axis is just one of many pathways involved in T-cell costimulation. While the different costimulatory pathways converge in their actions to augment or attenuate T-cell activation, they appear to differ in the relative importance they play in naïve, memory, and effector T-cells. In the EAE model, disruption of the activating costimulatory receptors OX-40, ICOS, DR3, or CD28 all ameliorate disease pathology (Sporici et al. 2001; Carboni et al. 2003; Meylan et al. 2008), while disruption of the inhibitory costimulatory receptors PD-1 or CTLA-4 both exacerbate the disease (Karandikar et al. 1996; Salama et al. 2003). However, in adoptive transfer EAE models, it appears that blockade of different costimulatory pathways has differential effects on disease mediated by memory versus effector T-cells (Elyaman et al. 2008).

Although many different costimulatory molecules have been reported to be present on infiltrating leukocytes in MS brains, the relative roles they play in MS pathogenesis has yet to be fully elucidated. Interestingly, at the time of this writing, three of the strongest genetic associations with the risk of developing MS are in genes with costimulatory functions: CD58, CD6, and CD226 (De Jager 2009a) (Hafler et al. 2009). CD58 is a ligand for the CD2 costimulatory receptor present on T-cells, and a polymorphism in the CD58 gene has been associated with an altered risk of MS susceptibility at a level of genomewide significance (OR = 0.78; $P = 4.13 \times 10^{-10}$) (De Jager 2009). In both EBV cell lines and whole PBMCs, expression of CD58 mRNA is higher in individuals with the protective allele at the CD58 locus, and in in vitro models of regulatory T-cell suppression, costimulation through the CD2 pathway has been shown to favor more suppression than costimulation through the CD28 pathway (De Jager et al. 2009b). Together, these findings suggest how an alteration in the balance of costimulatory signals could influence susceptibility to MS.

4.3   Costimulation as a Therapeutic Target

Advances in our understanding of the actions of costimulatory molecules in MS pathophysiology will allow proper evaluation and use of emerging molecular tools for the treatment of the disease. A soluble Ig-CTLA-4 fusion protein with the ability to inhibit B7 mediated costimulation has been shown to be effective in the treatment of rheumatoid arthritis and is now undergoing clinical trails for patients with MS

(Genovese et al. 2005; Viglietta et al. 2008). Several other therapeutics targeting costimulatory pathways, such as a CD58-Ig fusion protein, are in different stages of development for various autoimmune diseases. The advent of such agents promises to provide new tools to influence the course of MS. However, the proper use of these tools will benefit from greater understanding of the specific roles of different costimulatory pathways in disease pathogenesis.

5   T-Cell Polarization

The evidence of T-cell activation in MS pathophysiology does not indicate the functional consequences of that activation. An early observation in T-cell biology was that the immune system is capable of mounting different patterns of inflammatory response depending on the type of T-cells involved. For example, some activated T-cells mobilize macrophages to mediate delayed type hypersensitivity reactions, while others induce B-cell isotype class switching and IgE production. The type of reaction directed by T-cells is dictated by cues in the inflammatory environment in which a naïve T-cell first encounters antigen, but after initial activation, each T-cell tends to continue to react to subsequent encounters with antigen in the same manner.

The above observations led Mosmann and Coffman to propose the theory of T-cell polarization (Mosmann and Coffman 1989). In this theory, signals in the inflammatory environment at the time of initial encounter with antigen direct a T-cell to adopt a distinct phenotype of cytokine secretion which (1) can direct a specific pattern of inflammatory response, (2) can act in an autocrine fashion to reinforce and perpetuate the T-cell's polarized state, (3) can act in a paracrine fashion to induce other T-cells in proximity to adopt the same type of polarization, and (4) can antagonize the activity of T-cells with a different polarity. Polarization has been studied most extensively in the context of CD4$^+$ T helper (Th) cells. There is evidence that CD8$^+$ cells can also undergo polarizing differentiation, but that process is less well defined, and for the purposes of this chapter, we will restrict our discussion to Th polarization.

5.1   Th1/Th2/Th17 Cells

The initial formulation of the theory of polarization proposed that CD4$^+$ T-cells exist on a single axis of Th1 and Th2 polarities. Th1 cells are characterized by high levels of IFNγ secretion and function to induce macrophage activation and to coordinate immune responses to intracellular pathogens. Th2 cells are characterized by the production of IL-4, IL-5, and IL-13 and mediate responses to helminthes and other extracellular parasites. Th1 cells are induced when a naïve T-cell is activated in the presence of IL-12. Th2 cells are induced by activation in the presence of IL-4 (Abbas et al. 1996). In the Th1/Th2 model, all physiologic and pathologic immune responses were characterized as being driven by either a Th1 or Th2 polarized T-cell: clearance of intracellular pathogens and tissue destructive autoimmunity were ascribed to a Th1 polarization; attack of parasites and allergic reactions were ascribed to a Th2 polarization.

In recent years, a third pathogenic T-cell polarity has been described: the Th17 cell, named for its secretion of the proinflammatory cytokine IL-17a. The role of Th17 cells was recognized through studies of EAE mice. Knockout mice for IL-12 or the p35 component of the IL-12 receptor, failed to develop Th1 cells, and yet they still could be induced to have robust encephalomyelitis. In contrast, mice deficient in IL-23, the p19 component of the IL-23 receptor, or the p40 component common to both the IL-12 and IL-23 receptors, were all protected from encephalomyelitis, even though the mice deficient in IL-23 and IL-23R-p19 had normal Th1 responses (Becher et al. 2002; Cua et al. 2003). These findings suggested that EAE was not being mediated by Th1 cells. It was later discovered that the effect of IL-23 and IL-23 receptor knock-outs was a deficiency in specialized $CD4^+$ IL-17 secreting cells, that were named Th17 cells (Harrington et al. 2005; Park et al. 2005), and adoptive transfer of myelin reactive Th17 cells expanded in culture with IL-23 resulted in severe EAE (Langrish et al. 2005).

5.2   Characterization of Th17 Cells

The biology of Th17 cells has been further explored recently. In addition to IL-17, more precisely termed IL-17A, Th17 cells are characterized as secreting IL-17F, IL-21, and IL-22 (Liang et al. 2006; Nurieva et al. 2007). IL-17A is able to act on multiple cell types to initiate a cascade of proinflammatory cytokines, chemokines, and other inflammatory mediators with resultant neutrophil recruitment (Yao et al. 1995; Bettelli et al. 2008). It has also been shown to promote blood-brain barrier disruption and CNS inflammation (Kebir et al. 2007). Although Th17 cells were discovered by their dependence on IL-23, IL-23 does not induce the differentiation naïve $CD4^+$ cells into Th17 cells, but rather it appears to play a role in stabilization of their Th17 phenotype and enhancement of their pathogenic potential (Aggarwal et al. 2003; McGeachy and Cua 2007). In mice, the induction of Th17 cells from naïve T-cells requires the combined action of TGFβ and either of the proinflammatory cytokines IL-6 or IL-21 (Bettelli et al. 2006; Veldhoen et al. 2006; Korn et al. 2007). In humans, Th17 induction also requires the combination of TGFβ and similar proinflammatory cytokines; however, IL-21 appears to be a stronger inducer than IL-6 (Manel et al. 2008; Volpe et al. 2008; Yang et al. 2008). IL-21 also acts as a potent growth factor for differentiated Th17 cells, amplifying Th17 responses in both mice and humans (Korn et al. 2007). The differentiation of Th17 cells has been shown to be mediated by the transcription factor RORγt in mice (Ivanov et al. 2006), and in humans, the ortholog transcription factor, RORC, is restricted to IL-17 producing cells (Annunziato et al. 2007) and is induced by Th17 polarizing conditions (Yang et al. 2008).

5.3 Th1 Cells in MS

Prior to the discovery of Th17 cells, MS was considered a purely Th1-mediated disease. Evidence that has supported a Th1 pathology in MS has included the increased expression of IL-12 mRNA in acute MS lesions (Windhagen et al. 1995), a Th1 biased pattern of chemokine receptors on T-cells in the CSF and lesions (Sørensen et al. 1999), and increased secretion of IL-12 and IFNγ by PBMCs from MS patients after T-cell activation (Balashov et al. 1997). The idea of a Th1 bias has been further supported by an early attempt to treat MS patients with recombinant IFNγ, which resulted in exacerbations of disease (Panitch et al. 1987). In contrast, treatment with glatiramer acetate, an approved treatment for MS that reduces the mean relapse rate by 30% (Johnson et al. 2001), results in a shift from Th1 to Th2 profile in T-cells from treated patients, and this shift has been proposed to be its mechanism of therapeutic effect (Duda et al. 2000; Neuhaus et al. 2001).

5.4 IL-17 and Th17 Cells in Multiple Sclerosis

What is the evidence that Th17 cells may play a role in the pathophysiology of MS? Even before the recognition of Th17 cells as a distinct lineage, it was known that MS patients have elevated levels IL-17 mRNA in their CSF and blood mononuclear cells (Matusevicius et al. 1999), and a genomewide expression analysis had identified IL-17, TGFβ, and IL-6 as up-regulated in MS lesions (Lock et al. 2002). These findings have been replicated and extended more recently with in situ hybridization and immunohistochemistry showing IL-17 expressing T-cells in perivascular regions and in acute lesions (Tzartos et al. 2008). Markovic-Plese and colleagues reported extracting T-cells, from the brain of a patient who died with aggressive RRMS, showing a proliferative response to myelin derived peptides with up-regulation of RORC and Th17 cytokines (Montes et al. 2008). In MS patients, dendritic cells express higher levels of IL-23 and T-cells express higher levels of IL-17 compared with cells from healthy controls (Vaknin-Dembinsky et al. 2006).

5.5 Th1 Versus Th17

Is it possible that both Th1 and Th17 cells could mediate disease pathogenesis? One of the original concepts of T-cell polarization was that cells of a given polarity would antagonize the actions of other polarities, and there is some evidence of such reciprocal antagonism between Th1 and Th17 cells. The Th1 related cytokines IFNγ and IL-12, as well as the Th2 cytokine IL-4, can antagonize Th17 differentiation and inhibit IL-23 driven proliferation of Th17 cells (Harrington et al. 2005; Park et al. 2005). Furthermore expression of the Th1 transcription factor T-bet destabilizes the Th17 phenotype (Mathur et al. 2006). Conversely, IL-17 and other Th17 cytokines do not appear to affect Th1 differentiation directly, although in the presence of APCs, IL-17 and IL-23 have been reported to antagonize Th1 polarization in an in vitro polarization model (Nakae et al. 2007). The relationship between these cell types may be further complicated by the observation that Th17 polarized cells can transform into Th1 cells in vivo (Lee et al. 2009).

5.6   Th1 Versus Th17 in EAE

Both Th1 and Th17 cells can both participate the pathology of EAE. In the EAE model, both Th1 and Th17 polarized T-cells have been isolated from encephalitic tissue. Governman and colleagues have reported the ratio of pathogenic Th1/Th17 cells varies depending on the antigenic stimulus, and they have correlated this ratio with phenotype of the disease, with Th1 dominant responses associated with preferential spinal cord involvement and Th17 dominant responses correlating with preferential brain pathology (Stromnes et al. 2008). In adoptive transfer models of EAE, myelin specific T-cells polarized to either a Th1 or a Th17 phenotype are both capable of transferring disease, but result in distinct histopathology with macrophages more prominent in Th1 mediated disease and neutrophil infiltration in disease transferred by Th17 cells (Kroenke et al. 2008). Kuchroo and colleagues have reported a shift of Th polarization in the temporal course of EAE, with early involvement of Th17 cells giving way to Th1 predominance prior to the resolution of encephalitis (Dardalhon et al. 2008). They propose that this shift in polarization might be related to the natural course and recovery from an attack of EAE. Th17 cells are less susceptible to suppression by regulatory T-cells (Treg) than Th1 cells, and infiltrating T-cells isolated from early stages of EAE lesions are less suppressible than those isolated from later stages (Korn et al. 2007).

5.7   Th1 Versus Th17 in MS

In patients with MS, the relative contributions of Th17 and Th1 polarized cells are just beginning to be investigated. Various studies suggest that, as in different EAE models, the relative dominance of Th17 cells may vary with the temporal course and with different clinical variants of MS and related diseases. The percentage of T-cells costaining for IL-17 is reported to be higher in early active plaques compared with chronic active or inactive plaques (Tzartos et al. 2008). Similarly, in vitro stimulated PBMCs derived from MS patients produce higher levels of IL-17 if they come from patients within 2 years of diagnosis, compared with those from patients with longer standing disease (Graber et al. 2008). In the same set of experiments, PBMCs from patients with transverse myelitis produced even higher levels of IL-17 than those from MS patients. In evaluating the spinal fluid of Japanese patients,

IL-17 levels were found to be higher in patients with the optico-spinal variant compared with conventional MS patients (Ishizu et al. 2005).

6   FoxP3⁺ Regulatory T-Cells

Clonal deletion of self-reactive T-cells in the thymus and induction of T-cell anergy alone do not explain the maintenance of immunologic self-tolerance, as potentially pathogenic autoreactive T-cells are present in the periphery of healthy individuals (Ota et al. 1990; Pette et al. 1990). Thus, other regulatory mechanisms exist to prevent autoreactive T-cells from causing immune disorders. Active suppression by Treg plays a key role in the control of self-antigen-reactive T-cells and the induction of peripheral tolerance in vivo (Shevach et al. 2001; Sakaguchi et al. 2008). Seminal experiments performed by Sakaguchi and colleagues have shown that depletion of a subset of T-cells results in the onset of systemic autoimmune diseases in mice (Sakaguchi et al. 2008). The regulatory T-cell subset was characterized by the transcription factor FoxP3 and the surface immunophenotype of CD4⁺CD25⁺. Furthermore, cotransfer of these cells with CD4⁺CD25⁻ cells prevents the development of experimentally induced autoimmune diseases such as colitis, gastritis, insulin-dependent autoimmune diabetes, and thyroiditis (Sakaguchi et al. 1985; Read et al. 2000).

6.1   *Cell Surface Characterization of Human Tregs*

Our group and others described a population of CD4⁺CD25^{high} Treg in human peripheral blood and thymus (Baecher-Allan et al. 2001; Dieckmann et al. 2001; Dieckmann et al. 2002). Human CD4⁺CD25^{high}FoxP3⁺ natural Treg, similar to the mouse CD4⁺CD25⁺ suppressor cells, are anergic to in vitro antigenic stimulation and strongly suppress the proliferation of responder T-cells upon coculture. Although both murine and human CD4⁺CD25^{high}FoxP3⁺ Tregs similarly suppress the activation of CD4 responder T-cells (Tresp) in a cell-contact-dependent manner and express the Treg-specific lineage specification factor FoxP3, the human Treg population is by far more heterogeneous than that of the mouse, as gauged by both cell surface phenotype and functional capability. Human blood, isolated from an outbred population in a pathogenic environment, contains up to 30% CD4⁺CD25⁺ cells; only the 2–4% of the cells with the highest CD25 expression can be considered regulatory (Baecher-Allan et al. 2001). Problematically, there is no consensus as to where the boundary lies between CD25 high and CD25 intermediate expression, which has hindered both experimental reproducibility and clinical analysis of patients' blood. This is particularly concerning in patients where inflammatory conditions can lead to an increase in CD25 expression by activated T-cells. The majority of human Treg can, however, be differentiated from recently activated effector cells by CD62L (L-selectin) expression because CD4$^+$CD25^{high}FoxP3$^+$ Tregs are predominantly CD62L$^+$ (>95%). Additionally, human Treg isolated from adult peripheral blood or tonsils are predominantly CD45RO$^+$, CD45RA$^-$, and CD45RB^{low} (Dieckmann et al. 2001). For the most part, CD4$^+$CD25^{high}FoxP3$^+$ Tregs express a highly differentiated central memory phenotype and share many of the cell surface markers expressed on chronically activated CD4+ T-cells.

The Treg population is also expresses a number of different immunomodulatory surface determinants, including MHC class II, CD95 (Fas), glucocorticoid-induced tumor necrosis factor receptor family-related protein, and CTLA-4 that are not Treg specific. The most specific marker for Treg to date is the nuclear transcription factor FoxP3, the expression of which correlates with suppressive ability. In humans, FoxP3 is expressed strongly in CD4$^+$CD25^{high} T-cells and at low levels in activated CD4$^+$ T-cells (Wang et al. 2007), although FoxP3 expression has also been reported in a small fraction of the CD25$^-$, CD25 intermediate, and CD8$^+$ T-cell subsets. Indeed, data from FoxP3-GFP reporter mice has shown that CD25$^-$FoxP3$^-$GFP$^+$ cells exhibit the same suppressive capacity as CD25$^+$FoxP3$^-$GFP$^+$ cells in vitro (Fontenot et al. 2005). This evidence, paired with the absence of a unique Treg cell surface marker, suggests the possibility that the CD4$^+$CD25^{high} subset is merely enriched for human Treg and does not contain the entire Treg population.

Discovery of an alternative marker to CD25 would greatly enhance the identification and purification of human Treg. The best-accepted alternative is lack of cell surface CD127 (IL-7 receptor). Foxp3 expression and suppressive capacity have been detected in CD4+ T-cells expressing low levels of CD127 (Liu et al. 2006; Seddiki et al. 2006). Certainly, low CD127 expression may be useful in isolating FoxP3+ cells from the CD25$^+$ DN2 and DN3 populations of the thymus. In the periphery, however, the possibility remains that CD127 expression does not discriminate between adaptive Treg cells, such as the Tr1 and Th3 subsets, activated T-cells, and thymically derived Tregs.

6.2   Strength of Signal and Human Treg Suppression

Human Tregs must be activated through their TCR in order to be functionally suppressive (Dieckmann et al. 2002; Jonuleit et al. 2002), although once activated, these cells do not need to be viable in coculture in order to mediate Tresp suppression. The effective outcome of Tresp suppression is also dictated by the quality of T-cell stimulation, as the strength of stimuli applied to the Tresp population has a strong influence on whether suppression or proliferation will occur during Treg/Tresp coculture. Tresp activated in the presence of strong costimulation are refractory to Treg suppression, as are Tresp supplemented with growth-promoting cytokines (Baecher-Allan et al. 2001; Baecher-Allan et al. 2002). Furthermore, increasing the strength of TCR signal increases the resistance of Tresp to regulation (Baecher-Allan et al. 2002). These data suggest that human Tregs can only suppress Tresp activated with low signal strength. In highly inflammatory environments, where

Tresp are activated with high signal strength, human Tregs are unable to prevent Tresp proliferation and cytokine secretion.

6.3 Treg Frequency in Patients with MS

Autoreactive T-cells isolated from patients with autoimmune disease have a lower threshold of activation than Tresp in normal subjects (Reijonen et al. 2002). As discussed previously in this chapter, both the activation status of the Tresp and Treg can impact suppression ex vivo; we therefore sought to characterize Tregs in patients with MS as compared to healthy controls. Our group hypothesized that a lower threshold of Treg activation could reduce effector function or cause deficiency in Treg generation in MS patients. We first compared frequency of CD4$^+$CD25^{high} Treg derived from a group of untreated patients who have relapsing/ remitting MS with those from age-matched healthy control subjects (Viglietta et al. 2004). We were unable to detect any differences in Treg frequency between patients and healthy controls in either the CD4$^+$CD25^{high} or the bulk CD25$^+$ populations. Vandenbark and coworkers, however, showed decreases in FoxP3 levels in Tregs isolated from patients with MS and found that this decrease correlated with Treg loss of function (Huan et al. 2005). Additionally, Venken et al. reported that relapsing-remitting (but not secondary progressive) MS patients express lower levels of FoxP3 than healthy controls (Venken et al. 2006). It is not yet clear from these studies whether lower FoxP3 expression in MS patients is due to a reduced frequency of FoxP3 cells in the CD4$^+$CD25^{high} T-cell population or due to decreased FoxP3 expression at a cellular level.

6.4 Impaired Treg Function in Patients with MS

We isolated highly purified CD4$^+$CD25^{high} Tregs and CD4$^+$CD25$^-$ Tresp from untreated patients with relapsing/remitting MS and age-matched healthy control subjects to assay Treg function as well as frequency. Responder cells from both MS patients and healthy individuals responded similarly in a dose-dependent fashion to varying concentrations of plate-bound αCD3 monoclonal antibody (mAb). As expected, Tregs isolated from both groups were anergic to stimulation at all doses of plate-bound αCD3. This anergy indicates that CD4$^+$CD25^{high} T-cells isolated from patients with MS are not activated CD25$^+$ T-cells; such cells would not exhibit regulatory activity but rather enhance proliferation.

To quantify regulatory function, we cocultured Tregs with autologous Tresp at multiple ratios (responder:suppressor ratios of 1:1, 1:1/2, 1:1/4, and 1:1/8). Consistent with previous reports, Tregs isolated from our healthy controls consistently suppressed proliferation at a 1:1 ratio, and increasing the ratio of responder/ suppressor T-cells reduced suppression. In striking contrast, we found that Tregs isolated from the circulation of patients with MS, although normal in frequency, poorly inhibited Tresp proliferation. In these cultures, the secretion of the Th1 cytokine IFN-γ by Tresp was suppressed by healthy Tregs but not by Tregs derived from patients with MS. The suppressive cytokine IL-10 was secreted in these cultures, predominantly by the CD4⁺CD25⁻ Tresp, but we could not determine a role for IL-10 in mediating suppression by CD4⁺CD25ʰⁱᵍʰ Tregs. Additionally, blocking IL-10 or TGF-β did not result in loss of suppressor function in these cells.

The loss of Treg functionality in MS patients could be due to a decrease in CD4+CD25hi Treg performance or an increase in inhibition resistance by activated Tresp. Thus, we performed mixing experiments in which patient and control Treg were cocultured with the autologous and the converse Tresp isolated from either healthy subjects or patients with MS. Treg from patients with MS could not suppress the proliferative response of target responder T-cells from either patients or healthy controls (suppression 23%). In the reciprocal experiments, Treg from healthy controls suppressed the proliferative response of Treg isolated from both controls and patients with MS (suppression 78%). These data indicate that patients with MS do suffer a regulatory defect, which is due to impaired Treg function, but we have not yet determined the mechanism of this functional defect.

6.5   CD62L Expression on CD4⁺CD25hi Regulatory T-Cells

Although there were no differences in the frequency of CD4⁺CD25ʰⁱᵍʰ T-cells or in their proliferative or cytokine secretion in response to different stimuli between healthy subjects and patients with MS, it was important to determine whether an increase in the frequency of activated CD4⁺ T-cells in the circulation was diluting the regulatory CD4⁺CD25ʰⁱᵍʰ T-cells. Therefore, we used CD62L expression to further purify regulatory from the activated T-cells because L-selectin expression is down-regulated upon activation. We isolated CD4⁺CD25ʰⁱᵍʰCD62L⁺ and total CD4⁺CD25ʰⁱᵍʰCD62L⁺ Treg from healthy subjects and patients with MS. Whereas in the healthy controls both populations were able to suppress the proliferative response to anti-CD3 stimulation, the Tregs isolated from patients with MS, although further depleted of the potentially activated CD62L⁻ T-cells, were still unable to inhibit the proliferation of the Tresp population. These data strongly confirm a defect in the highly purified regulatory subset in patients with MS.

7   Tr1 Regulatory T-Cells

Type 1 regulatory (Tr1) cells comprise a distinct T-cell lineage that exerts regulatory effects through secretion of IL-10. IL-10 is a potent immunosuppressive cytokine that inhibits the activation and effector function of multiple cell types, although in certain contexts it can also act as a growth factor for activated CD8 cells (Groux et al. 1998; Moore et al. 2001). IL-10 deficient mice (IL-10$^{-/-}$) spontaneously develop inflammatory bowel disease due to loss of attenuation of responses to intestinal flora (Kuhn et al. 1993). Although IL-10 can be secreted transiently or at low levels by multiple cell types, such as Th2 cells, sustained expression of high levels of IL-10 defines the Tr1 lineage. Tr1 cells are also characterized by the absence of expression of IL-4 and low or absent levels of IL-2 (Groux et al. 1997). Tr1 cells do not express FoxP3, although it can be induced transiently after activation, and they express levels of CD25 comparable to that of other activated cells (Vieira et al. 2004).

Unlike FoxP3$^+$ Treg, Tr1 lineage fate is not determined in the thymus, but rather Tr1 cells are induced from naïve cells by activation in the setting of the appropriate conditioning signals. Several stimuli have been shown to result in the induction of Tr1 cells in vitro. Activating T-cells with CD3/CD28 crosslinking in the presence of IL-10 and IFNα, or with the combination of dexamethasone and 1-,25-(OH)^2 Vitamin D3, both result in Tr1 induction (Levings et al. 2001; Barrat et al. 2002). The latter combination has been of interest as low levels of Vitamin D have been identified as an environmental risk factor for MS (Munger et al. 2004). Tr1 cells can also be induced by crosslinking CD3 with the alternative costimulatory receptor CD46 (Kemper et al. 2003). Finally, the combination of IL-27 and TGFβ also induces Tr1 cells, and this combination is thought to be the mechanism through which FoxP3+ Treg modified dendritic cells induce Tr1 formation (Awasthi et al. 2007).

7.1 Tr1 Cells in MS

The involvement of IL-10 secreting cells in MS is suggested by the variation of frequency and extent of IL-10 production from peripheral mononuclear cells with the clinical course. Patients with early stable relapsing remitting MS or a history of optic neuritis have high numbers of IL-10 producing cells in their peripheral blood (Navikas et al. 1995). However, a decreased production of IL-10, associated with a significantly increased production of IL-12p40, is detected in patients with secondary progressive MS (Balashov et al. 1997; van Boxel-Dezaire et al. 1999; Soldan et al. 2004). It has also been reported that among patients with secondary progressive MS, lower IL-10 production correlates with higher disability and MRI lesion load (Petereit et al. 2003).

In a more direct test of the Tr1 pathway in MS, Tr1 cells were induced from naïve CD4$^+$ T-cells with CD3$^-$CD46 crosslinking in MS patients and controls (Astier et al. 2006). While no difference was observed in the proliferation of cells, the differentiation of IL-10 secreting Tr1 cells was significantly impaired in MS patients. Increasing strength of stimulation by stronger TCR stimulation or enhanced IL-2 concentrations did not restore IL-10 production. The deficit was specific to IL-10 production as there was no difference in the production of IFNγ by the two groups. Together with evidence of a FoxP3⁺ Treg suppression deficit, these data demonstrate that MS is associated with deficits of multiple regulatory pathways.

8 Regulatory T-Cells and Th17 Cells

With our increasing understanding of the developmental pathways of regulatory and proinflammatory T-cell lineages, it has become evident that Th17 cells are interrelated with Treg. Activation of naïve CD4⁺ T-cells in the presence of TGFβ results in the expression of FoxP3 and induction of a regulatory phenotype. Conversely, if IL-6 or IL-21 is added to TGFβ, RORC is expressed and pathogenic Th17 cells are induced. TGFβ has well-described antiinflammatory effects and has traditionally been thought of as a regulatory cytokine; a view supported by the development of systemic autoimmunity in TGFβ deficient mice (Shull et al. 1992; Kulkarni et al. 1993). However, TGFβ is also essential to Th17 development, as Th17 cells and Th17 mediated disease cannot be induced in mice lacking a component of the TGFβ receptor in their T-cells, showing a common requirement for both proinflammatory Th17 cells and antiinflammatory Tregs (Veldhoen et al. 2006). The reciprocal relationship of these two cell types is highlighted by the observation that the lineage defining transcription factors FoxP3 and RORC physically bind to each other in the nucleus and antagonize each other's activity (Ichiyama et al. 2008; Zhou et al. 2008). The relationship between these cell types is yet more complicated as it has been demonstrated FoxP3⁺ Tregs are able to substitute for the function of TGFβ in Th17 differentiation: activation in the presence of IL-6 and Tregs induces naïve T-cells to differentiate into Th17 cells, and in the presence of high amounts of IL-6, activated Tregs induce their own transformation into Th17 cells (Xu et al. 2007). Recently, we have also documented a subset of FoxP3⁺ human T-cell clones with a mixed phenotype: alternately and reversibly exhibiting the contact mediated suppression of Tregs, or producing the high levels of IL-17 characteristic of Th17 cells depending on the stimulus conditions (Beriou et al. 2009).

The induction of both Th17 cells and FoxP3⁺ Tregs is inhibited by addition of the cytokine IL-27. IL-27Rα deficient mice have enhanced susceptibility to EAE and generate more Th17 cells, and exogenous IL-27 inhibits both in vitro Th17 induction and IL-6 mediated proliferation of Th17 cells (Batten et al. 2006; Stumhofer et al. 2006). IL-27 also inhibits TGFβ induction of FoxP3⁺ Tregs, and instead the combination of IL-27 with TGFβ results in the generation of Tr1 cells (Awasthi et al. 2007). The combination of IL-27 and TGFβ appears to mediate the phenomena of "infectious tolerance," in which Treg conditioned dendritic cells induce naïve T-cells to become Tr1 cells. We are just beginning to understand how the relative predominance of highly pathogenic Th17 cells and regulatory FoxP3⁺ Tregs and Tr1 cells can be shifted by the balance of inductive signals, but it is becoming increasingly apparent that pro and antiinflammatory pathways are more closely interrelated than previously recognized.

9 T-Cell–CNS Interactions

Comprehensive understanding of the roles of T-cells in MS pathophysiology must take into account the mechanisms by which they specifically affect and are affected by their local environment: the CNS. The CNS has been regarded as an immune-privileged site with unique anatomic, cellular, and molecular features that limit both physiologic and pathophysiologic inflammation. The CNS lacks an intrinsic lymphatic drainage system and has a blood-brain barrier composed of tight endothelial junctions, which limit routine contact between the CNS parenchyma and the cells and proteins of the peripheral immune system. CNS astrocytes actively suppress immune activation by a number of mechanisms. Peripheral monocytes cocultured with human astrocytes produce less TNFα in response to activating stimuli and instead are induced to express IL-10. Astrocyte conditioning of monocytes also results in reduced T-cell proliferation and IFNγ secretion in in vitro assays of primary human cells (Kostianovsky et al. 2008). In animal models, rat T-cells cocultured with astrocytes assumed a regulatory phenotype and are able to suppress EAE upon adoptive transfer (Trajkovic et al. 2004).

9.1 Intrinsic Inflammatory Cells of the CNS

In addition to regulating the activity of the peripheral immune system, the CNS also has intrinsic immune properties that are able to mediate inflammatory responses and induce tissue injury. Microglia are resident APCs in the CNS ontogenically derived from myeloid progenator cells that are able to present antigen to T-cells and produce proinflammatory cytokines such as TNFα. Recent studies have shown that microglia are capable of mediating neural tissue damage in response to inflammatory signals. For example, lipopolysaccharide (LPS), a molecular component of gram negative bacteria, has been shown to induce neuronal and axonal loss both in vitro and in vivo due to activation of microglia through Toll-like receptor 4 (Lehnardt et al. 2003). In another study, peripheral administration of LPS to female rats resulted in a 240% increase in the density of activated microglia in the dentate gyrus of the hippocampus, which correlated with a 35% decrease in hippocampal neurogenesis (Monje et al. 2003). Coculture of neural progenitor cells with the LPS-stimulated but not resting microglia inhibited neurogenesis in vitro by approximately 50%, an effect that was mediated by microglial secretion of IL-6.

9.2 A Model of Reciprocal T-Cell-CNS Interactions: TIM-3

An insight into the reciprocal interactions between T-cells and the different resident cells of the CNS can be gained by looking at pleiotropic effects of the TIM-3 pathway in MS. TIM-3 is a type I transmembrane receptor, which was initially described for its expression on differentiated Th1 cells (Monney et al. 2002). Engagement of TIM-3 with its ligand, galectin-9, results in antagonism of Th1 effector functions (Zhu et al. 2005). When T-cells from CSF samples of MS patients and controls were examined, it was found that although T-cells from MS patients secrete higher levels of the Th1 cytokine IFNγ, they express lower levels of TIM-3, suggesting that the TIM-3 pathway is dysregulated in MS (Koguchi et al. 2006). Galectin-9 is known to be expressed on human astrocytes in response to inflammatory stimuli (Yoshida et al. 2001), suggesting that the galectin-9/TIM-3 pathway may be a CNS mediated mechanism for negative feedback of Th1 responses.

However, when human brain tissue was stained for TIM-3, it was discovered that TIM-3 expression is not restricted to T-cells, but rather it is also found on white matter microglial cells. It has been confirmed at both the RNA and protein level that TIM-3 is constitutively expressed on microglia from the white matter but not the gray matter, and it is present at higher levels in the inflammatory environment of MS. TIM-3 expression has also been described on other myeloid derived cells such as monocytes; however, its function is quite different on these cells than it is on T-cells. While TIM-3 is a negative regulator of Th1 cells, on myeloid cells such as monocytes or microglia, TIM-3 engagement results enhanced activation with secretion of increased amounts of the proinflammatory cytokine TNFα (Anderson et al. 2007). Therefore, while activation of the TIM-3 pathway would down-regulate the initial T-cell mediated inflammation, the subsequent coexpression of galectin-9 on astrocytes and TIM-3 on microglia, would be expected to result in ongoing microglial activation and tissue destruction.

10 Conclusion

T-cells are central regulators of both adaptive immunity and specific tolerance. By virtue of their diverse repertoire of TCRs, T-cells can respond specifically to a range of different antigens, and upon antigenic encounter, they can produce a variety of either proinflammatory or antiinflammatory effects. In MS, there is evidence of both enhanced activation of effector T-cells and defective function of Treg. In recent years, much has been learned about the different subtypes of T-cells as well as the factors that mediate their differentiation, activation, and reciprocal interactions. Understanding of these principals of T-cell biology is providing new tools to explore the mechanisms through which the interactions of T-cells with other T-cells, with other cells of the peripheral immune system, and with the intrinsic cells of the CNS converge in the pathogenesis and progression of MS.

References

Abbas AK, Murphy KM, et al (1996) Functional diversity of helper T lymphocytes. Nature 383(6603):787–793

Aggarwal S, Ghilardi N, et al (2003) Interleukin-23 promotes a distinct CD4 T cell activation state characterized by the production of interleukin-17. J Biol Chem 278(3):1910–1914

Anderson AC, Anderson DE, et al (2007) Promotion of tissue inflammation by the immune receptor Tim-3 expressed on innate immune cells. Science 318(5853):1141–1143

Anderson DE, Sharpe AH, et al (1999) The B7-CD28/CTLA-4 costimulatory pathways in autoimmune disease of the central nervous system. Curr Opin Immunol 11(6):677–683

Annunziato F, Cosmi L, et al (2007) Phenotypic and functional features of human Th17 cells. J Exp Med 204(8):1849–1861

Anthony DC, Miller KM, et al (1998) Matrix metalloproteinase expression in an experimentally-induced DTH model of multiple sclerosis in the rat CNS. J Neuroimmunol 87(1–2):62–72

Astier A, Meiffren G, et al (2006) Alterations in CD46-mediated Tr1 regulatory T cells in patients with multiple sclerosis. J Clin Invest 116(12):3252–3257

Awasthi A, Carrier Y, et al (2007) A dominant function for interleukin 27 in generating interleukin 10-producing anti-inflammatory T cells. Nat Immunol 8(12):1380–1389

Babbe H, Roers A, et al (2000) Clonal expansions of CD8(+) T cells dominate the T cell infiltrate in active multiple sclerosis lesions as shown by micromanipulation and single cell polymerase chain reaction. J Exp Med 192(3):393–404

Babinski J (1885) Recherches sur l'anatomie pathologique de la sclérose en plaques et etude comparative des diverses varietes de la scléroses de la moelle. Arch Physiol 2–6:186–207

Baecher-Allan C, Brown JA, et al (2001) CD4+CD25high regulatory cells in human peripheral blood. J Immunol 167(3):1245–1253

Baecher-Allan C, Viglietta V, et al (2002) Inhibition of human CD4(+)CD25(+high) regulatory T cell function. J Immunol 169(11):6210–6217

Balashov KE, Smith DR, et al (1997) Increased interleukin 12 production in progressive multiple sclerosis: induction by activated CD4+ T cells via CD40 ligand. Proc Natl Acad Sci U S A 94(2):599–603

Barrat FJ, Cua DJ, et al (2002) In vitro generation of interleukin 10-producing regulatory CD4(+) T cells is induced by immunosuppressive drugs and inhibited by T helper type 1 (Th1)- and Th2-inducing cytokines. J Exp Med 195(5):603–616

Batten M, Li J, et al (2006) Interleukin 27 limits autoimmune encephalomyelitis by suppressing the development of interleukin 17-producing T cells. Nat Immunol 7(9):929–936

Becher B, Durell BG, et al (2002) Experimental autoimmune encephalitis and inflammation in the absence of interleukin-12. J Clin Invest 110(4):493–497

Beriou G, Costantino CM, et al (2009) IL-17 producing human peripheral regulatory T cells retain suppressive function. Blood 113(18):4240–4249

Bettelli E, Carrier Y, et al (2006) Reciprocal developmental pathways for the generation of pathogenic effector TH17 and regulatory T cells. Nature 441(7090):235–238

Bettelli E, Korn T, et al (2008) Induction and effector functions of T(H)17 cells. Nature 453(7198):1051–1057

Bo L, Peterson JW, et al (1996) Distribution of immunoglobulin superfamily members ICAM-1, -2, -3, and the beta 2 integrin LFA-1 in multiple sclerosis lesions. J Neuropathol Exp Neurol 55(10):1060–1072

Carboni S, Aboul-Enein F, et al (2003) CD134 plays a crucial role in the pathogenesis of EAE and is upregulated in the CNS of patients with multiple sclerosis. J Neuroimmunol 145(1–2):1–11

Charcot J (1868) Histologic de la sclérose en plaque. Gaz Hôp 41:554–556

International Multiple Sclerosis Genetics Consortium, Hafler DA, et al (2007) Risk alleles for multiple sclerosis identified by a genomewide study. N Engl J Med 357(9):851–862

Cua D, Sherlock J, et al (2003) Interleukin-23 rather than interleukin-12 is the critical cytokine for autoimmune inflammation of the brain. Nature 421(6924):744–748

Cuzner ML, Gveric D, et al (1996) The expression of tissue-type plasminogen activator, matrix metalloproteases and endogenous inhibitors in the central nervous system in multiple sclerosis: comparison of stages in lesion evolution. J Neuropathol Exp Neurol 55(12):1194–1204

Dardalhon V, Korn T, et al (2008) Role of Th1 and Th17 cells in organ-specific autoimmunity. J Autoimmun 31(3):252–256

De Jager PL, Hafler DA (2007) New therapeutic approaches for multiple sclerosis. Annu Rev Med 58:417–432

De Jager PL, Jia X, et al (2009a) Meta-analysis of genome scans and replication identify CD6, IRF8 and TNFRSF1A as new multiple sclerosis susceptibility loci. Nat Genet 14:14

De Jager PL, Baecher-Allan C, et al (2009b) The role of the CD58 locus in multiple sclerosis. Proc Natl Acad Sci USA 106(13):5264–5269

Dieckmann D, Bruett CH, et al (2002) Human CD4(+)CD25(+) regulatory contact-dependent T cells induce interleukin 10-producing, contact-independent type 1-like regulatory T cells [corrected]. J Exp Med 196(2):247–253

Dieckmann D, Plottner H, et al (2001) Ex vivo isolation and characterization of CD4(+)CD25(+) T cells with regulatory properties from human blood. J Exp Med 193(11):1303–1310

Duda PW, Schmied MC, et al (2000) Glatiramer acetate (Copaxone) induces degenerate, Th2-polarized immune responses in patients with multiple sclerosis. J Clin Invest 105(7):967–976

Elovaara I, Ukkonen M, et al (2000) Adhesion molecules in multiple sclerosis: relation to subtypes of disease and methylprednisolone therapy. Arch Neurol 57(4):546–551

Elyaman W, Kivisäkk P, et al (2008) Distinct functions of autoreactive memory and effector CD4+ T cells in experimental autoimmune encephalomyelitis. Am J Pathol 173(2):411–422

Engelhardt B, Ransohoff R (2005) The ins and outs of T-lymphocyte trafficking to the CNS: anatomical sites and molecular mechanisms. Trends Immunol 26(9):485–495

Fontenot JD, Rasmussen JP, et al (2005) Regulatory T cell lineage specification by the forkhead transcription factor foxp3. Immunity 22(3):329–341

Genovese MC, Becker JC, et al (2005) Abatacept for rheumatoid arthritis refractory to tumor necrosis factor alpha inhibition. N Engl J Med 353(11):1114–1123

Giovannoni G, Cutter GR, et al (2006) Infectious causes of multiple sclerosis. Lancet Neurol 5(10):887–894

Graber JJ, Allie SR, et al (2008) Interleukin-17 in transverse myelitis and multiple sclerosis. J Neuroimmunol 196(1–2):124–132

Groux H, Bigler M, et al (1998) Inhibitory and stimulatory effects of IL-10 on human CD8+ T cells. J Immunol 160(7):3188–3193

Groux H, O'Garra A, et al (1997) A CD4+ T-cell subset inhibits antigen-specific T-cell responses and prevents colitis. Nature 389(6652):737–742

Hafler DA (2004) Multiple sclerosis. J Clin Invest 113(6):788–794

Hafler DA, Fox DA, et al (1985) In vivo activated T lymphocytes in the peripheral blood and cerebrospinal fluid of patients with multiple sclerosis. N Engl J Med 312(22):1405–1411

Hafler JP, Maier LM, et al (2009) CD226 Gly307Ser association with multiple autoimmune diseases. Genes Immun 10(1):5–10

Harding FA, McArthur JG, et al (1992) CD28-mediated signalling co-stimulates murine T-cells and prevents induction of anergy in T-cell clones. Nature 356(6370):607–609

Harrington L, Hatton R, et al (2005) Interleukin 17-producing CD4+ effector T cells develop via a lineage distinct from the T helper type 1 and 2 lineages. Nat Immunol 6(11):1123–1132

Hauser SL, Bhan AK, et al (1986) Immunohistochemical analysis of the cellular infiltrate in multiple sclerosis lesions. Ann Neurol 19(6):578–587

Huan J, Culbertson N, et al (2005) Decreased FOXP3 levels in multiple sclerosis patients. J Neurosci Res 81(1):45–52

Ichiyama K, Yoshida H, et al (2008) Foxp3 inhibits RORgammat-mediated IL-17A mRNA transcription through direct interaction with RORgammat. J Biol Chem 283(25):17003–17008

Ishizu T, Osoegawa M, et al (2005) Intrathecal activation of the IL-17/IL-8 axis in opticospinal multiple sclerosis. Brain 128(Pt 5):988–1002

Ivanov I, McKenzie BS, et al (2006) The orphan nuclear receptor RORgammat directs the differentiation program of proinflammatory IL-17+ T helper cells. Cell 126(6):1121–1133

Jenkins MK, Taylor PS, et al (1991) CD28 delivers a costimulatory signal involved in antigen-specific IL-2 production by human T cells. J Immunol 147(8):2461–2466

Johnson KP, Brooks BR, et al (2001) Copolymer 1 reduces relapse rate and improves disability in relapsing-remitting multiple sclerosis: results of a phase III multicenter, double-blind, placebo-controlled trial. 1995. Neurology 57(12 Suppl 5):S16–S24

Jonuleit H, Schmitt E, et al (2002) Infectious tolerance: human CD25(+) regulatory T cells convey suppressor activity to conventional CD4(+) T helper cells. J Exp Med 196(2):255–260

Kabat EA, Glusman M, et al (1948) Quantitative estimation of the albumin and gamma globulin in normal and pathologic cerebrospinal fluid by immunochemical methods. Am J Med 4(5): 653–662

Karandikar NJ, Vanderlugt CL, et al (1996) CTLA-4: a negative regulator of autoimmune disease. J Exp Med 184(2):783–788

Kebir H, Kreymborg K, et al (2007) Human TH17 lymphocytes promote blood-brain barrier disruption and central nervous system inflammation. Nat Med 13(10):1173–1175

Keir ME, Sharpe AH (2005) The B7/CD28 costimulatory family in autoimmunity. Immunol Rev 204:128–143

Kemper C, Chan AC, et al (2003) Activation of human CD4+ cells with CD3 and CD46 induces a T-regulatory cell 1 phenotype. Nature 421(6921):388–392

Kieseier BC, Seifert T, et al (1999) Matrix metalloproteinases in inflammatory demyelination: targets for treatment. Neurology 53(1):20–25

Koguchi K, Anderson D, et al (2006) Dysregulated T cell expression of TIM3 in multiple sclerosis. J Exp Med 203(6):1413–1418

Korn T, Bettelli E, et al (2007) IL-21 initiates an alternative pathway to induce proinflammatory T(H)17 cells. Nature 448(7152):484–487

Korn T, Reddy J, et al (2007) Myelin-specific regulatory T cells accumulate in the CNS but fail to control autoimmune inflammation. Nat Med 13(4):423–431

Kostianovsky AM, Maier LM, et al (2008) Astrocytic regulation of human monocytic/microglial activation. J Immunol 181(8):5425–5432

Kroenke MA, Carlson TJ, et al (2008) IL-12- and IL-23-modulated T cells induce distinct types of EAE based on histology, CNS chemokine profile, and response to cytokine inhibition. J Exp Med 205(7):1535–1541

Krummel MF, Allison JP (1996) CTLA-4 engagement inhibits IL-2 accumulation and cell cycle progression upon activation of resting T cells. J Exp Med 183(6):2533–2540

Kuhn R, Lohler J, et al (1993) Interleukin-10-deficient mice develop chronic enterocolitis. Cell 75(2):263–274

Kulkarni AB, Huh CG, et al (1993) Transforming growth factor beta 1 null mutation in mice causes excessive inflammatory response and early death. Proc Natl Acad Sci U S A 90(2):770–774

Lafferty KJ, Woolnough J (1977) The origin and mechanism of the allograft reaction. Immunol Rev 35:231–262

Langrish CL, Chen Y, et al (2005) IL-23 drives a pathogenic T cell population that induces autoimmune inflammation. J Exp Med 201(2):233–240

Lassmann H, Raine CS, et al (1998) Immunopathology of multiple sclerosis: report on an international meeting held at the Institute of Neurology of the University of Vienna. J Neuroimmunol 86(2):213–217

Lee YK, Turner H, et al (2009) Late developmental plasticity in the T helper 17 lineage. Immunity 30(1):92–107

Lehnardt S, Massillon L, et al (2003) Activation of innate immunity in the CNS triggers neurodegeneration trough a Toll-like receptor 4-dependent pathway. Proc Natl Acad Sci U S A 100(14):8514–8519

Leppert D, Waubant E, et al (1995) T cell gelatinases mediate basement membrane transmigration in vitro. J Immunol 154(9):4379–4389

Levings MK, Sangregorio R, et al (2001) IFN-alpha and IL-10 induce the differentiation of human type 1 T regulatory cells. J Immunol 166(9):5530–5539

Liang SC, Tan XY, et al (2006) Interleukin (IL)-22 and IL-17 are coexpressed by Th17 cells and cooperatively enhance expression of antimicrobial peptides. J Exp Med 203(10):2271–2279

Liu W, Putnam AL, et al (2006) CD127 expression inversely correlates with FoxP3 and suppressive function of human CD4+ T reg cells. J Exp Med 203(7):1701–1711

Lock C, Hermans G, et al (2002) Gene-microarray analysis of multiple sclerosis lesions yields new targets validated in autoimmune encephalomyelitis. Nat Med 8(5):500–508

Maeda A, Sobel RA (1996) Matrix metalloproteinases in the normal human central nervous system, microglial nodules, and multiple sclerosis lesions. J Neuropathol Exp Neurol 55(3):300–309

Manel N, Unutmaz D, et al (2008) The differentiation of human T(H)-17 cells requires transforming growth factor-beta and induction of the nuclear receptor RORgammat. Nat Immunol 9(6): 641–649

Markovic-Plese S, Cortese I, et al (2001) CD4+CD28- costimulation-independent T cells in multiple sclerosis. J Clin Invest 108(8):1185–1194

Mathur AN, Chang HC, et al (2006) T-bet is a critical determinant in the instability of the IL-17-secreting T-helper phenotype. Blood 108(5):1595–1601

Matusevicius D, Kivisäkk P, et al (1999) Interleukin-17 mRNA expression in blood and CSF mononuclear cells is augmented in multiple sclerosis. Mult Scler 5(2):101–104

McGeachy MJ, Cua DJ (2007) The link between IL-23 and Th17 cell-mediated immune pathologies. Semin Immunol 19(6):372–376

Meylan F, Davidson TS, et al (2008) The TNF-family receptor DR3 is essential for diverse T-cell-mediated inflammatory diseases. Immunity 29(1):79–89

Miller SD, Vanderlugt CL, et al (1995) Blockade of CD28/B7-1 interaction prevents epitope spreading and clinical relapses of murine EAE. Immunity 3(6):739–745

Monje ML, Toda H, et al (2003) Inflammatory blockade restores adult hippocampal neurogenesis. Science 302(5651):1760–1765

Monney L, Sabatos CA, et al (2002) Th1-specific cell surface protein Tim-3 regulates macrophage activation and severity of an autoimmune disease. Nature 415(6871):536–541

Montes M, Zhang X, et al (2008) Oligoclonal myelin-reactive T-cell infiltrates derived from multiple sclerosis lesions are enriched in Th17 cells. Clin Immunol

Moore KW, de Waal Malefyt R, et al (2001) Interleukin-10 and the interleukin-10 receptor. Annu Rev Immunol 19:683–765

Mosmann TR, Coffman RL (1989) TH1 and TH2 cells: different patterns of lymphokine secretion lead to different functional properties. Annu Rev Immunol 7:145–173

Munger KL, Zhang SM, et al (2004) Vitamin D intake and incidence of multiple sclerosis. Neurology 62(1):60–65

Nakae S, Iwakura Y, et al (2007) Phenotypic differences between Th1 and Th17 cells and negative regulation of Th1 cell differentiation by IL-17. J Leukoc Biol 81(5):1258–1268

Navikas V, Link J, et al (1995) Increased mRNA expression of IL-10 in mononuclear cells in multiple sclerosis and optic neuritis. Scand J Immunol 41(2):171–178

Neuhaus O, Farina C, et al (2001) Mechanisms of action of glatiramer acetate in multiple sclerosis. Neurology 56(6):702–708

Nurieva R, Yang X, et al (2007) Essential autocrine regulation by IL-21 in the generation of inflammatory T cells. Nature 448(7152):480–483

Oliveira EM, Bar-Or A, et al (2003) CTLA-4 dysregulation in the activation of myelin basic protein reactive T cells may distinguish patients with multiple sclerosis from healthy controls. J Autoimmun 20(1):71–81

Ota K, Matsui M, et al (1990) T-cell recognition of an immunodominant myelin basic protein epitope in multiple sclerosis. Nature 346(6280):183–187

Panitch HS, Hirsch RL, et al (1987) Exacerbations of multiple sclerosis in patients treated with gamma interferon. Lancet 1(8538):893–895

Park H, Li Z, et al (2005) A distinct lineage of CD4 T cells regulates tissue inflammation by producing interleukin 17. Nat Immunol 6(11):1133–1141

Pettereit HF, Pukrop R, et al (2003) Low interleukin-10 production is associated with higher disability and MRI lesion load in secondary progressive multiple sclerosis. J Neurol Sci 206(2):209–214

Pette M, Fujita K, et al (1990) Myelin basic protein-specific T lymphocyte lines from MS patients and healthy individuals. Neurology 40(11):1770–1776

Prineas J (1975) Pathology of the early lesion in multiple sclerosis. Hum Pathol 6(5):531–554

Prineas JW, Wright RG (1978) Macrophages, lymphocytes, and plasma cells in the perivascular compartment in chronic multiple sclerosis. Lab Invest 38(4):409–421

Raine CS (1994) The Dale E. McFarlin Memorial Lecture: the immunology of the multiple sclerosis lesion. Ann Neurol 36(Suppl):S61–S72

Read S, Malmstrom V, et al (2000) Cytotoxic T lymphocyte-associated antigen 4 plays an essential role in the function of CD25(+)CD4(+) regulatory cells that control intestinal inflammation. J Exp Med 192(2):295–302

Reijonen H, Novak EJ, et al (2002) Detection of GAD65-specific T-cells by major histocompatibility complex class II tetramers in type 1 diabetic patients and at-risk subjects. Diabetes 51(5):1375–1382

Rivers TM, Schwentker FF (1935) Encephalomyelitis accompanied by myelin destruction experimentally produced in monkeys. J Exp Med 61:689–702

Sakaguchi S, Fukuma K, et al (1985) Organ-specific autoimmune diseases induced in mice by elimination of T cell subset. I. Evidence for the active participation of T cells in natural self-tolerance; deficit of a T cell subset as a possible cause of autoimmune disease. J Exp Med 161(1):72–87

Sakaguchi S, Yamaguchi T, et al (2008) Regulatory T cells and immune tolerance. Cell 133(5):775–787

Salama AD, Chitnis T, et al (2003) Critical role of the programmed death-1 (PD-1) pathway in regulation of experimental autoimmune encephalomyelitis. J Exp Med 198(1):71–78

Scholz C, Patton KT, et al (1998) Expansion of autoreactive T cells in multiple sclerosis is independent of exogenous B7 costimulation. J Immunol 160(3):1532–1538

Seddiki N, Santner-Nanan B, et al (2006) Expression of interleukin (IL)-2 and IL-7 receptors discriminates between human regulatory and activated T cells. J Exp Med 203(7):1693–1700

Shevach EM, McHugh RS, et al (2001) Control of T-cell activation by CD4+ CD25+ suppressor T cells. Immunol Rev 182:58–67

Shull MM, Ormsby I, et al (1992) Targeted disruption of the mouse transforming growth factor-beta 1 gene results in multifocal inflammatory disease. Nature 359(6397):693–699

Sobel RA, Mitchell ME, et al (1990) Intercellular adhesion molecule-1 (ICAM-1) in cellular immune reactions in the human central nervous system. Am J Pathol 136(6):1309–1316

Soldan SS, Alvarez Retuerto AI, et al (2004) Dysregulation of IL-10 and IL-12p40 in secondary progressive multiple sclerosis. J Neuroimmunol 146(1–2):209–215

Sørensen TL, Tani M, et al (1999) Expression of specific chemokines and chemokine receptors in the central nervous system of multiple sclerosis patients. J Clin Invest 103(6):807–815

Sporici RA, Beswick RL, et al (2001) ICOS ligand costimulation is required for T-cell encephalitogenicity. Clin Immunol 100(3):277–288

Springer TA (1994) Traffic signals for lymphocyte recirculation and leukocyte emigration: the multistep paradigm. Cell 76(2):301–314

Stromnes I, Cerretti L, et al (2008) Differential regulation of central nervous system autoimmunity by T(H)1 and T(H)17 cells. Nat Med 14(3):337–342

Stumhofer J, Laurence A, et al (2006) Interleukin 27 negatively regulates the development of interleukin 17-producing T helper cells during chronic inflammation of the central nervous system. Nat Immunol 7(9):937–945

Trajkovic V, Vuckovic O, et al (2004) Astrocyte-induced regulatory T cells mitigate CNS autoimmunity. Glia 47(2):168–179

Traugott U, Reinherz EL, et al (1983) Multiple sclerosis: distribution of T cell subsets within active chronic lesions. Science 219(4582):308–310

Tzartos JS, Friese MA, et al (2008) Interleukin-17 production in central nervous system-infiltrating T cells and glial cells is associated with active disease in multiple sclerosis. Am J Pathol 172(1):146–155

Vaknin-Dembinsky A, Balashov K, et al (2006) IL-23 is increased in dendritic cells in multiple sclerosis and down-regulation of IL-23 by antisense oligos increases dendritic cell IL-10 production. J Immunol 176(12):7768–7774 van Boxel-Dezaire AH, Hoff SC, et al (1999) Decreased interleukin-10 and increased interleukin-12p40 mRNA are associated with disease activity and characterize different disease stages in multiple sclerosis. Ann Neurol 45(6):695–703

Veldhoen M, Hocking RJ, et al (2006) TGFbeta in the context of an inflammatory cytokine milieu supports de novo differentiation of IL-17-producing T cells. Immunity 24(2):179–189

Venken K, Hellings N, et al (2006) Secondary progressive in contrast to relapsing-remitting multiple sclerosis patients show a normal CD4+CD25+ regulatory T-cell function and FOXP3 expression. J Neurosci Res 83(8):1432–1446

Vieira PL, Christensen JR, et al (2004) IL-10-secreting regulatory T cells do not express Foxp3 but have comparable regulatory function to naturally occurring CD4+CD25+ regulatory T cells. J Immunol 172(10):5986–5993

Viglietta V, Baecher-Allan C, et al (2004) Loss of functional suppression by CD4+CD25+ regulatory T cells in patients with multiple sclerosis. J Exp Med 199(7):971–979

Viglietta V, Bourcier K, et al (2008) CTLA4Ig treatment in patients with multiple sclerosis: an open-label, phase 1 clinical trial. Neurology 71(12):917–924

Volpe E, Servant N, et al (2008) A critical function for transforming growth factor-beta, interleukin 23 and proinflammatory cytokines in driving and modulating human T(H)-17 responses. Nat Immunol 9(6):650–657

Walunas TL, Lenschow DJ, et al (1994) CTLA-4 can function as a negative regulator of T-cell activation. Immunity 1(5):405–413

Wang J, Ioan-Facsinay A, et al (2007) Transient expression of FOXP3 in human activated non-regulatory CD4+ T cells. Eur J Immunol 37(1):129–138

Washington R, Burton J, et al (1994) Expression of immunologically relevant endothelial cell activation antigens on isolated central nervous system microvessels from patients with multiple sclerosis. Ann Neurol 35(1):89–97

Windhagen A, Newcombe J, et al (1995) Expression of costimulatory molecules B7–1 (CD80), B7–2 (CD86), and interleukin 12 cytokine in multiple sclerosis lesions. J Exp Med 182(6):1985–1996

Wucherpfennig KW, Newcombe J, et al (1992a) Gamma delta T-cell receptor repertoire in acute multiple sclerosis lesions. Proc Natl Acad Sci U S A 89(10):4588–4592

Wucherpfennig KW, Newcombe J, et al (1992b). T cell receptor V alpha-V beta repertoire and cytokine gene expression in active multiple sclerosis lesions. J Exp Med 175(4):993–1002

Xu L, Kitani A, et al (2007) Cutting edge: regulatory T cells induce CD4+CD25-Foxp3- T cells or are self-induced to become Th17 cells in the absence of exogenous TGF-beta. J Immunol 178(11):6725–6729

Yang L, Anderson D, et al (2008) IL-21 and TGF-beta are required for differentiation of human T(H)17 cells. Nature 454(7202):350–352

Yao Z, Painter SL, et al (1995) Human IL-17: a novel cytokine derived from T cells. J Immunol 155(12):5483–5486

Yeo TW, De Jager PL, et al (2007) A second major histocompatibility complex susceptibility locus for multiple sclerosis. Ann Neurol 61(3):228–236

Yong VW, Krekoski CA, et al (1998) Matrix metalloproteinases and diseases of the CNS. Trends Neurosci 21(2):75–80

Yoshida H, Imaizumi T, et al (2001) Interleukin-1beta stimulates galectin-9 expression in human astrocytes. Neuroreport 12(17):3755–3758

Zamvil S, Nelson P, et al (1985) T-cell clones specific for myelin basic protein induce chronic relapsing paralysis and demyelination. Nature 317(6035):355–358

Zhang GX, Baker CM, et al (2000) Chemokines and chemokine receptors in the pathogenesis of multiple sclerosis. Mult Scler 6(1):3–13

Zhang J, Markovic-Plese S, et al (1994) Increased frequency of interleukin 2-responsive T cells specific for myelin basic protein and proteolipid protein in peripheral blood and cerebrospinal fluid of patients with multiple sclerosis. J Exp Med 179(3):973–984

Zhou L, Lopes J, et al (2008) TGF-beta-induced Foxp3 inhibits T(H)17 cell differentiation by antagonizing RORgammat function. Nature 453(7192):236–240

Zhu C, Anderson AC, et al (2005) The Tim-3 ligand galectin-9 negatively regulates T helper type 1 immunity. Nat Immunol 6(12):1245–1252

B Cells and Antibodies in MS

Markus Reindl, Bettina Kuenz, and Thomas Berger

Abstract Increasing research activities on humoral immune responses involved in the immunopathogenesis of multiple sclerosis (MS) led to a revival of the importance of B cells and antibodies in MS. B cells seem now to play various immunopathogenetic roles in the initiation and propagation of inflammatory demyelinating processes at different disease stages of MS. The biological activities of antibodies in MS is, in general, still less known, although it emerges that antibodies are specifically involved in demyelination or, at least, mirror tissue destruction in the central nervous system. Finally, there is growing evidence that treatments, which specifically target B cells and/or antibodies, are effective in MS and its variants neuromyelitis optica (NMO). This chapter therefore aims to summarize the present knowledge and to outline future directions about the role of B cells and antibodies in research and therapy of MS and NMO.

1 Introduction

Since the seminal finding of elevated immunoglobulins (Ig) in the cerebrospinal fluid (CSF) of more than 90% of multiple sclerosis (MS) patients (Kabat et al. 1948), the detection of oligoclonal Ig is an important diagnostic marker in MS (Freedman et al. 2005). Since then, many studies have tried to identify both, autoantibodies directed against central nervous system (CNS) antigens and antibodies to infectious agents, in the serum and CSF of MS patients, but none of these studies convincingly demonstrated an MS specific antibody response (Reindl et al. 2006). Despite the fact that the antigenic targets of the intrathecal antibody response in MS are still unidentified there is accumulating evidence

M. Reindl(✉), B. Kuenz and T. Berger
Neuroimmunological and Multiple Sclerosis Clinic and Research Unit,
Clinical Department of Neurology, Innsbruck Medical University,
Anichstrasse 35, A-6020 Innsbruck, Austria
e-mail: Markus.Reindl@i-med.ac.at

Results Probl Cell Differ, doi:10.1007/400_2008_16
© Springer-Verlag Berlin Heidelberg 2009

from immunological, pathological, and therapeutic studies that B cells are key components in the pathophysiology of MS. In this chapter we aim to summarize the present knowledge on the role of B cells and antibodies in MS and to outline future directions in related research and MS therapy.

2 The Role of CSF B Cells in CNS Inflammation

The early differentiation of pro- and pre-B cell precursors to naïve immature B cells from hematopoetic stem cells occurs in the bone marrow. Naïve immature B cells then migrate from the bone marrow to secondary lymphoid tissues, mainly the spleen, in order to emerge as CD19+ CD20+ surface(s)IgM–sIgD+ mature naïve B cells. In line with an antigen-driven germinal center reaction in secondary lymphatic organs (MacLennan 1994) and after stimulation by T-cells and dendritic cells, B cells become activated and differentiate to memory B cells (CD19+ CD20+ CD27+ CD138–) and antibody-secreting effector cells, plasma blasts (CD19+ CD27++ CD138+ CD38++ HLA-DR++), and plasma cells (CD19–CD27++ CD138+ CD38+ HLA-DR–), which subsequently can enter the circulation (Shapiro-Shelef and Calame 2005). Whereas plasma blasts are short-lived and disappear quickly after the removal of their challenging antigens (Bernasconi et al. 2002), plasma cells are long-living for several months to years in specific survival niches (e.g., bone marrow, inflamed tissue). Plasma cells release antibodies, thus assuring humoral immunity, but are also potent to generate autoimmunity (Arce et al. 2002; Manz et al. 2002, 2005). The maturation and migration of B cells is summarized in Fig. 1.

CD19+ B cells accumulate in the CSF during CNS inflammation, but are largely absent in noninflammatory conditions (Cepok et al. 2001; Kuenz et al. 2008). The majority of these B cells are CD27+ antigen-experienced memory B cells (Cepok et al. 2005a, 2006; Corcione et al. 2004), which are, irrespective of antigen presence, considered to persist in the CNS compartment for years. The main effector B cell subset in the CSF of MS patients are short-lived plasma blasts (CD19+ CD27++ CD138+), which in an infectious condition usually occur quickly in CSF following antigen contact, but disappear also quickly after antigen removal, thus indicating in the condition of MS a sustained local antigen exposition to the immune system (Cepok et al. 2005a; Winges et al. 2007). Although any prevalence of specific CSF immune cells in different MS disease courses could not be proven, a predominance of B cells was associated with more rapid disease progression (Cepok et al. 2001). Recently we could demonstrate that the extent of CSF B cells correlate with brain inflammation measured by MRI and with other inflammatory CSF parameters such as intrathecal IgM and IgG production and intrathecal matrix metalloproteinase (MMP)-9 (Kuenz et al. 2008). These findings were only seen in clinically isolated syndromes (CIS) and relapsing–remitting (RR) MS, but not in chronic progressive MS, thus indicating an important role of CSF B cells in acute brain inflammation.

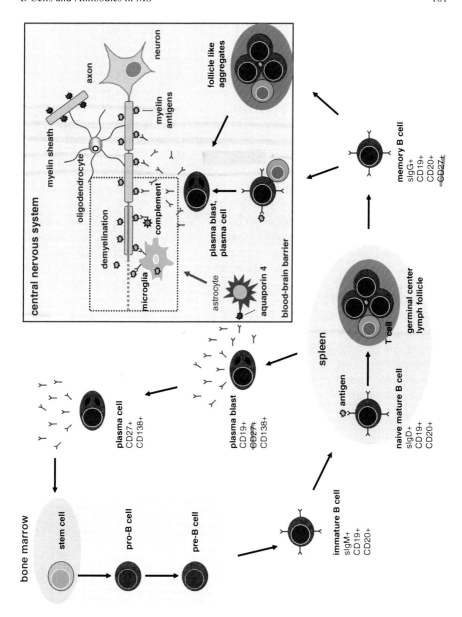

Fig. 1 Maturation and migration of B cells to the CNS. B-cell development starts in the bone marrow and is continued in the spleen, where mature naïve B cells differentiate in an antigen-driven germinal center reaction to memory B cells or plasmablasts which then appear in peripheral blood. Plasmablasts can then enter the bone marrow as long-lived plasma cells. Memory B cells can enter the CNS and differentiate into antibody-secreting cells in response to antigen outside of follicles or in meningeal follicle-like aggregates. Plasmablasts that find appropriate survival conditions in the CNS develop to long-lived plasma cells that cause oligoclonal bands. B cells and antibodies and complement can contribute to BBB damage, demyelination and axonal injury.

All these findings are, however, not specific for MS. In acute infectious or chronic inflammatory neurological diseases such as acute meningoradiculitis, caused by *Borrelia burgdorferi,* CSF cells may comprise up to 30% of B cells. During remission cytokine levels normalize and plasma cells disappear, followed by a long-lasting antigen-specific B cell response within the CNS (Cepok et al. 2003). Further, Cepok et al. could demonstrate the persistence of CSF CD19+ B cells in neuroborreliosis for at least 1.5 years after disease onset and treatment, whereas CD19+ CD138+ plasma blasts disappeared from CSF within weeks after initial diagnostic lumbar puncture. By contrast, in MS patients levels of B cells and plasma blasts remained constantly elevated during the course of the disease (observation period > 2.5 years) (Cepok et al. 2005a). With regard to chronic neuroinflammation caused by human immunodeficiency virus (HIV) infection, the incidence of CSF B cells was shown to be consistent during early and late stages of HIV infection. However, plasma blasts prevailed in the CSF during early infection and correlated with HIV RNA copy numbers in CSF, thus indicating an HIV-triggered CSF B-cell response with plasma blasts serving as the main virus-related B-cell subset (Cepok et al. 2007).

In conclusion, several studies indicate that the migration and persistence of B cells in the CSF compartment is an early step in CNS inflammation, which seems to be controlled by persistent antigenic stimulation. The crucial role of B cells for CNS inflammation was also confirmed by recent studies, which demonstrated a highly significant reduction of Gd-enhancing brain MRI lesions in MS patients treated with the B cell depleting anti-CD20 antibody rituximab (Bar-Or et al. 2008; Cross et al. 2006; Hauser et al. 2008).

3 The Role of B Cells in CNS Pathology

The characterization of B cell populations in MS plaques revealed an accumulation of clonally expanded B lymphocytes (Baranzini et al. 1999; Colombo et al. 2000; Owens et al. 1998, 2003) and neuropathological studies identified a key role of B cells, antibodies and complement in demyelination (Breij et al. 2008; Lucchinetti et al. 1996; Storch et al. 1998).

Besides their role in acute demyelination, B cells could also propagate important disease progression in MS (Dal Bianco et al. 2008; Magliozzi et al. 2007; Uccelli et al. 2005). Whereas in early RRMS new focal white matter lesions develop following new waves of inflammation, which enter the brain from the peripheral immune system and cause major blood brain barrier leakage, the inflammatory reaction in progressive MS seems to be compartmentalized within the CNS tissue. This is reflected by the formation of aberrant lymphatic tissue within the connective tissue compartments in the brain, the meninges and the perivascular spaces (Serafini et al. 2004). These meningeal ectopic follicular B cell structures might be crucial for sustained pathogenic B cell responses (Uccelli et al. 2005). Lymphoid follicles with germinal centers are critical sites where antigen-activated B cells undergo somatic hypermutation and affinity maturation (Shapiro-Shelef and Calame 2005). Lymphoid follicle-like aggregates have also been identified in the meninges of animals affected by experimental autoimmune encephalomyelitis (EAE) and in the cerebral meninges of patients with secondary chronic progressive MS, but not in RRMS and primary chronic progressive MS (Magliozzi et al. 2004, 2007; Serafini et al. 2004). Consistent with these results, a recent study reported plasma cells (CD19–CD138+) and centroblasts (CD19+ CD38+ CD27+) in the CSF of MS patients (Corcione et al. 2004). Centroblasts are found exclusively in the germinal centers of secondary lymphoid structures and their presence in the CSF of MS patients thus indicates the occurrence of a germinal centre reaction in the CNS with consecutive spreading of these cells into the CSF. These intrameningeal follicles in secondary chronic progressive MS have been shown to contain B cells, T cells, plasma cells and a network of follicular dendritic cells producing homing lymphoid chemokines, such as CXCL-13 (Serafini et al. 2004). CXCL-13, also known as B-lymphocyte chemoattractant (BLC) or B cell-attracting chemokine 1 (BCA-1), is a member of the CXC subtype of the chemokine superfamily constitutively expressed in the B cell follicles of secondary lymphoid organs (Gunn et al. 1998). CXCL-13 plays a crucial role in germinal centre organization, recruitment of B cells into the lymphatic tissue and for follicle development by triggering a positive feedback loop (Allen et al. 2004; Yu et al. 2002). Via a stimulatory cytokine loop CXCL-13 induces upregulation of lymphotoxin $\alpha1\beta2$ followed by an enhanced development of follicular dendritic cells with subsequent production of CXCL-13 in turn (Ansel et al. 2000). Since the ectopic expression of CXCL-13 is considered to promote the formation of lymphoid-like structures (Luther et al. 2003) its expression in MS lesion might have similar effects in this disease (Krumbholz et al. 2005; Serafini et al. 2004). Further, in recent study a strong correlation between CXCL-13 levels and inflammatory activity in MS, reflected by an upregulation of CXCL-13 in actively demyelinating MS lesions and increased CXCL-13 concentrations with intrathecal IgG synthesis and the presence of B cells, T cells, and plasma blasts in the CSF of MS patients was observed (Krumbholz et al. 2006; Kuenz et al. 2008).

All these data indicate that a chronic intrathecal B cell response may develop behind an intact (or repaired) blood brain barrier, become independent from peripheral immune control and drive chronic MS disease progression by cortical demyelination and diffuse axonal injury in the so-called "normal appearing white matter."

In conclusion B cells could contribute to MS disease pathogenesis in various ways (Fig. 2):

(1) B cells can serve as antigen-presenting cells for autoreactive T cells. This possible function is supported by the observation that the epitope specificity of antibodies generated during EAE, encephalitogenic T cell epitopes, and human immunodominant T and B cell epitopes are often overlapping (Lindert et al. 1999; Wang and Fujinami 1997; Wucherpfennig et al. 1997).

(2) Production of proinflammatory and regulatory cytokines in chronic progressive MS.

(3) The production of myelin-specific antibodies and the destruction of myelin within plaques (see also Sect. 4).

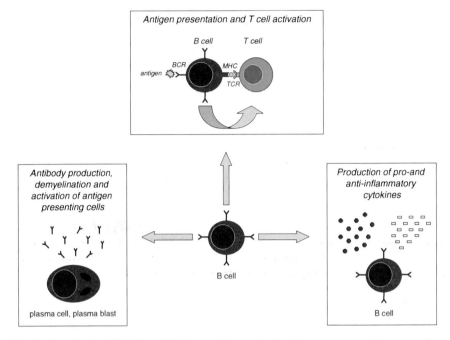

Fig. 2 Contribution of B cells to MS disease pathogenesis. B cells can serve as antigen-presenting cells for autoreactive T cells, secrete proinflammatory and regulatory cytokines and secrete myelin-specific antibodies and thus be involved in the destruction of myelin within MS plaques

4   The Role of B Cells in MS: Therapeutic Studies

Pathological studies indicated that antibodies play an important role in immunopathogenesis in a major subset of MS patients (Lucchinetti et al. 2000) thus providing a rationale for treatment strategies which target pathogenic humoral immune responses, such as intravenous immunoglobulins or plasma exchange. Indeed, some case series and small, but well-designed controlled studies demonstrated that at least subgroups of MS patients with either severe relapses or other more rare idiopathic inflammatory demyelinating CNS diseases recovered promptly after plasma exchange (Keegan et al. 2005, 2002; Moldenhauer et al. 2005; Weinshenker et al. 1999).

Rituximab, a chimeric (mouse/human) monoclonal antibody, depletes the CD20+ B cell subpopulation. As shown in Fig. 1 CD20 is a B-cell surface antigen with restricted expression on pre-B and mature B cells, but lacking on stem cells and on B cells differentiating into plasma cells. Cell lysis of CD20+ B lymphocytes is supposed to be mediated by rituximab via a combination of complement-dependent and antibody-dependent, cell-mediated cytotoxicity and the promotion of apoptosis (Reff et al. 1994). Hauser et al. reported in a phase II trial a reduction of inflammatory brain MRI lesions and clinical relapses for 48 weeks in patients with RRMS after a single course of rituximab. CD20+ peripheral B cells were rapidly and completely depleted after rituximab administration, but median immunoglobulin levels were not significantly reduced due to absence of the surface antigen on plasma cells (Hauser et al. 2008). Another recent 72-week, open-label, phase I trial in relapsing remitting MS patients showed a considerable reduction concerning relapse rate and the number of Gd-enhancing or T2 lesions after twofold treatment with rituximab at an interval of 6 months (Bar-Or et al. 2008). These preliminary impressive effects of rituximab on disease activity raises the question how B-cell depletion mediates these effects. Since rituximab does not target mature plasma cells, which are the main producers of (so far in MS undefined) pathological antibodies, it is rather unlikely that the reduction of presumed pathologic antibodies alone can explain the rapid effects of this treatment. It seems more likely that rituximab is targeting the development of lesions in acute disease including antigen presentation and activation of T cells. Thus, these studies indicate that the biological role of B cells in MS is much broader as anticipated and includes not only the production of presumed pathogenic antibodies, but also the antigen presentation and activation of B cells and the production of inflammatory and regulatory cytokines (Fig. 2).

5 Specificity of the Antibody Response in MS

5.1 Intrathecal IgG Production and OCBs

The qualitative and quantitative measurement of intrathecal IgG production is an established paraclinical method in the diagnosis of MS. Quantitative measurement of IgG production by the IgG index has a sensitivity of about 73% for MS (Freedman et al. 2005), whereas the qualitative detection of intrathecal IgG synthesis by isoelectric focusing offers a sensitivity for MS higher than 95% (Kabat et al. 1948; McDonald et al. 2001). Although the precise nature of oligoclonal bands (OCB) in MS is still unclear and no MS specific antibody response has been identified so far, the presence of elevated IgG and OCB indicates a localized B-cell expansion within the brain (Reiber 1998). Further, serial investigations demonstrated persistence or even accumulation of OCB in the CSF of MS patients, reflecting a continuous intrathecal humoral immune response (Link and Huang 2006). However, both quantitative and qualitative IgG synthesis are not specific for MS but are also found in other inflammatory, especially infectious, neurological diseases such as neuroborreliosis (Freedman et al. 2005; Hansen et al. 1990). Although the majority of CSF specialists doubt the methods (e.g., isoelectric focusing) for qualitative detection of oligoclonal IgM bands several studies argue for an association of intrathecal IgM with a worse disease course, comprising a shorter time period to the next relapse, an earlier disease conversion to secondary chronic progressive MS and a higher disability score (Sharief et al. 1990; Sharief and Thompson 1991; Villar et al. 2002, 2003).

5.2 OCB and EBV Antigens

Epstein–Barr virus (EBV), a ubiquitous B-lymphotrophic herpesvirus, is also considered to be associated with MS, albeit positive proof of its involvement in the immunopathogenesis of the disease is still missing. Several sero-epidemiological studies consistently demonstrated a higher EBV seropositivity rate in MS patients than in controls (99% versus 90–95%) and in children with MS compared to age-matched controls (83–99% versus 42–72%), respectively (Ascherio and Munger 2007; Haahr and Hollsberg 2006; Pohl et al. 2006; Wandinger et al. 2000). Moreover, infectious mononucleosis and elevated serum anti-EBV antibody titers, particularly IgG antibodies to EBNA-1 and EBNA complex, have been shown to be associated with an increased risk of developing MS (DeLorenze et al. 2006; Haahr and Hollsberg 2006; Levin et al. 2005; Thacker et al. 2006), thus indicating an indirect role of EBV in activation of underlying disease processes and MS development. A raised humoral immune response to EBV and oligoclonal IgG binding to EBV proteins, EBNA-1 and BRRF2, was also detected in CSF of MS patients (Bray et al. 1992; Cepok et al. 2005b). Consistent with abnormal virus-targeting immune responses, an increased EBV-specific CD4+ and CD8+ T-cell response was found in CSF and peripheral blood of MS patients (Cepok et al. 2005b; Hollsberg et al. 2003; Lunemann et al. 2006). Recently, Serafini et al. showed an accumulation of EBV-infected B cells as well as plasma cells in postmortem brain tissue from MS cases with different clinical disease courses, though ectopic B-cell follicle-like structures within the cerebral meninges of some secondary chronic progressive MS cases constituted the major sites of EBV persistence (Serafini et al. 2007). In addition, activation of CD8+ T-cells, which are mediating cytotoxicity towards plasma cells and mainly restricted to accumulation spots of EBV-infected cells, was demonstrated in the same study (Serafini et al. 2007). Apart from the fact that these data and their reproducibility provoked controversial discussions, these findings could path the way to new insights, not only in MS pathogenesis, but also in the role of (EBV transformed) B cells in MS.

5.3 Anti-Myelin Antibodies

Several CNS-antigens have been described as targets for autoantibodies in MS, but their role in disease pathogenesis is still obscure (Reindl et al. 2006). One exception is myelin–oligodendrocyte–glycoprotein (MOG), a CNS specific antigen, which is exclusively localized on the surface of myelin sheaths and oligodendrocytes (Brunner et al. 1989). Antibodies to MOG cause demyelination in vitro (Kerlero de Rosbo et al. 1990), in vivo (Linington et al. 1988) and were found in active MS brain lesions (Genain et al. 1999). Based on the important role of anti-MOG antibodies in experimental models of MS, the presence of anti-MOG antibodies in the CSF and serum of MS patients has been extensively studied (Reindl et al. 2006). Results were controversial, with frequencies ranging from 0 to 80% in MS patients and 0 to 60% in healthy controls, mainly due to the use of different antigen preparations and detection methods. Most of these studies used soluble, refolded MOG produced in *Escherichia coli* lacking the correct membrane topology and glycosylation. However, several recent studies have shown that these factors are crucial for the pathogenic activity of MOG antibodies (Marta et al. 2005). Thus it would be helpful to develop immunoassays recognizing antibodies to conformational epitopes on surface-exposed and glycosylated MOG. Indeed, promising data from recent studies seem to detect increased serum levels of these antibodies in MS patients using mammalian cells expressing human MOG on their surface (Gaertner et al. 2004; Lalive et al. 2006; Zhou et al. 2006). The authors identified elevated titers of anti-MOG antibody in sera of a subset of MS patients, which were cytotoxic in vitro and induced MS-like disease when injected in susceptible rats, thus supporting a pathogenic role of anti-MOG antibodies in MS (Zhou et al. 2006). However, it remains to be elucidated if the detection of anti-MOG antibodies directed against the native protein really reflects an autoantibody driven immunopathogenic process in MS patients and if these patients consequently respond to a specifically targeted therapy. Most recently a novel tetramer radioimmunoassay was reported to be another improvement in the detection of conformation-dependent anti-MOG antibodies (O'Connor et al. 2007). The authors developed a sensitive assay based on self-assembling radio-labeled tetramers, which were more sensitive for anti-MOG autoantibody detection than other methods. Using this assay MOG-specific autoantibodies were identified in a subgroup of (juvenile) patients with acute demyelinating encephalomyelitis (ADEM) but only rarely in adult-onset MS cases.

Apart from their possible pathogenetic role, the presence of anti-myelin antibodies in MS have been suggested to be a valuable biological marker for prognosis and disease progression in MS. A clinical significance of serum anti-myelin antibodies in MS patients has been first reported by a study from our department (Berger et al. 2003). In this study we demonstrated that the presence of serum IgM anti-MOG and anti-myelin basic protein (MBP) antibodies predicts the risk of conversion to clinically definite MS in patients with a CIS. Since this publication the predictive role of anti-MOG and anti-myelin basic protein (MBP) antibodies has been published in eight different CIS cohorts with correlations ranging from highly significant (Berger et al. 2003; Greeve et al. 2007; Tomassini et al. 2007), significant in subanalyses (Kuhle et al. 2007a; Rauer et al. 2006) to not significant at all (Kuhle et al. 2007b; Lim et al. 2005; Pelayo et al. 2007). It is important to note that these controversial results were all obtained with the same detection method for antimyelin antibodies, i.e., immunoblotting. Thus, these different results may primarily reflect differences in study cohorts rather than methodical variations. Recently we have analyzed by ELISA whether the presence of serum anti-MOG antibodies was associated with an increased risk of MS in healthy young adults (Wang et al. 2008). We found that the presence of serum anti-MOG antibodies was associated with an increase in the risk of developing MS although this association was attenuated and no longer significant after adjustment for titers of antibodies to EBV (EBNA-1). Further, we found a significant correlation between EBNA antibodies and MOG antibodies. Although it has been documented that EBV antibodies may recognize, and possibly react with MBP

(Bray et al. 1992), a possible association between antibody responses to EBV antigens and MOG has now to be assessed in future studies. In this study we have examined the association between MOG antibody status and risk of MS in a prospective way, due to the availability of serum samples before the onset of MS. Our findings provide moderate support for the notion that MOG antibodies might have a place in the prediction and prognosis of MS. This observation, together with the finding that circulating titers of EBNA antibodies are strongly associated with risk of MS (Levin et al. 2005), might however also reflect a general activation of peripheral B cells in MS.

5.4   Anti-aquaporin-4 Antibody

NMO is an inflammatory demyelinating disease with generally poor prognosis, selectively targeting optic nerves and spinal cord (Wingerchuk et al. 1999). Neuropathologically NMO seems to be the most prominent clinical presentation of antibody mediated demyelination (Lucchinetti et al. 2002). Early accurate diagnosis is important because NMO is associated with a substantial higher mortality and morbidity compared to MS. Treatment approaches differ considerably between NMO and MS, because approved disease modifying therapies used in MS are not effective in NMO.

Recently, specific serum IgG autoantibodies (NMO-IgG) was detected in NMO patients using indirect immunofluorescence staining (Lennon et al. 2004). The targeted binding site of this antibody was identified as the aquaporin-4 (AQP4) water channel, located at astrocytic foot processes attached to the blood–brain barrier (Lennon et al. 2005). Aquaporin-4 is diffusely expressed throughout the healthy brain, but selectively lost in early lesions of NMO (Misu et al. 2007; Roemer et al. 2007). These data indicate that aquaporin-4-specific antibodies are induced by an unknown peripheral stimulus with consecutive transmigration through the damaged blood–brain barrier and trigger a specific inflammatory response within the CNS (Ishizu et al. 2005). Lennon et al. demonstrated seropositivity for NMO-IgG antibodies in 73% of patients with definite NMO and in 46% of patients at high risk for NMO (Lennon et al. 2004). The high specificity (>90%) of NMO-IgG suggests these antibodies to be useful biomarkers in the diagnosis of NMO (Weinshenker et al. 2006a; Wingerchuk et al. 2006). Recently, these findings were confirmed by several other laboratories using different assays and cohorts of patients (Jarius et al. 2007; Marignier et al. 2008; Matsuoka et al. 2007; Paul et al. 2007; Takahashi et al. 2007). In addition, seropositivity for NMO-IgG at the initial occurrence of longitudinally extensive transverse myelitis might predict a future relapse of myelitis or optic neuritis (Weinshenker et al. 2006b). Thus NMO-IgG antibodies could serve not only as diagnostic but also prognostic markers providing a specific distinction of NMO from MS as well as other neurological diseases with consequent stratification for NMO specific treatment, such as plasma exchange (Keegan et al. 2002) or rituximab (Cree et al. 2005).

6 Outlook

MS has long been considered as T-cell mediated autoimmune disease of the CNS. However, there is accumulating evidence from immunological, pathological and therapeutic studies that B cells are key components in the pathophysiology of MS. B cells seem to play various roles in the initiation and propagation of inflammatory demyelinating processes at different disease stages of MS. Recent therapeutic trials indicated that monoclonal antibodies that specifically target B cells (e.g., rituximab) are effective in MS and its variant NMO. Future studies are now necessary to identify markers, such as CSF B-cell profiles, which could be helpful in identifying patients who might benefit from these treatments. Further, it would be very important to identify MRI correlates of cellular responses in the CSF and in MS lesions in future studies.

Several studies indicated increased antibody responses to EBV in MS and recent findings indicated the presence of EBV-infected B cells as well as plasma cells in postmortem brain tissue from MS cases. Additional studies are now necessary to reproduce these important findings and to investigate whether circulating titers of EBNA antibodies really reflect infection by EBV or rather a general activation of peripheral B cells in MS.

Although antibodies to various autoantigens are frequently found in serum and CSF of MS patients, it remains, with the exception of NMO-IgG, to be proven whether these autoantibodies are involved in the initiation and/or propagation of the pathology of MS or whether they develop as an epiphenomenon in response to tissue damage. In the first case detection of pathogenic autoantibodies would be an important step to identify patients who then might respond to a specifically tailored therapy. In the latter case the presence of autoantibodies might serve as biological markers for prognosis and disease monitoring (with regard to, e.g., disease progression and, indirectly, treatment responses) in the management of MS patients. In both cases the development of assays that detect specific antibodies in MS patients, like it has been proven for NMO-IgG, is urgently needed. Future studies should focus on antibodies directed to native, correctly-folded CNS antigens and a combination of different antigens in multiplex assays might also improve the specificity and sensitivity of these assays. Future studies should be performed in prospective MS cohorts and include MRI correlates of inflammation and progression.

References

Allen CD et al. (2004) Germinal center dark and light zone organization is mediated by CXCR4 and CXCR5. Nat Immunol 5:943–952

Ansel KM et al. (2000) A chemokine-driven positive feedback loop organizes lymphoid follicles. Nature 406:309–314

Arce S et al. (2002)wThe role of long-lived plasma cells in autoimmunity. Immunobiology 206:558–562

Ascherio A, Munger KL (2007) Environmental risk factors for multiple sclerosis. Part I: the role of infection. Ann Neurol 61:288–299

Bar-Or A et al. (2008) Rituximab in relapsing-remitting multiple sclerosis: a 72-week, open-label, phase I trial. Ann Neurol 63:395–400

Baranzini SE et al. (1999) B cell repertoire diversity and clonal expansion in multiple sclerosis brain lesions. J Immunol 163:5133–5144

Berger T et al. (2003) Antimyelin antibodies as a predictor of clinically definite multiple sclerosis after a first demyelinating event. N Engl J Med 349:139–145

Bernasconi NL et al. (2002) Maintenance of serological memory by polyclonal activation of human memory B cells. Science 298:2199–2202

Bray PF et al. (1992) Antibodies against Epstein-Barr nuclear antigen (EBNA) in multiple sclerosis CSF, and two pentapeptide sequence identities between EBNA and myelin basic protein. Neurology 42:1798–1804

Breij EC et al. (2008) Homogeneity of active demyelinating lesions in established multiple sclerosis. Ann Neurol 63:16–25

Brunner C et al. (1989) Differential ultrastructural localization of myelin basic protein, myelin/ oligodendroglial glycoprotein, and 2′,3′-cyclic nucleotide 3′-phosphodiesterase in the CNS of adult rats. J Neurochem 52:296–2304

Cepok S et al. (2001) Patterns of cerebrospinal fluid pathology correlate with disease progression in multiple sclerosis. Brain 124:2169–2176

Cepok S et al. (2003) The immune response at onset and during recovery from Borrelia burgdorferi meningoradiculitis. Arch Neurol 60:849–855

Cepok S et al. (2005a) Short-lived plasma blasts are the main B cell effector subset during the course of multiple sclerosis. Brain 128:1667–1676

Cepok S et al. (2005b) Identification of Epstein-Barr virus proteins as putative targets of the immune response in multiple sclerosis. J Clin Invest 115:1352–1360

Cepok S et al. (2006) Accumulation of class switched IgD-IgM- memory B cells in the cerebrospinal fluid during neuroinflammation. J Neuroimmunol 180:33–39

Cepok S et al. (2007) Viral load determines the B-cell response in the cerebrospinal fluid during human immunodeficiency virus infection. Ann Neurol 62:458–467

Colombo M et al. (2000) Accumulation of clonally related B lymphocytes in the cerebrospinal fluid of multiple sclerosis patients. J Immunol 164:2782–2789

Corcione A et al. (2004) Recapitulation of B cell differentiation in the central nervous system of patients with multiple sclerosis. Proc Natl Acad Sci U S A 101:11064–11069

Cree BA et al. (2005) An open label study of the effects of rituximab in neuromyelitis optica. Neurology 64:1270–1272

Cross AH et al. (2006) Rituximab reduces B cells and T cells in cerebrospinal fluid of multiple sclerosis patients. J Neuroimmunol 180:63–70

Dal Bianco A et al. (2008) Multiple sclerosis and Alzheimer's disease. Ann Neurol 63:174–183

DeLorenze GN et al. (2006) Epstein-Barr virus and multiple sclerosis: evidence of association from a prospective study with long-term follow-up. Arch Neurol 63:839–844

Freedman MS et al. (2005) Recommended standard of cerebrospinal fluid analysis in the diagnosis of multiple sclerosis: a consensus statement. Arch Neurol 62:865–870

Gaertner S et al. (2004) Antibodies against glycosylated native MOG are elevated in patients with multiple sclerosis. Neurology 63:2381–2383

Genain CP et al. (1999) Identification of autoantibodies associated with myelin damage in multiple sclerosis. Nat Med 5:170–175

Greeve I et al. (2007) Anti-myelin antibodies in clinically isolated syndrome indicate the risk of multiple sclerosis in a Swiss cohort. Acta Neurol Scand 116:207–210

Gunn MD et al. (1998) A B-cell-homing chemokine made in lymphoid follicles activates Burkitt's lymphoma receptor-1. Nature 391:799–803

Haahr S, Hollsberg P (2006) Multiple sclerosis is linked to Epstein-Barr virus infection. Rev Med Virol 16:297–310

Hansen K et al. (1990) Oligoclonal Borrelia burgdorferi-specific IgG antibodies in cerebrospinal fluid in Lyme neuroborreliosis. J Infect Dis 161:1194–1202

Hauser SL et al. (2008) B-cell depletion with rituximab in relapsing-remitting multiple sclerosis. N Engl J Med 358:676–688

Hollsberg P et al. (2003) Altered CD8+ T cell responses to selected Epstein-Barr virus immunodominant epitopes in patients with multiple sclerosis. Clin Exp Immunol 132:137–143

Ishizu T et al. (2005) Intrathecal activation of the IL-17/IL-8 axis in opticospinal multiple sclerosis. Brain 128:988–1002

Jarius S et al. (2007) NMO-IgG in the diagnosis of neuromyelitis optica. Neurology 68:1076–1077

Kabat EA et al. (1948) Quantitative estimation of the albumin and gamma-globulin in normal and pathological cerebrospinal fluid by immunochemical methods. Am J Med 4:653–662

Keegan M et al. (2002) Plasma exchange for severe attacks of CNS demyelination: predictors of response. Neurology 58:143–146

Keegan M et al. (2005) Relation between humoral pathological changes in multiple sclerosis and response to therapeutic plasma exchange. Lancet 366:579–582

Kerlero de Rosbo N et al. (1990) Demyelination induced in aggregating brain cell cultures by a monoclonal antibody against myelin/oligodendrocyte glycoprotein. J Neurochem 55:583–587

Krumbholz M et al. (2005) BAFF is produced by astrocytes and up-regulated in multiple sclerosis lesions and primary central nervous system lymphoma. J Exp Med 201:195–200

Krumbholz M et al. (2006) Chemokines in multiple sclerosis: CXCL12 and CXCL13 up-regulation is differentially linked to CNS immune cell recruitment. Brain 129:200–211

Kuenz B et al. (2008) Cerebrospinal fluid B cells correlate with early brain inflammation in multiple sclerosis. PloSONE 3(7):e2559

Kuhle J et al (2007a) Antimyelin antibodies in clinically isolated syndromes correlate with inflammation in MRI and CSF. J Neurol 254:160–168

Kuhle J et al. (2007b) Lack of association between antimyelin antibodies and progression to multiple sclerosis. N Engl J Med 356:371–378

Lalive PH et al. (2006) Antibodies to native myelin oligodendrocyte glycoprotein are serologic markers of early inflammation in multiple sclerosis. Proc Natl Acad Sci U S A 103:2280–2285

Lennon VA et al. (2004) A serum autoantibody marker of neuromyelitis optica: distinction from multiple sclerosis. Lancet 364:2106–2112

Lennon VA et al. (2005) IgG marker of optic-spinal multiple sclerosis binds to the aquaporin-4 water channel. J Exp Med 202:473–477

Levin LI et al. (2005) Temporal relationship between elevation of epstein-barr virus antibody titers and initial onset of neurological symptoms in multiple sclerosis. JAMA 293:2496–2500

Lim ET et al. (2005) Anti-myelin antibodies do not allow earlier diagnosis of multiple sclerosis. Mult Scler 11:492–494

Lindert RB et al. (1999) Multiple sclerosis: B- and T-cell responses to the extracellular domain of the myelin oligodendrocyte glycoprotein. Brain 122(Pt 11):2089–2100

Linington C et al. (1988) Augmentation of demyelination in rat acute allergic encephalomyelitis by circulating mouse monoclonal antibodies directed against a myelin/oligodendrocyte glycoprotein. Am J Pathol 130:443–454

Link H, Huang YM (2006) Oligoclonal bands in multiple sclerosis cerebrospinal fluid: an update on methodology and clinical usefulness. J Neuroimmunol 180:17–28

Lucchinetti CF et al. (1996) Distinct patterns of multiple sclerosis pathology indicates heterogeneity on pathogenesis. Brain Pathol 6:259–274

Lucchinetti C et al. (2000) Heterogeneity of multiple sclerosis lesions: implications for the pathogenesis of demyelination. Ann Neurol 47:707–717

Lucchinetti CF et al. (2002) A role for humoral mechanisms in the pathogenesis of Devic's neuromyelitis optica. Brain 125:1450–1461

Lunemann JD et al. (2006) Increased frequency and broadened specificity of latent EBV nuclear antigen-1-specific T cells in multiple sclerosis. Brain 129:1493–1506

Luther SA et al. (2003) Overlapping roles of CXCL13, interleukin 7 receptor alpha, and CCR7 ligands in lymph node development. J Exp Med 197:1191–1198

MacLennan IC (1994) Germinal centers. Annu Rev Immunol 12:117–139

Magliozzi R et al. (2004) Intracerebral expression of CXCL13 and BAFF is accompanied by formation of lymphoid follicle-like structures in the meninges of mice with relapsing experimental autoimmune encephalomyelitis. J Neuroimmunol 148:11–23

Magliozzi R et al. (2007) Meningeal B-cell follicles in secondary progressive multiple sclerosis associate with early onset of disease and severe cortical pathology. Brain 130:1089–1104

Manz RA et al. (2002) Humoral immunity and long-lived plasma cells. Curr Opin Immunol 14:517–521

Manz RA et al. (2005) Maintenance of serum antibody levels. Annu Rev Immunol 23:367–386

Marignier R et al. (2008) NMO-IgG and Devic's neuromyelitis optica: a French experience. Mult Scler 14:440–445

Marta CB et al. (2005) Pathogenic myelin oligodendrocyte glycoprotein antibodies recognize glycosylated epitopes and perturb oligodendrocyte physiology. Proc Natl Acad Sci U S A 102:13992–13997

Matsuoka T et al. (2007) Heterogeneity of aquaporin-4 autoimmunity and spinal cord lesions in multiple sclerosis in Japanese. Brain 130:1206–1223

McDonald WI et al. (2001) Recommended diagnostic criteria for multiple sclerosis: guidelines from the International Panel on the diagnosis of multiple sclerosis. Ann Neurol 50:121–127

Misu T et al. (2007) Loss of aquaporin 4 in lesions of neuromyelitis optica: distinction from multiple sclerosis. Brain 130:1224–1234

Moldenhauer A et al. (2005) Immunoadsorption patients with multiple sclerosis: an open-label pilot study. Eur J Clin Invest 35:523–530

O'Connor KC et al. (2007) Self-antigen tetramers discriminate between myelin autoantibodies to native or denatured protein. Nat Med 13:211–217

Owens GP et al. (1998) Restricted use of VH4 germline segments in an acute multiple sclerosis brain. Ann Neurol 43:236–243

Owens GP et al. (2003) Single-cell repertoire analysis demonstrates that clonal expansion is a prominent feature of the B cell response in multiple sclerosis cerebrospinal fluid. J Immunol 171:2725–2733

Paul F et al. (2007) Antibody to aquaporin 4 in the diagnosis of neuromyelitis optica. PLoS Med 4:e133

Pelayo R et al. (2007) Antimyelin antibodies with no progression to multiple sclerosis. N Engl J Med 356:426–428

Pohl D et al. (2006) High seroprevalence of Epstein-Barr virus in children with multiple sclerosis. Neurology 67:2063–2065

Rauer S et al. (2006) Antimyelin antibodies and the risk of relapse in patients with a primary demyelinating event. J Neurol Neurosurg Psych 77:739–742

Reff ME et al. (1994) Depletion of B cells in vivo by a chimeric mouse human monoclonal antibody to CD20. Blood 83:435–445

Reiber H (1998) Cerebrospinal fluid–physiology, analysis and interpretation of protein patterns for diagnosis of neurological diseases. Mult Scler 4:99–107

Reindl M et al. (2006) Antibodies as biological markers for pathophysiological processes in MS. J Neuroimmunol 180:50–62

Roemer SF et al. (2007) Pattern-specific loss of aquaporin-4 immunoreactivity distinguishes neuromyelitis optica from multiple sclerosis. Brain 130:1194–1205

Serafini B et al. (2004) Detection of ectopic B-cell follicles with germinal centers in the meninges of patients with secondary progressive multiple sclerosis. Brain Pathol 14:164–174

Serafini B et al. (2007) Dysregulated Epstein-Barr virus infection in the multiple sclerosis brain. J Exp Med 204:2899–2912

Shapiro-Shelef M, Calame K (2005) Regulation of plasma-cell development. Nat Rev Immunol 5:230–242

Sharief MK, Thompson EJ (1991) Intrathecal immunoglobulin M synthesis in multiple sclerosis. Relationship with clinical and cerebrospinal fluid parameters. Brain 114(Pt 1A):181–195

Sharief MK et al. (1990) Intrathecal synthesis of IgM in neurological diseases: a comparison between detection of oligoclonal bands and quantitative estimation. J Neurol Sci 96:131–142

Storch MK et al. (1998) Multiple sclerosis: in situ evidence for antibody- and complement-mediated demyelination. Ann Neurol 43:465–471

Takahashi T et al. (2007) Anti-aquaporin-4 antibody is involved in the pathogenesis of NMO: a study on antibody titre. Brain 130:1235–1243

Thacker EL et al. (2006) Infectious mononucleosis and risk for multiple sclerosis: a meta-analysis. Ann Neurol 59:499–503

Tomassini V et al. (2007) Anti-myelin antibodies predict the clinical outcome after a first episode suggestive of MS. Mult Scler 13:1086–1094

Uccelli A et al. (2005) Unveiling the enigma of the CNS as a B-cell fostering environment. Trends Immunol 26:254–259

Villar LM et al. (2002) Intrathecal IgM synthesis predicts the onset of new relapses and a worse disease course in MS. Neurology 59:555–559

Villar LM et al. (2003) Intrathecal IgM synthesis is a prognostic factor in multiple sclerosis. Ann Neurol 53:222–226

Wandinger K et al. (2000) Association between clinical disease activity and Epstein-Barr virus reactivation in MS. Neurology 55:178–184

Wang LY, Fujinami RS (1997) Enhancement of EAE and induction of autoantibodies to T-cell epitopes in mice infected with a recombinant vaccinia virus encoding myelin proteolipid protein. J Neuroimmunol 75:75–83

Wang H et al. (2008) Myelin oligodendrocyte glycoprotein antibodies and multiple sclerosis in healthy young adults. Neurology 71:1142–1146

Weinshenker BG et al. (1999) A randomized trial of plasma exchange in acute central nervous system inflammatory demyelinating disease. Ann Neurol 46:878–886

Weinshenker BG et al. (2006a) NMO-IgG: a specific biomarker for neuromyelitis optica. Dis Markers 22:197–206

Weinshenker BG et al. (2006b) Neuromyelitis optica IgG predicts relapse after longitudinally extensive transverse myelitis. Ann Neurol 59:566–569

Wingerchuk DM et al. (1999) The clinical course of neuromyelitis optica (Devic's syndrome). Neurology 53:1107–1114

Wingerchuk DM et al. (2006) Revised diagnostic criteria for neuromyelitis optica. Neurology 66:1485–1489

Winges KM et al. (2007) Analysis of multiple sclerosis cerebrospinal fluid reveals a continuum of clonally related antibody-secreting cells that are predominantly plasma blasts. J Neuroimmunol 192:226–234

Wucherpfennig KW et al. (1997) Recognition of the immunodominant myelin basic protein peptide by autoantibodies and HLA-DR2-restricted T cell clones from multiple sclerosis patients. Identity of key contact residues in the B-cell and T-cell epitopes. J Clin Invest 100:1114–1122

Yu P et al. (2002) B cells control the migration of a subset of dendritic cells into B cell follicles via CXC chemokine ligand 13 in a lymphotoxin-dependent fashion. J Immunol 168:5117–5123

Zhou D et al. (2006) Identification of a pathogenic antibody response to native myelin oligodendrocyte glycoprotein in multiple sclerosis. Proc Natl Acad Sci U S A 103:19057–19062

Cooperation of B Cells and T Cells in the Pathogenesis of Multiple Sclerosis

Martin S. Weber and Bernhard Hemmer

Abstract B cells and T cells are two major players in the pathogenesis of multiple sclerosis (MS) and cooperate at various check points. B cells, besides serving as a source for antibody-secreting plasma cells, are efficient antigen presenting cells for processing of intact myelin antigen and subsequent activation and pro-inflammatory differentiation of T cells. This notion is supported by the immediate clinical benefit of therapeutic B cell depletion in MS, presumably abrogating development of encephalitogenic T cells. However, different B cell subsets strongly vary in their respective effect on T cell differentiation which may relate to B cell phenotype, activation status, antigen specificity and the immunological environment where a B cell encounters a naïve T cell in. In this regard, some B cells also have anti-inflammatory properties producing regulatory cytokines and facilitating development and maintenance of other immunomodulatory immune cells, such as regulatory T cells. Reciprocally, differentiated T cells influence T cell polarizing B cell properties establishing a positive feedback loop of joint pro- or anti-inflammatory B and T cell developments. Further, under the control of activated T helper cells, antigen-primed B cells can switch immunoglobulin isotype, terminally commit to the plasma cell pathway or enter the germinal center reaction to memory B Cell development. Taken together, B cells and T cells thus closely support one another to participate in the pathogenesis of MS in an inflammatory but also in a regulatory manner.

1 The Role of B Cells in Multiple Sclerosis: Effects on T Cells

B cells may have various roles in the pathogenesis of multiple sclerosis (MS), serving as a source of plasma cells that secrete auto-reactive antibodies, and also as antigen presenting cells for activation of encephalitogenic T cells. Data indicate that

M.S. Weber and B. Hemmer (✉)
Department of Neurology, Technische Universität München, Ismaningerstrasse 22, 81675, Munich, Germany
e-mail: hemmer@lrz.tu-muenchen.de

Results Probl Cell Differ, DOI 10.1007/400_2009_21

myelin-specific antibodies promote demyelination in MS and experimental auto-immune encephalomyelitis (EAE), the animal model of MS, while the role of B cells themselves in CNS auto-immune disease is less clear. Like dendritic cells, B cells are professional antigen presenting cells as defined by a constitutive expression of a major histocompatibility complex (MHC) class II. Compared to other antigen presenting cell populations, antigen-specific B cells are very competent in the presentation of protein antigen when their receptor recognizes the same antigen as the responding T cells. As processing and presentation of CNS protein antigen are required for initiation of CNS auto-immune disease (Slavin et al. 2001), B cells and, in particular, B cells specific for CNS auto-antigen may have an important role as antigen presenting cells for the activation of myelin-specific T cells in MS. On the other hand, B cells may have an anti-inflammatory role in MS pathogenesis modulating effector function of other immune cells, e.g., dendritic cells or FoxP3$^+$ regulatory T cells. Which of these diverse roles a respective B cell may play in the pathogenesis of MS appears to relate to its maturation and activation status, antigen specificity as well as the immunological environment.

1.1   B Cells as the Source for Antibody-Secreting Plasma Cells

B cell-derived myelin-specific antibodies may be differentially important in the pathogenesis of certain subtypes and at different stages of MS (Reindl et al. 1999; Baranzini et al. 1999; Berger and Reindl 2000; Qin et al. 2003). Antibodies are found in large number in active MS lesions and in areas of myelin breakdown in EAE (Prineas and Connell 1978; Genain et al. 1999). Whether these antibodies truly play a pathogenic role in CNS auto-immune disease is still discussed controversially. Whereas it is clear that antibodies against myelin alone are not capable of initiating inflammation in brain and spinal cord, certain EAE studies suggest that they might facilitate CNS demyelination (Genain et al. 1999; Linington and Lassmann 1987; Schluesener et al. 1987). Experiments in B cell deficient mice indicate that B cells and antibodies may be required for EAE induced by the extra-cellular domain of human myelin oligodendrocyte glycoprotein (MOG), one suspected auto-antigen in MS pathogenesis, whereas they appear not to be involved in EAE induced by its short encephalitogenic MOG peptide (Lyons et al. 1999, 2002). Further, not all myelin-specific antibody responses are thought to be pathogenic which can be determined by their antigen recognition. In general, pathogenic MOG-specific antibodies recognize conformation-dependent MOG epitopes that are only available within the intact protein, whereas antibodies that recognize linear determinants are non-pathogenic (von Budingen et al. 2002, 2004; Bourquin et al. 2003; Marta et al. 2005; Zhou et al. 2006). In a recent publication, the development of non-conformational myelin-specific antibodies has been associated with an increased risk for patients that have experienced a first attack of CNS demyelination, known as a clinically isolated syndrome (CIS) to develop clinically definite MS (CDMS)(Berger et al. 2003; Greeve et al. 2007). However, these findings could not be reproduced by other groups (Lim et al. 2005; Kuhle et al. 2007).

Humoral responses are probably more relevant in neuromyelitis optica (NMO), also known as Devic's syndrome, another severe CNS demyelinating disorder that preferentially affects optic nerves and spinal cord (Mandler et al. 1993; Cree et al. 2002). In a significant proportion of individuals with NMO, a serum antibody named NMO IgG has been detected, which was determined to recognize the water channel protein, aquaporin-4. More recently, a murine model of NMO has been described (Krishnamoorthy et al. 2006; Bettelli et al. 2006a). MOG knock-in mice containing the rearranged Ig heavy chain variable gene for a pathogenic MOG-specific antibody (8.18-C.5) were crossed with MOG-specific T cell receptor transgenic mice, of which approximately 40% are affected by optic neuritis (Krishnamoorthy et al. 2006; Bettelli et al. 2003). Among the progeny, around 60% spontaneously developed a severe form of EAE in which meningeal and parenchymal lesions were restricted to the spinal cord and optic nerves (Krishnamoorthy et al. 2006; Bettelli et al. 2006a). In these mice, MOG-specific B cells efficiently presented rMOG to MOG p35–55-specific T cells, but these mice also revealed elevated levels of MOG-specific IgG. It remains to be elucidated whether inflammation and demyelination in the spinal cord and optic nerves of these mice is related to enhanced anti-MOG antibody secretion or activation of encephalitogenic T cells by myelin-specific B cells. Although this murine model shares certain clinical properties with human NMO, it needs to be emphasized that the auto-reactive antibody response in these mice is solely directed against MOG. In NMO however, growing evidence suggests that the antibody directed against aquaporin-4 displays not only a robust biomarker, but may indeed play a crucial pathogenic role in the development of NMO (Jarius et al. 2008; Hinson et al. 2008).

1.2 B Cells as Antigen Presenting Cells for Activation of T Cells

Pro-inflammatory CD4+ Th1 and Th17 cells are thought to have a central role in the pathogenesis of MS and its model EAE (Zamvil and Steinman 1990, 2003; Steinman and Zamvil 2005). Activation of myelin-specific T cells requires recognition of myelin antigen in association with MHC class II molecules expressed on antigen presenting cells. Several types of APC may participate in T cell activation in MS pathogenesis which can be generally divided into resident (CNS) and non-resident (bone marrow-derived) antigen presenting cell. Among resident CNS antigen presenting cells, astrocytes and parenchymal microglia could participate in antigen presentation (Fontana et al. 1984; Soos et al. 1998), its contribution to in vivo activation of T cells is however still under debate (Stuve et al. 2002). Non-resident bone marrow-derived antigen presenting cells however appear to have a more important role in initiation of CNS demyelinating disease. B cells, perivascular macrophages and dendritic cells are found in EAE and active MS lesions. In EAE, it has been demonstrated that MHC class II-restricted antigen presentation by dendritic cells is sufficient to induce CNS demyelinating disease. While these data suggest that dendritic cells alone are capable to activate T cells, they do not exclude that other antigen presenting cells may significantly contribute to MHC II-restricted activation of T cells. Other data suggest that Ag-specific B cells may have a key role particularly in presentation of protein antigen in EAE and MS. While dendritic cells are the most potent antigen presenting cells for presentation of peptide antigen, antigen-specific B cells appear to be at least similarly efficient in endocytic processing and presentation of protein antigen (van der Veen et al. 1992; Constant et al. 1995a, b). In this regard, it has been demonstrated that antigen-specific B cells are essential for T cell expansion in lymph nodes and for systemic T cell responses to low antigen concentrations (Rivera et al. 2001). Furthermore, exclusive antigen presentation by B cells resulted in substantial expansion of both CD4[+] T cells and B cells (Rodriguez-Pinto and Moreno 2005), indicating that similar to dendritic cells, B cell antigen presentation alone is capable to activate T cells in vivo. A recent study further elucidated this interplay between B cells and T cells in their reciprocal activation (Harp et al. 2008). The authors demonstrated in detail that activation of B cells by a combination of the T cell products CD40L and IL-4, but not by unspecific stimuli such as a toll-like receptor ligand and IL-2 rendered B cells capable to activate T cells in a myelin antigen specific manner. These data indicate that T cell-mediated activation of B cells is crucial for their capability to process and present antigen to T cells and that both populations thereby stimulate each other in a reciprocal manner. In the non-obese diabetic mouse model of insulin-dependent diabetes mellitus, B cell deficient mice failed to develop T cell-mediated diabetes despite normal number of T cells. Thus, B lymphocytes appear to play a crucial role for the initial development and/or activation of auto-reactive T cells in non-obese diabetic mice (Serreze et al. 1996). While several studies indicate that B cell deficient mice are generally susceptible to EAE induced by short encephalitogenic peptide (Dittel et al. 2000; Wolf et al. 1996), it remains to be elucidated whether myelin-specific B cells contribute to processing and presenting protein Ag to myelin-specific T cells.

Interestingly, different B cell subsets appear to preferentially facilitate development of pro- or anti-inflammatory T cell subsets. For other antigen presenting cells, such as dendritic cells, this has been well established. Various subsets of dendritic cells are known to orchestrate T cell priming and differentiation as antigen presenting cells by differential release of cytokines and expression of costimulatory molecules. In the case of B cells, two distinct subsets, peritoneal B1 and conventional follicular B2 cells, can be distinguished. They differ in respect to lineage, location, gene expression, antibody repertoire, proliferative responses and immunoglobulin secretion. B1 cells are the predominant population of B cells in the peritoneal cavity, rare in spleen and lymph nodes (Hardy 2006). B1 cells, presumably through potent antigen presentation, are thought to play a role in auto-immunity (Sato et al. 2004). In this regard, a recent publication demonstrated that B1 cells used as APC indeed induced Th1/Th17 cells whereas conventional B2 cells promoted development of induced regulatory T cells with suppressive capacity. Interestingly, blockade of CD86, a costimulatory molecule primarily associated with Th2 responses (Ranger et al. 1996), enhanced the ability of B1 cells to induce regulatory T cells (Zhong et al. 2007). Taken together, these findings indicate that B1 and B2 cells, similar to different subsets of dendritic cells, play distinct roles in promotion of selective T cell lineages due to differential expression of costimulatory molecules as well as release of polarizing cytokines. Data presented below, further suggest that once polarized, effector T cells also influence differentiation of B cells establishing a positive feedback loop of pro- or anti-inflammatory differentiation of B and T cells.

1.3   B Cells as Therapeutic Target in Multiple Sclerosis: Abrogation of T Cell Activation?

Anti-CD20 (rituximab, Rituxan®) is an approved treatment of non-Hodgkins lymphoma and has shown promising results in treatment of rheumatoid arthritis (Okamoto and Kamatani 2004), in systemic lupus erythematosis (SLE) (van Vollenhoven et al. 2004; Vigna-Perez et al. 2006; Torrente-Segarra et al. 2009), and in treatment of IgM-associated peripheral neuropathy (Pestronk et al. 2003). It is currently elucidated whether the beneficial effect of anti-CD20 treatment in those conditions translates to amelioration of CNS neurological diseases with possible auto-immune B cell involvement. A small open-label pilot study evaluating B cell depletion in NMO suggested that B cell depletion was well tolerated. Seven out of eight patients receiving rituximab experienced substantial improvement of neurologic function over 1 year (Cree et al. 2005). In a follow-up study, rituximab treatment significantly reduced the annualized relapse rate and improved or stabilized disability in 20 out of 25 NMO patients (Jacob et al. 2008). In relapsing–remitting MS, a recent placebo-controlled phase II trial revealed that MS patient receiving rituximab exerted a substantial reduction in development of newly emerging inflammatory brain lesions. Despite the short period of the trial, the same study at least suggested a reduction in clinical attack frequency (Hauser et al. 2007, 2008). Another placebo-controlled phase II/III trial is currently testing rituximab-mediated depletion in primary progressive MS.

Mechanistically, it remains to be elucidated elimination of which B cell function may correlate best with a potential benefit of B cell depletion in CNS auto-immune disease. It can be speculated that rituximab immediately abolishes cellular B cell functions, such as B cell antigen presentation, whereas myelin-specific antibody titers may decrease with a substantial delay due to the fact that antibody-secreting plasma cells no longer express CD20 (Cragg et al. 2005). Based on this assumption it appears likely that the immediate benefit of B cell depletion in T cell-mediated auto-immune disease may indeed relate to an impaired activation of T cells. Detailed mechanistic studies are about to more thoroughly investigate the impact of B cell depletion on T cell activation and differentiation. Similarly, only limited information is available regarding the content of B cells in various compartments, e.g., the central nervous system, following systemic B cell depletion. In a smaller study with 16 relapsing–remitting MS patients who had not responded optimally to standard immunomodulatory therapies, rituximab treatment reduced the number of B cells within the cerebrospinal fluid at 6 months post-treatment (Cross et al. 2006).

To a lesser degree, the number of CNS T cells also declined. Although suggestive, it remains to be determined whether this reduction of cerebrospinal fluid leukocytes reflects the ability of rituximab to deplete B cells and reduce the prevalence of T cells within parenchymal inflammatory lesions themselves. Taken together, rituximab-mediated B cell depletion appears to mediate a clinical benefit in treatment of CNS auto-immune disease that waits to be understood mechanistically in a more thorough manner.

1.4 B Cells with Immunomodulatory Properties

The data mentioned in the sections above primarily suggest that B cells may play a pro-inflammatory role in the pathogenesis of CNS auto-immune disease. In addition, B cells, and primarily naïve B cells may however also serve as regulatory cells in auto-immunity. In EAE, B cell deficient mice did not recover from acute EAE like wild-type mice, suggesting that B cells had a regulatory role preventing progression to chronic and relapsing EAE (Wolf et al. 1996). Mice that contained B cells that could not produce IL-10 did not recover from EAE either. Interestingly, when reconstituted with IL-10 competent B cells, these mice recovered from EAE, demonstrating that B cell IL-10 production conferred protection from development of chronic EAE (Fillatreau et al. 2002). The seminal publication by Harris et al. more generally defined two "effector" subsets of B cells, pro-inflammatory B effector 1 cells and anti-inflammatory B effector 2 cells. Their development was shown to primarily depend on the cytokine environment in which naïve B cells are stimulated during their primary encounter with antigen. Specifically, naïve B cells were co-cultured with T cells that had been polarized into Th1 and Th2 cells in the presence of B cell antigen. The B cells activated with the Th1 effectors were named B effector 1 cells (Be1 cells), and the B cells that were co-cultured with the Th2 cells in the primary cultures were named B effector 2 cells (Be2 cells). When re-stimulated, these Be1 and Be2 cells produced a substantially different cytokine signature. While IL-2, IL-6 and IL-10 were made by both Be1 and Be2 cells upon restimulation, IL-4 was produced exclusively by the Be2 cells, but not by Be1-stimulated cells. Conversely, IFN-γ was produced by Be1 cells, whereas only minimal amounts were detected in supernatants from re-stimulated Be2 cells. Functionally, these effector B cell subsets differentially regulated development of naïve CD4+ T cells into Th1 and Th2 cells. While Be1 cells supported differentiation into Th1 cells through production of polarizing IFN-γ, Be2 cells facilitated generation of Th2 cells through provision of IL-4. Taken together, effector B cells and T cells regulated each other in a reciprocal manner. Development of anti-inflammatory Th2 cells is associated with differentiation of naïve B cells into Be2 cells which support development of Th2 cells in a positive feedback loop (Harris et al. 2000). Such Be2 cells with a regulatory capacity have now been further described in other experimental models of arthritis (Mauri et al. 2003), inflammatory bowel disease, lupus (Brummel and Lenert 2005) and parasitic infections

(Mangan et al. 2004; Gillan et al. 2005). These findings may explain why in ulcerative colitis B cell depleting rituximab failed to exert a clinical benefit and rather led to an exacerbation of disease (Goetz et al. 2007). Very recently it has been described that B cells with regulatory capacity may also facilitate development of regulatory T cells. In this report (Shah and Qiao 2008), resting, but not activated B cells expanded a population of CD4$^+$CD25$^+$FoxP3$^+$ regulatory T cells. This effect correlated with the ability of resting B cells to produce TGF-β, which was diminished upon B cell activation. Thus, while B cells may serve as APC and secrete antibody, supporting a pathogenic role for myelin-specific B cells, it may be important to further elucidate the regulatory role of B cells in CNS auto-immune disease and how this role may be therapeutically exhausted.

2 T Cells Regulate B Cell Immunity

In the sections above, we focused on the ability of B cells to initiate and/or modulate T cell priming and differentiation. However, in the pathogenesis of MS, B and T cell interaction is not one directional but rather a dynamic and tremendously complex interplay. As mentioned above, polarized T cells can influence B cell maturation and differentiation into a pro-inflammatory but also an anti-inflammatory phenotype in a positive feedback mechanism. Specifically, antigen-specific, activated T helper (Th) cells are recruited into the T cell zones of secondary lymphoid tissue in order to support the development of antigen-primed B cells. Under the control of Th cells, antigen-primed B cells can switch immunoglobulin isotype, terminally commit to the plasma cell pathway or enter the germinal center (GC) reaction to memory B cell development.

2.1 Antigen-Specific B Cell Activation Requires T Helper Cells

B cells are capable of responding to some highly repetitive epitopes – mainly parts of bacterial pathogens – in a T cell-independent manner. However, most antigens require T cell help to sufficiently augment B cell responsiveness. In a first step, antigen-specific B cells internalize, process and present antigen. Upon antigen encounter, B cells up-regulate costimulatory molecules (Benschop et al. 1999) such as CD40, CD80 and CD86 as well as constitutively expressed MHC class II (Wagle et al. 2000) and thereby strongly gain capability to activate T cells. Differentiated Th cells which recognize the processed peptide presented by the corresponding B cell in the context of MHC class II, express elevated levels of CD40-ligand. Resulting B cell CD40–T cell CD40 ligand interaction together with enhanced secretion of T cell cytokines, such as IL-4 activates B cells to proliferate and further differentiate. For this interaction, Th cells and B cells do not need to recognize the same epitope as long as they share recognition of an identical antigen. This mechanism ensures that specific B cell antigen recognition alone is not sufficient to mediate development of antibody-secreting plasma cells or memory B cells and thereby prevents a false activation also of potentially auto-reactive B cells. In return, this also means, that in the pathogenesis of B cell-mediated auto-immune disease or inflammatory conditions in which B cells are pathogenetically involved, such as MS, an affected individual necessarily contains self-reactive B and T cells.

2.2 Polarized T Helper Cells Facilitate Isotype Switching

Prior to antigen recognition and Th cell encounter, B cells express immunoglobulin (Ig)M on their membrane. Through gene rearrangement, B cells and B cell-derived plasma cells can switch expression and secretion of antibody to various isotypes, which ultimately encode for antibody effector function. As mentioned above, together with CD40 ligand, IL-4 strongly activates B cells to proliferate and differentiate into terminally committed plasma cells. In addition, Th2 cell-derived IL-4 preferentially promotes isotype switching towards IgG1, IgG4 and IgE (Ishizaka et al. 1990). TGF-β, an immunomodulatory cytokine centrally involved in development and expansion of CD4+CD25+FoxP3+ regulatory T cells (Bettelli et al. 2006b), enhances preferential synthesis of IgG2b and IgA. While pro-inflammatory CD4+ Th1 cells hardly contribute to initial B cell activation, they facilitate antibody-switching towards the IgG2a and IgG3 isotype presumably through secretion of IFN-γ. In MS pathogenesis, it is still under debate whether the entire antibody response against myelin antigens has prognostic or even immuno-pathological significance (Reindl et al. 1999; Berger et al. 2003; Lim et al. 2005; Kuhle et al. 2007; Karni et al. 1999). It thus remains to be elucidated whether preferential expression of a specific isotype may be a desirable therapeutic goal at all. However, in consistency with the above, it is worth mentioning that the treatment of MS with glatiramer acetate, an approved treatment presumably effective through development of Th2 cells (Neuhaus et al. 2000) and CD4+CD25+FoxP3+ regulatory T cells (Jee et al. 2007), is associated with a preferential occurrence of glatiramer acetate-specific antibodies of an IgG4 isotype (Farina et al. 2002).

2.3 T Cells Participate in the Formation of Germinal Centers

Upon Th cell-mediated activation of B cells, some antigen-specific B cells migrate back into follicles in lymph node and spleen where they clonally expand, now called secondary B cell follicle. This secondary follicle then polarizes into a region of rapidly dividing B cells (centroblasts) and an opposing region of non-cycling B cells (centrocytes) establishing a dynamic environment called germinal center. Traditionally, GCs have been functionally divided into two corresponding zones, the dark zone with proliferation and somatic hypermutation, and the light zone, the area for B cell selection and differentiation (Tarlinton 1998). Migration between these zones is an important component of normal GC function, allowing rounds of mutation and selection to occur which ultimately results in affinity maturation of the membrane bound Ig. B cells which survived this selection process leave the GC to develop into antibody-secreting plasma cells or memory B cells. Antigen-specific Th cells are recruited into this GC reaction (Farina et al. 2002). Whether T cells control development of high affinity B cells is not entirely established, but Th cells interfere with B cell selection within the GC. In this regard, it has been demonstrated that disruption of costimulatory molecule interaction between GC T and B cells hinders affinity based B cell selection (Han et al. 1995). Further, a recent report indicates that in an auto-immune model, T cell-derived IL-17 enhanced GC formation and auto-antibody formation (Hsu et al. 2008), suggesting that in MS pro-inflammatory T cell differentiation may synergistically accelerate affinity maturation of B cells and B cell-derived plasma cells.

References

Baranzini SE et al (1999) B cell repertoire diversity and clonal expansion in multiple sclerosis brain lesions. J Immunol 163:5133–5144

Benschop RJ, Melamed D, Nemazee D, Cambier JC (1999) Distinct signal thresholds for the unique antigen receptor-linked gene expression programs in mature and immature B cells. J Exp Med 190:749–756

Berger T, Reindl M (2000) Immunopathogenic and clinical relevance of antibodies against myelin oligodendrocyte glycoprotein (MOG) in multiple sclerosis. J Neural Transm Suppl 60:351–360

Berger T et al (2003) Antimyelin antibodies as a predictor of clinically definite multiple sclerosis after a first demyelinating event. N Engl J Med 349:139–145

Bettelli E et al (2003) Myelin oligodendrocyte glycoprotein-specific T cell receptor transgenic mice develop spontaneous autoimmune optic neuritis. J Exp Med 197:1073–1081

Bettelli E, Baeten D, Jager A, Sobel RA, Kuchroo VK (2006a) Myelin oligodendrocyte glycoprotein-specific T and B cells cooperate to induce a Devic-like disease in mice. J Clin Invest 116:2393–2402

Bettelli E et al (2006b) Reciprocal developmental pathways for the generation of pathogenic effector TH17 and regulatory T cells. Nature 441:235–238

Bourquin C et al (2003) Selective unresponsiveness to conformational B cell epitopes of the myelin oligodendrocyte glycoprotein in H-2b mice. J Immunol 171:455–461

Brummel R, Lenert P (2005) Activation of marginal zone B cells from lupus mice with type A(D) CpG-oligodeoxynucleotides. J Immunol 174:2429–2434

Constant S, Schweitzer N, West J, Ranney P, Bottomly K (1995a) B lymphocytes can be competent antigen-presenting cells for priming CD4+ T cells to protein antigens in vivo. J Immunol 155:3734–3741

Constant S et al (1995b) Peptide and protein antigens require distinct antigen-presenting cell subsets for the priming of CD4+ T cells. J Immunol 154:4915–4923

Cragg MS, Walshe CA, Ivanov AO, Glennie MJ (2005) The biology of CD20 and its potential as a target for mAb therapy. Curr Dir Autoimmun 8:140–174

Cree BA, Goodin DS, Hauser SL (2002) Neuromyelitis optica. Semin Neurol 22:105–122

Cree BA et al (2005) An open label study of the effects of rituximab in neuromyelitis optica. Neurology 64:1270–1272

Cross AH, Stark JL, Lauber J, Ramsbottom MJ, Lyons JA (2006) Rituximab reduces B cells and T cells in cerebrospinal fluid of multiple sclerosis patients. J Neuroimmunol 180:63–70

Dittel BN, Urbania TH, Janeway CA Jr (2000) Relapsing and remitting experimental autoimmune encephalomyelitis in B cell deficient mice. J Autoimmun 14:311–318

Farina C et al (2002) Treatment with glatiramer acetate induces specific IgG4 antibodies in multiple sclerosis patients. J Neuroimmunol 123:188–192

Fillatreau S, Sweenie CH, McGeachy MJ, Gray D, Anderton SM (2002) B cells regulate autoimmunity by provision of IL-10. Nat Immunol 3:944–950

Fontana A, Fierz W, Wekerle H (1984) Astrocytes present myelin basic protein to encephalitogenic T-cell lines. Nature 307:273–276

Genain CP, Cannella B, Hauser SL, Raine CS (1999) Identification of autoantibodies associated with myelin damage in multiple sclerosis. Nat Med 5:170–175

Gillan V, Lawrence RA, Devaney E (2005) B cells play a regulatory role in mice infected with the L3 of Brugia pahangi. Int Immunol 17:373–382

Goetz M, Atreya R, Ghalibafian M, Galle PR, Neurath MF (2007) Exacerbation of ulcerative colitis after rituximab salvage therapy. Inflamm Bowel Dis 13:1365–1368

Greeve I et al (2007) Anti-myelin antibodies in clinically isolated syndrome indicate the risk of multiple sclerosis in a Swiss cohort. Acta Neurol Scand 116:207–210

Han S et al (1995) Cellular interaction in germinal centers. Roles of CD40 ligand and B7-2 in established germinal centers. J Immunol 155:556–567

Hardy RR (2006) B-1 B cell development. J Immunol 177:2749–2754

Harp CT, Lovett-Racke AE, Racke MK, Frohman EM, Monson NL (2008) Impact of myelin-specific antigen presenting B cells on T cell activation in multiple sclerosis. Clin Immunol 128:382–391

Harris DP et al (2000) Reciprocal regulation of polarized cytokine production by effector B and T cells. Nat Immunol 1:475–482

Hauser S et al (2007) A phase II randomized, placebo-controlled, multicenter trial of rituximab in adults with relapsing remitting multiple sclerosis (RRMS). Neurology 68(Suppl):A99–A100

Hauser SL et al (2008) B-cell depletion with rituximab in relapsing-remitting multiple sclerosis. N Engl J Med 358:676–688

Hinson SR et al (2008) Aquaporin-4-binding autoantibodies in patients with neuromyelitis optica impair glutamate transport by down-regulating EAAT2. J Exp Med 205:2473–2481

Hsu HC et al (2008) Interleukin 17-producing T helper cells and interleukin 17 orchestrate autoreactive germinal center development in autoimmune BXD2 mice. Nat Immunol 9:166–175

Ishizaka A et al (1990) The inductive effect of interleukin-4 on IgG4 and IgE synthesis in human peripheral blood lymphocytes. Clin Exp Immunol 79:392–396

Jacob A et al (2008) Treatment of neuromyelitis optica with rituximab: retrospective analysis of 25 patients. Arch Neurol 65:1443–1448

Jarius S et al (2008) Antibody to aquaporin-4 in the long-term course of neuromyelitis optica. Brain 131:3072–3080

Jee Y et al (2007) CD4(+) CD25(+) regulatory T cells contribute to the therapeutic effects of glatiramer acetate in experimental autoimmune encephalomyelitis. Clin Immunol 125:34–42

Karni A, Bakimer-Kleiner R, Abramsky O, Ben-Nun A (1999) Elevated levels of antibody to myelin oligodendrocyte glycoprotein is not specific for patients with multiple sclerosis. Arch Neurol 56:311–315

Krishnamoorthy G, Lassmann H, Wekerle H, Holz A (2006) Spontaneous opticospinal encephalomyelitis in a double-transgenic mouse model of autoimmune T cell/B cell cooperation. J Clin Invest 116:2385–2392

Kuhle J et al (2007) Lack of association between antimyelin antibodies and progression to multiple sclerosis. N Engl J Med 356:371–378

Lim ET et al (2005) Anti-myelin antibodies do not allow earlier diagnosis of multiple sclerosis. Mult Scler 11:492–494

Linington C, Lassmann H (1987) Antibody responses in chronic relapsing experimental allergic encephalomyelitis: correlation of serum demyelinating activity with antibody titre to the myelin/oligodendrocyte glycoprotein (MOG). J Neuroimmunol 17:61–69

Lyons JA, San M, Happ MP, Cross AH (1999) B cells are critical to induction of experimental allergic encephalomyelitis by protein but not by a short encephalitogenic peptide. Eur J Immunol 29:3432–3439

Lyons JA, Ramsbottom MJ, Cross AH (2002) Critical role of antigen-specific antibody in experimental autoimmune encephalomyelitis induced by recombinant myelin oligodendrocyte glycoprotein. Eur J Immunol 32:1905–1913

Mandler RN, Davis LE, Jeffery DR, Kornfeld M (1993) Devic's neuromyelitis optica: a clinicopathological study of 8 patients. Ann Neurol 34:162–168

Mangan NE et al (2004) Helminth infection protects mice from anaphylaxis via IL-10-producing B cells. J Immunol 173:6346–6356

Marta CB, Oliver AR, Sweet RA, Pfeiffer SE, Ruddle NH (2005) Pathogenic myelin oligodendrocyte glycoprotein antibodies recognize glycosylated epitopes and perturb oligodendrocyte physiology. Proc Natl Acad Sci USA 102:13992–13997

Mauri C, Gray D, Mushtaq N, Londei M (2003) Prevention of arthritis by interleukin 10-producing B cells. J Exp Med 197:489–501

Neuhaus O et al (2000) Multiple sclerosis: comparison of copolymer-1- reactive T cell lines from treated and untreated subjects reveals cytokine shift from T helper 1 to T helper 2 cells. Proc Natl Acad Sci USA 97:7452–7457

Okamoto H, Kamatani N (2004) Rituximab for rheumatoid arthritis. N Engl J Med 351:1909 author reply 1909

Pestronk A et al (2003) Treatment of IgM antibody associated polyneuropathies using rituximab. J Neurol Neurosurg Psychiatry 74:485–489

Prineas JW, Connell F (1978) The fine structure of chronically active multiple sclerosis plaques. Neurology 28:68–75

Qin Y et al (2003) Intrathecal B-cell clonal expansion, an early sign of humoral immunity, in the cerebrospinal fluid of patients with clinically isolated syndrome suggestive of multiple sclerosis. Lab Invest 83:1081–1088

Ranger AM, Das MP, Kuchroo VK, Glimcher LH (1996) B7-2 (CD86) is essential for the development of IL-4-producing T cells. Int Immunol 8:1549–1560

Reindl M et al (1999) Antibodies against the myelin oligodendrocyte glycoprotein and the myelin basic protein in multiple sclerosis and other neurological diseases: a comparative study. Brain 122(Pt 11):2047–2056

Rivera A, Chen CC, Ron N, Dougherty JP, Ron Y (2001) Role of B cells as antigen-presenting cells in vivo revisited: antigen-specific B cells are essential for T cell expansion in lymph nodes and for systemic T cell responses to low antigen concentrations. Int Immunol 13:1583–1593

Rodriguez-Pinto D, Moreno J (2005) B cells can prime naive CD4+ T cells in vivo in the absence of other professional antigen-presenting cells in a CD154-CD40-dependent manner. Eur J Immunol 35:1097–1105

Sato T et al (2004) Aberrant B1 cell migration into the thymus results in activation of CD4 T cells through its potent antigen-presenting activity in the development of murine lupus. Eur J Immunol 34:3346–3358

Schluesener HJ, Sobel RA, Linington C, Weiner HL (1987) A monoclonal antibody against a myelin oligodendrocyte glycoprotein induces relapses and demyelination in central nervous system autoimmune disease. J Immunol 139:4016–4021

Serreze DV et al (1996) B lymphocytes are essential for the initiation of T cell-mediated autoimmune diabetes: analysis of a new "speed congenic" stock of NOD.Ig mu null mice. J Exp Med 184:2049–2053

Shah S, Qiao L (2008) Resting B cells expand a CD4(+) CD25(+) Foxp3(+) Treg population via TGF-beta3. Eur J Immunol 38:2488–2498

Slavin AJ et al (2001) Requirement for endocytic antigen processing and influence of invariant chain and H-2M deficiencies in CNS autoimmunity. J Clin Invest 108:1133–1139

Soos JM et al (1998) Astrocytes express elements of the class II endocytic pathway and process central nervous system autoantigen for presentation to encephalitogenic T cells. J Immunol 161:5959–5966

Steinman L, Zamvil SS (2005) Virtues and pitfalls of EAE for the development of therapies for multiple sclerosis. Trends Immunol 26:565–571

Stuve O et al (2002) The role of the MHC class II transactivator in class II expression and antigen presentation by astrocytes and in susceptibility to central nervous system autoimmune disease. J Immunol 169:6720–6732

Tarlinton D (1998) Germinal centers: form and function. Curr Opin Immunol 10:245–251

Torrente Segarra V, Lisbona-Perez M, Rotes-Sala D, Castro-Oreiro S, Carbonell-Abello J (2009) Clinical, biological and ultrasonographic remission in a patient with musculoskeletal systemic lupus erythematosus with rituximab. Lupus 18:270–272

van der Veen RC, Trotter JL, Kapp JA (1992) Immune processing of proteolipid protein by subsets of antigen-presenting spleen cells. J Neuroimmunol 38:139–146

van Vollenhoven RF et al (2004) Biopsy-verified response of severe lupus nephritis to treatment with rituximab (anti-CD20 monoclonal antibody) plus cyclophosphamide after biopsy-documented failure to respond to cyclophosphamide alone. Scand J Rheumatol 33:423–427

Vigna-Perez M et al (2006) Clinical and immunological effects of Rituximab in patients with lupus nephritis refractory to conventional therapy: a pilot study. Arthritis Res Ther 8:R83

von Budingen HC et al (2002) Molecular characterization of antibody specificities against myelin/ oligodendrocyte glycoprotein in autoimmune demyelination. Proc Natl Acad Sci USA 99:8207–8212

von Budingen HC et al (2004) Frontline: epitope recognition on the myelin/oligodendrocyte glycoprotein differentially influences disease phenotype and antibody effector functions in autoimmune demyelination. Eur J Immunol 34:2072–2083

Wagle NM et al (2000) B-lymphocyte signaling receptors and the control of class-II antigen processing. Curr Top Microbiol Immunol 245:101–126

Wolf SD, Dittel BN, Hardardottir F, Janeway CA Jr (1996) Experimental autoimmune encephalomyelitis induction in genetically B cell-deficient mice. J Exp Med 184:2271–2278

Zamvil SS, Steinman L (1990) The T lymphocyte in experimental allergic encephalomyelitis. Annu Rev Immunol 8:579–621

Zamvil SS, Steinman L (2003) Diverse targets for intervention during inflammatory and neurodegenerative phases of multiple sclerosis. Neuron 38:685–688

Zhong X et al (2007) Reciprocal generation of Th1/Th17 and T(reg) cells by B1 and B2 B cells. Eur J Immunol 37:2400–2404

Zhou D et al (2006) Identification of a pathogenic antibody response to native myelin oligodendrocyte glycoprotein in multiple sclerosis. Proc Natl Acad Sci USA 103:19057–19062

Role of NK Cells and Invariant NKT Cells in Multiple Sclerosis

Kaori Sakuishi, Sachiko Miyake, and Takashi Yamamura

Abstract Natural killer (NK) cells and invariant natural killer T (iNKT) cells are two distinctive lymphocyte populations, each possessing its own unique features. Although NK cells are innate lymphocytes with cytotoxic property, they play an immunoregulatory role in the pathogenesis of autoimmune diseases. NKT cells are T cells expressing invariant TCR α-chains, which are known to bridge innate and adaptive arms of the immune system. Accumulating data now support active involvement of these cells in multiple sclerosis (MS). However, unlike professionally committed regulatory cells such as Foxp3+ regulatory T cells, NK, and iNKT cells have dual potential of acting as either protective or pathogenic lymphocytes depending on the disease setting, adding complexity to the interpretation of data obtained from human and rodent studies. They are potential therapeutic targets in MS, and further in-depth understanding of these cells will lead to designing new strategies to overcome the disabling disease MS.

1 Introduction

Over the past years, a growing number of evidence has indicated that multiple sclerosis (MS) is an autoimmune disease mediated by T cell immunity (Sospedra and Martin 2005). As described in detail in other chapters, pathogenesis of MS would actually involve autoreactive T cells that recognize the central nervous system (CNS) antigens. The target antigens include myelin basic protein (MBP) (Bielekova et al. 2000; Martin et al. 1991; Ota et al. 1990; Pette et al. 1990; Richert et al. 1989), myelin proteolipid protein (PLP) (Correale et al. 1995; Illes et al. 1999; Kondo et al. 1996; Ohashi et al. 1995; Pelfrey et al. 1993), and myelin oligodendrocyte glycoprotein (MOG) (Iglesias et al. 2001; Koehler et al. 2002; Mendel et al. 1995).

K. Sakuishi, S. Miyake, and T. Yamamura (✉)

Department of Immunology, National Institute of Neuroscience, National Center of Neurology and Psychiatry, 4–1–1 Ogawahigashi, KodairaTokyo, 187–8502, Japan

e-mail: yamamura@ncnp.go.jp

Results Probl Cell Differ, DOI 10.1007/400_2009_11

Although the dominant role of CD4⁺ T cells in MS has long been emphasized (Hafler 2004), more recent works indicate that CD8⁺ T cells (Huseby et al. 2001; Skulina et al. 2004) and B cells also play a critical role in the disease development, and actually comprise a proportion of the CNS infiltrating cells. CD8⁺ cells are reported to be predominant in the CNS lesions of MS, although compositions of cellular infiltrates vary greatly, depending on types and stages of this disease (Sospedra and Martin 2005). Now, the key question in MS lies in what disrupts the T cell and B cell immunological tolerance against the CNS antigens that are usually kept well secluded from the systemic immune system (Goodnow et al. 2005; Kyewski and Derbinski 2004; Walker and Abbas 2002). The relevance of this question is obvious because better understanding of the mechanism for the disruption of self-tolerance will lead to development of various new approaches to prevent the onset of MS and to control its further progression.

One of the distinctive and intriguing aspects of MS is that individual patients show various patterns in the longitudinal changes of its disease activity. While a large majority of the patients exhibit a relapsing and remitting course, some patients develop into or even start out as a progressive chronic illness (Sospedra and Martin 2005; Steinman 2001). Despite the vigorous efforts to control the activity of MS, currently available therapeutics do not halt the progression of disease in a majority of cases, although some patients do not exhibit any sign of worsening for a long period of time even without treatment.

To clarify the regulation of autoimmune responses, much efforts have been dedicated to investigate the role of specialized adaptive regulatory T cells, including CD4⁺ T cells expressing transcription factor Foxp3 (Miyara and Sakaguchi 2007), IL-10 producing T regulatory 1 (Tr1) cells (Roncarolo et al. 2006), and TGF-β producing Th3 cells (Awasthi et al. 2007; Baecher-Allan and Hafler 2006). However, recent publications provide evidence that cells of the innate immune system also have an unexpected potential to inhibit autoreactive CD4⁺ T cells from mediating autoimmune disease and to protect tissues from collateral damage by T cells reactive to exogenous pathogens (Carrol and Prodeus 1998; Fearon and Locksley 1996; Medzhitov and Janeway 1997; Shi et al. 2001). Natural killer (NK) cells and invariant natural killer T (iNKT) cells, the main focus of this review, are also now recognized as innate cells with immunoregulatory potentials. Although they sense external ligands with different receptors (TCR for iNKT cells and NK receptor for NK cells), they behave like innate cells when they need to rapidly respond to stimuli. Therefore, it was believed previously that both cell types would primarily function within the innate arms of immunity. However, recent works have provided evidence that they would actively regulate T cell responses, thereby influencing the adaptive immune system (Bendelac et al. 1997; Carrol and Prodeus 1998; Fearon and Locksley 1996; Medzhitov and Janeway 1997; Shi et al. 2001; Shi and Van Kaer 2006).

In summary, NK cells and iNKT cells are now considered as multipotent cells that work at the border of innate and adaptive immunity, to prevent the induction, propagation, and activation of autoimmune T cells. Here, we review the latest advances in the research of the regulatory NK and iNKT lymphocytes with regard to the pathogenesis of MS and discuss the possibilities that they may serve as an effective target for MS therapy.

2 NK Cells and MS

2.1 General Properties of NK Cells

Natural killer cells are evolutionary primitive lymphocytes that lack antigen-specific receptors. They were originally identified as lymphoid cells capable of lysing tumor cell lines in the absence of prior stimulation *in vivo* or *in vitro*, which was the basis of their denomination (Trinchieri 1989). Constituting about 10% of the lymphocyte in human peripheral blood mononuclear cells (PBMC), NK cells possess cytotoxic properties, directed against virus-infected cells, thus considered as an important part of the innate immune system. Their cytotoxic reaction is determined by collective signaling of an array of inhibitory and stimulatory receptors expressed on their surface (Kirwan and Burshtyn 2007) (Fig. 1). Inhibitory receptors, commonly referred to as killer inhibitory Ig-like receptors (KIRs), interact with shared allelic determinants of classical and non classical MHC class I. Hence, NK cells are kept in an inactivated state through contact with self MHC class I molecule expressed on healthy cells. For example, CD94/NKG2A heterodimer expressed on NK cell surface recognize HLA-class Ib molecule, HLA-E (Borrego et al. 2006; Lopez-Botet et al. 1997). On the contrary, stimulatory receptors on NK cell surface bind to NK stimulatory receptor ligand up-regulated on other cells upon undergoing cellular stress. The main activating receptors constitutively found on all NK cells in peripheral blood are NKG2D, 2B4, and the two of the three natural cytotoxicity receptors (NCRs), NKp30, and NKp46. One example of NK stimulatory receptor ligand is the protein encoded by retinoic acid early inducible gene (RAE-I), which was isolated from tumor lines. RAE-1 is also expressed on virus-infected cells (Backstrom et al. 2007), and binds to the stimulatory receptor expressed on NK cells, NKG2D (Diefenbach et al. 2000; Smyth et al. 2005). As an overall effect, NK cells would lyse target cells that have lost or express low amounts of MHC class I molecules, including tumor cells or cells infected by viruses such as certain Herpes viruses or Adenoviruses.

Once activated, NK cells display cytotoxic functions which is mediated by direct cell-to-cell contact as well as secretion of cytokines and chemokines. The cell contact pathways include perforin/granzyme (Warren and Smyth 1999), Fas/Fas-ligand (Screpanti et al. 2005), and TRAIL/TRAIL ligand interaction (Takeda et al. 2001). They also produce inflammatory cytokines such as IFN-γ, TGF-β, and GM-CSF. Despite these cytotoxic actions against tumor cells and virus infected cells, it is now well conceived that some NK cells could act as modulator of adaptive immunity and have the potential to eliminate self-reactive T cells.

Although the diversity of NK cells remained to be ambiguous some time ago, recent works have greatly contributed to clarifying their heterogeneity in phenotypes and functions. The majority of human NK cells in PBMC belong to CD56^{dim}CD16$^+$ cytolytic NK subset. These cells express homing markers for inflamed peripheral sites and carry perforin to rapidly mediate cytotoxicity. CD56^{bright} CD16$^-$ cells constitute a minor NK subset that lacks perforin but secrete large amounts of IFN-γ and

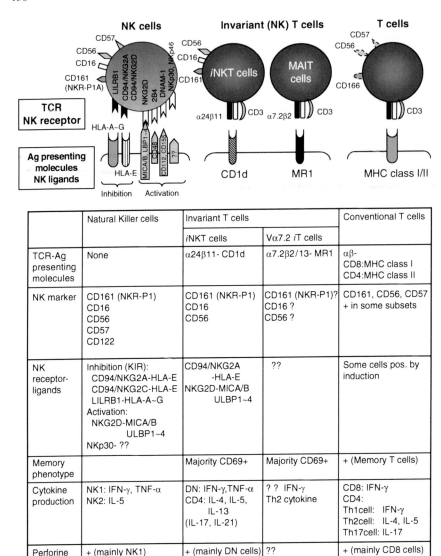

Fig. 1 Comparative features of human NK cells, invariant *i*NKT cells, and conventional T cells

	Natural Killer cells	Invariant T cells		Conventional T cells
		*i*NKT cells	Vα7.2 *i*T cells	
TCR-Ag presenting molecules	None	α24β11- CD1d	α7.2β2/13- MR1	αβ- CD8:MHC class I CD4:MHC class II
NK marker	CD161 (NKR-P1) CD16 CD56 CD57 CD122	CD161 (NKR-P1) CD16 CD56	CD161 (NKR-P1)? CD16 ? CD56 ?	CD161, CD56, CD57 + in some subsets

NK receptor-ligands	Inhibition (KIR):
	  CD94/NKG2A-HLA-E
	  CD94/NKG2C-HLA-E
	  LILRB1-HLA-A~G
	Activation:
	  NKG2D-MICA/B
	          ULBP1~4
	  NKp30- ??
		CD94/NKG2A
		  -HLA-E
		NKG2D-MICA/B
		  ULBP1~4
			??	Some cells pos. by induction

Memory phenotype		Majority CD69+	Majority CD69+	+ (Memory T cells)
Cytokine production	NK1: IFN-γ, TNF-α NK2: IL-5	DN: IFN-γ,TNF-α CD4: IL-4, IL-5, IL-13 (IL-17, IL-21)	? ? IFN-γ Th2 cytokine	CD8: IFN-γ CD4: Th1cell: IFN-γ Th2cell: IL-4, IL-5 Th17cell: IL-17
Perforine activation	+ (mainly NK1)	+ (mainly DN cells)	??	+ (mainly CD8 cells)
Frequency in PBMC	10 %	0.1 - 0.5 %	??	30-40 %

TNF-α upon activation. They are superior to CD56^{dim} cells in the regulatory functions that are mediated by these cytokines (Moretta et al. 2001). Moreover, they express surface markers such as CCR7 and CD62L that allow their homing to the lymph nodes, which results in the predominance of this NK cell subset in the secondary lymphoid organs.

Recent studies have shown that human NK cells are able to polarize *in vitro* into two functionally distinct subsets NK type 1 (NK1) or NK2 cells, analogous to T cell subsets Th1 or Th2. NK cells cultured in a condition favoring Th1 deviation (cultured with IL-12) would differentiate into NK1 cells producing IFN-γ and IL-10, whereas NK cells grown in a Th2 condition (cultured with IL-4) differentiate into NK2 cells producing IL-5 and IL-13 (Peritt et al. 1998). Although it was ambiguous whether the polarization actually occurs *in vivo*, an expansion of NK2 like cells producing IL-5 and IL-13 was observed in IFN-γ knockout mice (Hoshino et al. 1999), indicating that NK cells could functionally polarize into NK2-like cells in vivo.

Phenotypical analysis of NK cells in rodents has also identified a distinct population of NK cells that express CD11c, a prototypical dendritic cell (DC) marker. As the CD11c NK cells were shown to exhibit both NK and DC function, they are often referred to as "bitypic NK/DC cells" (Homann et al. 2002; Pillarisetty et al. 2005). CD11c molecule is known to be associated with integrin CD18 and form CD11c/CD18 complex. Although the precise function is not clear, CD11c is reportedly involved in binding of iC3b (Bilisland et al. 1994), adhesion to stimulated endothelium (Stacker and Springer 1991), and phagocytosis of apoptotic cells (Morelli et al. 2003). Bearing in mind that we have only very little knowledge of how these NK cell subsets are correlated to each other, we will next discuss on the recent progress which correlates the regulatory aspects of NK cells with the pathogenesis of MS.

2.2 NK Cell in MS

Despite the extensive studies in the past, there has been no simple uniform consensus regarding the role of NK cells in MS. Some of the earlier studies have found an inverse relationship between the number or the functional activity of circulating NK cells and the clinical or radiological activity of the patients with MS. NK cells isolated from MS patients were reported to be inefficient at cytotoxic killing and IFN-γ production (Benczur et al. 1980; Kastrukoff et al. 1998; Munschauer et al. 1995; Vranes et al. 1989). Furthermore, a longitudinal study showed that the functional activities of NK cells would decline during the relapse and then normalized during remission (Kastrukoff et al. 2003). On the contrary, several earlier studies failed to reveal any quantitative or qualitative difference between NK populations in MS patients versus controls (Hauser et al. 1981; Rauch et al. 1985; Rice et al. 1983; Santoli et al. 1981). The reason for these controversial findings remains to be unclear. However, it is of note that the criteria used to classify NK cells have been variable among the researchers and as a result the assays and protocols used to measure their functions and frequencies differ widely among the studies above mentioned. Moreover, because of difficulties in enrollment of patients, each of the studies might have examined the group of patients in different conditions. We also assume that they did not unify various confounding factors, some of which were not recognized when the study was conducted. Even duration of time between blood sampling and examination may affect the condition of NK cells (Takahashi et al. 2001).

In spite of the setbacks, the notion that NK cells have a significant role in reducing neuroinflammation and CNS injury stems from indirect evidences that were extracted from studies of an animal model experimental autoimmune encephalomyelitis (EAE) and from human clinical trials.

2.2.1 Protective Role of NK Cells in EAE

Monophasic EAE can be induced in C57BL/6 strain of mice (B6 mice) by immunizing the mice with an encephalitogenic myelin oligodendrocyte glycoprotein peptide (MOG_{35-55}). When NK cells were depleted in vivo by antibody specific for NK1.1 molecule (CD161), mice developed an aggravated form of EAE in terms of onset and clinical severity (Zhang et al. 1997). Furthermore, NK cell depletion was found to increase proliferation and production of Th1 cytokines by memory CD4[+] T cells in the recall response to MOG. Similarly, NK cell depletion augmented the severity of EAE induced in β_2-microglobulin$^{-/-}$ mice. As the mice are lacking expression of CD1d molecule necessary for NK1.1[+] T cell development, it was assumed that NK cells would play a regulatory role in a manner independent of NK1.1[+] T cells. Furthermore, co-transfer of whole splenocytes, but not of NK cell-depleted splenocytes, ameliorated EAE that was induced by adoptive transfer of MOG-specific T cells into Rag2$^{-/-}$ hosts. Taken together, it was concluded that NK cells play a regulatory role in EAE. Involvement of NK cells was also demonstrated in Lewis rat EAE model which can be induced by sensitization to MBP (Matsumoto et al. 1998). When NK cells were depleted by antibody specific for either NKR-P1 (analogous to NK1.1) or asialo GM1, the rats developed an aggravated form of EAE, characterized by higher maximal clinical scores and increased mortality rates. Subsequently, Swanborg et al. have shown that rat bone marrow-derived NK cells would exhibit potent inhibitory effects on proliferation of auto-reactive T cells (Smeltz et al. 1999), further strengthening the postulate that NK cells play a regulatory role in the CNS autoimmunity.

More recently, Huang et al. have reported that mice deficient in CX3CR1 (the fractalkine receptor) develop a more severe form of EAE (Huang et al. 2006). Compared with their littermates, CX3CR1$^{-/-}$ mice immunized with MOG_{35-55} would exhibit a higher incidence of CNS hemorrhage, leading to a higher mortality rate. Moreover, the survived mice failed to recover neurological functions after they reached the peak of EAE. Although the CX3CR1$^{-/-}$ mice developed more serious manifestations of EAE, recall responses to MOG_{35-55} and generation of encephalogenic T cells in the peripheral lymphoid organs were not augmented in the mice. Notable differences were found in the CNS infiltrating cells. Namely, NK1.1[+]CD3[-] cells were selectively depleted from mononuclear cells isolated from the spinal cord of the CX3CR1$^{-/-}$ mice, whereas they comprised 10–20% of the CNS infiltrates in wild-type mice and heterozygous CX3CR1$^{+/-}$ littermates. These findings led the authors to speculate that the exacerbated disease in CX3CR1$^{-/-}$ mice was due to a failure of regulatory NK cells to enter the target organ. In support of this, the majority of CNS-infiltrating NK cells in the littermate mice suffering from EAE expressed CX3CR1.

When NK cells were depleted in vivo by injecting anti-NK1.1 antibody, difference between CX3CR1$^{-/-}$ and the littermate CX3CR1$^{+/-}$ mice in the severity of EAE was no more evident. Of interest, soluble CX3CL1 was increased in the CNS of the EAE mice, and protein extracts from the CNS tissues showed a chemotactic activity for NK cells. It is of particular interest that a reduced number of circulating CX3CR1^{+} NK cells has recently been reported in patients with MS (Infante-Duarte et al. 2005), which would prompt further investigation to examine a possible correlate between EAE and MS with regard to NK cell-mediated immunoregulation.

2.2.2 Ex Vivo Analysis Revealed an Alteration of NK cells in MS

Given putative roles of NK cells in MS, one may ask if there is a significant correlation of NK cell functions and the disease activity of MS. By analyzing surface pheno-types and cytokine secretion profile of peripheral blood NK cells, we demonstrated in 2001 that NK cells from MS patients during clinical remission are characterized by a higher frequency of CD95^{+} cells as well as a higher expression level of IL-5, which represents a feature highly reminiscent of NK2 cells (Takahashi et al. 2001). The patients were selected from those who were not given any disease-modifying drugs, including corticosteroids. Remarkably, the NK2 cell-like feature, that is, a strong bias toward producing IL-5, was lost during the relapse of MS and regained after recovery. It was also found that NK2 cells induced *in vitro* from the peripheral blood of healthy subjects would inhibit the induction of Th1 cells, suggesting that the NK2 cells in vivo may also prohibit autoimmune effector T cells. Subsequently, we showed that when MS patients in remission are divided into two groups, accord-ing to the CD95^{+} NK cell frequency, memory T cells reactive to MBP are increased in patients who possess a higher number of CD95^{+} NK cells (Takahashi et al. 2004). Interestingly, NK cells from the "CD95 high patients" exhibited an ability to actively suppress the autoimmune T cells. These results allowed us to propose a model that CD95 low patients are enjoying very stable remission wherein an actual frequency of pathogenic autoimmune T cells is low, whereas CD95 high patients are in a more active state (which we call "smoldering state") wherein a higher number of autoreactive T cells are counter-regulated by NK cells (Fig. 2).

In a separate study, we found that CD11c expression on peripheral NK cells tends to correlate with temporal disease activity of MS (Aranami et al. 2006). Our study has revealed that surface CD11c expression on NK cells is significantly up-regulated in a proportion of patients with MS in remission, compared with healthy subjects or the rest of the patients. In the group of patients whose NK cells express higher levels of CD11c ("CD11c high patients"), IL-5 production from NK cells was significantly down-regulated and conversely, HLA-DR class II molecule was up-regulated. Accordingly, NK cells from "CD11c low patients" are NK2-biased, whereas those from "CD11c high patients" are not. NK cells from human PBMC would up-regulate expression of both CD11c and HLA-DR molecules after culture with IL-15 or a combination of IL-12 and IL-18 inflammatory cytokines commonly found in MS. Remarkably, the "CD11c high patients" tended to relapse significantly

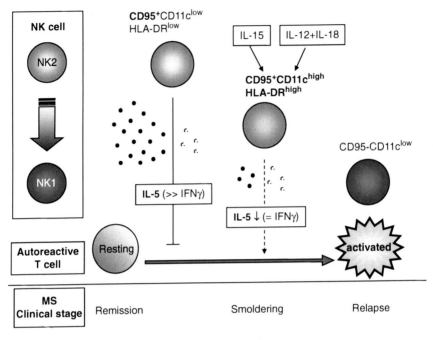

Fig. 2 Regulatory role of CD95+ NK 2 cell in MS remission

earlier than "CD11c low patients," indicating that "CD11c high patients" are clinically more active. We, therefore, propose that expression levels of CD11c on NK cells may serve as a good indicator of the disease activity (Fig. 2).

Another evidence for the role of NK cells in MS was obtained in the clinical trial of a new humanized monoclonal antibody against IL-2 receptor α-chain. In a recent phase II trial with the antibody (daclizumab), Bielekova et al. have noticed that an expansion of CD56^{bright} immunoregulatory NK cells and their increased perforin expression would highly correlate with the reduction of the disease activity (Bielekova et al. 2006). In fact, contrast enhanced lesion on brain MRI was significantly suppressed along with an expansion of circulating CD56^{bright} NK cells. NK cells isolated from patients being given daclizumab were found to exhibit cytotoxity towards autologous activated T cells, even without prestimulating NK cells with IL-2. These results raise a possibility that induced regulatory NK cells may at least partly mediate daclizumab effects on MS. In another study, an increase of CD56^{bright} NK cells was demonstrated in the blood of newly diagnosed patients with relapsing-remitting MS who were started on interferon-β treatment a few months ago (Saraste et al. 2007). This work also supports a role for induced regulatory NK cells in patients who respond to immunomodulatory therapy. Taking the available data together, we assume that NK cells harbor functional subpopulations that play a protective role in CNS autoimmunity. Regulatory NK cells could be CD56^{high}, CD95+, or CX3CR1+, although mutual relationship of the populations still remains unclear. Further attempts to find a way to selectively activate regulatory NK cells are warranted, because it will lead to developing a new treatment strategy for MS. It is known that NK cells show cytotoxic insults against CNS components in some *in vitro* conditions (Morse et al. 2001). To develop safe and effective drugs targeting NK cells, it is also important to know if regulatory NK cells could be selectively induced without augmenting cytotoxic NK cells that are potentially harmful for MS.

3 *i*NKT Cells in MS

3.1 What Is iNKT Cell?

3.1.1 General Properties of Invariant NKT (*i*NKT) Cells

Invariant NKT (*i*NKT) cells are a unique subset of lymphocytes that recognize a glycolipid antigen such as α-galactosylceramide (α-GC) (Kawano et al. 1997), that is bound to a monomorphic MHC class I-like molecule CD1d (Bendelac et al. 2007; Kronenberg 2005; Taniguchi et al. 2003). The term "NKT cells" was first introduced in mice to define a broader range of T cells that express the NK cell-associated marker NK1.1 (CD161) (Ballas and Rasmussen 1990; Fowlkes et al. 1987). The term "*i*NKT cells" defines a more limited population among NK1.1$^+$ T cells that express a single invariant α-chain (Vα14-Jα18 in mice and Vα24-Jα18 in humans) and respond to α-GC bound to CD1d (Dellabona et al. 1994; Exley et al. 1997; Koseki et al. 1991) (Fig. 1). The invariant α-chain is coupled with a noninvariant β-chain which selectively uses Vβ8.2, Vβ7, and Vβ2 gene segments in mice and Vβ11 (a molecule homologous to mice Vβ 8.2) in humans. It is currently known that mouse NK1.1$^+$ T cells (or NKT cells in the classic definition) are composed of *i*NKT cells, CD1d-restricted noninvariant T cells, conventional T cells that are not restricted by CD1d, and MAIT cells (see Sect. 4). On the other hand, there are a significant number of NK1.1-negative T cells that express the invariant Vα14-Jα18 TCR and react to α-GC/CD1d. In most of the current literatures, such T cells are also called *i*NKT cells.

$\quad$*i*NKT cells constitutively express memory/activated T cell phenotype and are capable of robustly producing pro- and anti-inflammatory cytokines within hours after TCR engagement. The cytokine burst following *i*NKT cell activation then triggers a maturation process of downstream cells, such as NK cells, DCs, B cells, and T cells, which leads to subsequent alteration of a broad range of adaptive immune responses. Although *i*NKT cells utilize TCR for sensing a specific antigen, the behavior of the cells in response to external stimuli resembles that of innate lymphocytes (Mempel et al. 2002). Owing to the swift responsiveness to external stimuli, it is thought that *i*NKT cells play an important role in bridging innate and adaptive arms of immune response.

$\quad$Another striking property of *i*NKT cells is to produce diverse combinations of cytokines, depending on how they are stimulated. Mouse *i*NKT cells can produce IFN-γ, IL-2, -5, -13, -17, -21, GM-CSF, TNF-α, and osteopontin after an optimal engagement of TCR (Yamamura et al. 2007). In fact, they can produce a broad range of pro- and anti-inflammatory cytokines upon stimulation with α-GC, a highly potent ligand for *i*NKT cells (Kawano et al. 1997). In contrast, cytokine production by *i*NKT cells is much more finely regulated under physiological environment, which could result in production of a set of Th2 cytokines (Sakuishi et al. 2007).

*i*NKT cells are segregated into CD4$^+$CD8$^-$ and CD4$^-$CD8$^-$ double negative (DN) subsets. It has been shown that each subset differs remarkably in their functional properties. In humans, about 40–60% of *i*NKT cells are CD4$^+$, and a large majority of the remaining cells are DN cells. Some *i*NKT cells express CD8α, but only very few cells co-express CD8β. The CD4$^+$ subset potently produces both Th1 and Th2 cytokines, whereas the DN population selectively produces the Th1 cytokines (IFN-γ and TNF-α) and preferentially up-regulates perforin in response to IL-2 or IL-12 (Gumperz et al. 2002; Lee et al. 2002). It is also known that the CD4$^+$ and DN *i*NKT cells differentially express chemokine receptors: CCR4 on CD4$^+$ cells and CCR1, CCR6, and CXCR6 on DN cells (Kim et al. 2002). These results suggest the presence of a functional dichotomy in *i*NKT cells.

3.1.2 iNKT Cells and Their Ligands

To evaluate the potential of *i*NKT cells to regulate autoimmune diseases, it is particularly important to understand how they recognize a glycolipid antigen bound to CD1d. The CD1d molecule, highly conserved among mammalian species (Exley et al. 2000), is primarily expressed on the cells of hematopoietic origin, including thymocytes, B cells, macrophages, and DCs, and could also be induced on T cells upon activation. The binding cleft of the CD1d molecule consists of two nonpolar lined grooves, which makes it ideal for the presentation of hydrophobic antigens such as glycolipids. In 1997, a marine sponge-derived glycosphingolipid, α-GC, was identified as a potent ligand for mouse *i*NKT cells (Kawano et al. 1997). It was subsequently found that α-GC is stimulatory for human *i*NKT cells as well (Brossay et al. 1998). Thereafter, a synthetic α-GC has been used extensively for research (Fig. 3). A widely supported view on the topology of TCR/ligand/CD1d is that the two lipid chains of α-GC would be inserted into the CD1d hydrophobic grooves and α-linked sugar moiety becomes accessible for the TCR of *i*NKT cells (McCarthy et al. 2007). More recently, crystal structure analysis has demonstrated that the invariant α-chain of the *i*NKT cells would selectively recognize the α-linked sugar of α-GC (Borg et al. 2007). It is of note that glycolipids with α-linked sugars such as α-GC could not be found in mammalian tissues, but are rather ubiquitously present in the environment. After LPS-negative α-proteobacteria extracts were found to contain glycosphingolipids stimulatory for *i*NKT cells, a growing number of bacterial lipid antigens has been shown to stimulate *i*NKT cells (Bendelac et al. 2007), including diacylglycerol glycolipid extracted from Borrelia burgdorferi (Kinjo et al. 2005). Given that the TCRs of *i*NKT cells recognize such pathogen-derived antigens, the lipid antigens may be an important initiator for triggering the immune response in bacterial and parasite infection. However, it has recently been demonstrated that *i*NKT cells are activated during infection without recognizing a bacteria component

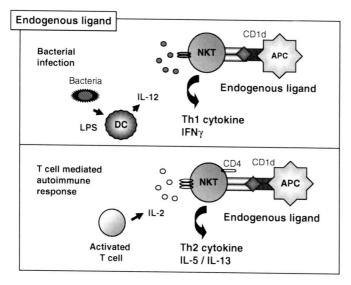

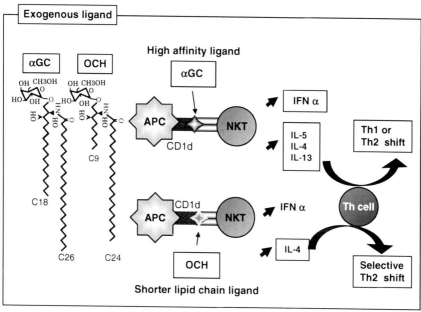

Fig. 3 Effects of lipid chain lengths in alpha-galactosylceramides on cyotokine release by natural killer T cells

via TCR (Brigl and Brenner 2004; Mattner et al. 2005). The antigen recognized by the TCR of *i*NKT cells is thought to be an endogenous ligand bound with CD1d, but not an exogenous microbial ligand. These studies also showed that the role for the bacterial LPS is to trigger production of IL-12 from DCs. Although *i*NKT cells exhibit little response to the endogenous ligand/CD1d iNKT cells expressed by DCs, the presence of excessive amount of IL-12 would remarkably augment the iNKT cell response to endogenous ligand, which leads to production of a large amount of IFNγ from iNKT cells. Thus, iNKT cells may act as crucial amplifiers of Th1 cells in the initial inflammatory response to the pathogens.

Of note, not only Th1 but Th2 cytokine response could also be amplified through a similar mechanism. We have recently revealed that in the presence of excessive IL-2, TCR recognition of putative endogenous ligand would trigger production of IL-5 and IL-13 from human CD4$^+$ iNKT cells (Sakuishi et al. 2007). These findings indicate that under physiological conditions, cytokine milieu would be decisive in directing iNKT cell responses towards Th1 or Th2, and are relevant for understanding the mechanism of how iNKT cells would regulate the adaptive immune response in vivo (Fig. 3).

Since α-anomeric glycolipids do not exist in mammalian tissues, a number of β-anomeric glycolipids have been evaluated for their possible role as an endogenous ligand for iNKT cells. The search has led to the identification of lysosomal glycolipid isoglobotrihexosylceramide (iGb3) as a putative endogenous ligand (Zhou et al. 2004; Mattner et al. 2005). However, it has recently been demonstrated that iNKT cells are normal in number and function in iGb3 synthetase deficient mice, despite of lacking endogenous iGb3 (Porubsky et al. 2007). Moreover, a highly sensitive HPLC assay has failed to detect the presence of iGb3 in various mouse tissues except for the dorsal root ganglion. Nor was iGb3 detected in any human tissue (Speak et al. 2007). Therefore, the search for endogenous ligand is still not over. Regarding the pathogenesis of MS, it is of key interest whether any myelin-derived lipid antigen may stimulate iNKT cells.

Another subject of growing interest is to use iNKT cell ligands as therapeutic agents for autoimmune diseases. The prototypical ligand α-GC showed some efficacy for autoimmune diseases (Hong et al. 2001). However, as it provokes production of a wide range of cytokines including proinflammatory ones, it may worsen some disease conditions. To overcome this problem, structurally altered analogs of α-GC were synthesized and their ability to inhibit the development of autoimmune disease has been examined. A work from our laboratory has demonstrated that an α-GC analog bearing a shorter sphingosine chain compared with α-GC (named as OCH) would selectively stimulate IL-4 production from iNKT cells, whereas α-GC stimulation induces both IL-4 and IFNγ (Miyamoto et al. 2001; Oki et al. 2004). Accordingly, OCH stimulation of iNKT cells favors a Th2 bias of immune response in vivo as compared with α-GC stimulation and showed better efficacy for treatment of various autoimmune disease models (Fig. 3) (see Sect. 3.3 as well).

3.2   Studies of iNKT Cells in MS

Using single-strand conformation polymorphism (SSCP), a method for examining the TCR repertoire, we have previously analyzed blood samples from subjects with MS as well as other neurological diseases (Illes et al. 2000). Expression of the invariant Vα24-Jα18 rearrangement, the invariant TCR α-chain expressed by human iNKT cells, was greatly reduced in the blood lymphocytes of the patients with MS, compared with those from healthy subjects. The reduction was not observed in the patients with other autoimmune/inflammatory neurological diseases. Interestingly, the Vα24-Jα18 TCR was only rarely found in the CNS lesions of MS but was often detected in the biopsy samples from chronic inflammatory demyelinating polyneuropathy (CIDP).

More recently, we have reanalyzed the frequency of iNKT cells in the peripheral blood of MS by using flow cytometry. A striking reduction of the total number of iNKT cells was confirmed in the peripheral blood of the patients with MS in a drug-free remission state (Araki et al. 2003). Interestingly, when CD4$^+$ and DN iNKT cells were analyzed separately, a remarkable iNKT cell reduction was found to reflect a great reduction of DN iNKT cells, that are known to preferentially produce proinflammatory cytokines (Gumperz et al. 2002; Lee et al. 2002). Moreover, we found that the CD4$^+$ iNKT cell lines from MS patients were significantly biased for Th2: they produced much more IL-4 than those from healthy subjects, although the production of IFN-γ was not altered significantly (Araki et al. 2003). Collectively, the changes found in iNKT cells (a reduction of DN and Th2 bias of CD4$^+$ iNKT cells) are thought to be beneficial for maintaining the remission state of MS.

It is also worthwhile to mention that the currently available drugs may exert their actions through targeting iNKT cells. Although the drug-free remission state of MS was associated with a great reduction of iNKT cells in the peripheral blood (Araki et al. 2003), patients who were continuously given a low dose oral corticosteroid showed a normal frequency of iNKT cells in the blood, indicating that oral corticosteroid treatment may restore the frequency of iNKT cells (Araki et al. 2004). Interestingly, the cytokine profile of DN NKT cells from the corticosteroid-treated MS showed a trend for Th2 bias. This may represent one of the mechanisms of the corticosteroid effects in MS and other autoimmune diseases.

In a recent longitudinal study, IFN-β treatment significantly increased the number of iNKT cells in the peripheral blood mononuclear cell within same patients (Gigli et al. 2007). Furthermore, iNKT cells of IFN-β treated individuals showed a dramatically improved secretion of INF-γ, IL-4, and IL-5 in response to α-GC stimulation compared with those isolated from the same individuals before IFN-β treatment. The study also showed up-regulation of key costimulatory molecules expressed by DCs in the IFN-β treated patients. Thus, immune regulatory effect of IFN-β therapy in MS may possibly mediate iNKT cells.

3.3 iNKT Cells as a Therapeutic Target in MS/EAE

Results of EAE studies give us clues to understanding the role of iNKT cells in the pathogenesis of MS. It is well known that SJL/J mice are very susceptible to induction of EAE and other autoimmune diseases. In this strain of mice, iNKT cells are reduced in number and defective in IL-4 production (Yoshimoto et al. 1995), allowing us to speculate that the iNKT cell defects may account for the autoimmune susceptible nature. On the contrary, transgenic overexpression of the invariant TCR of iNKT cells was found to protect NOD strain of mice from development of EAE. This EAE protection was associated with an inhibition of antigen-specific IFN-γ production but was independent of IL-4 (Mars et al. 2002). These results indicate an inverse correlation of iNKT cell numbers/functions with the susceptibility to EAE, raising a simple idea that expanding iNKT cells may be beneficial for treating patients with MS.

After α-GC was identified as a potent ligand for iNKT cells, several laboratories have examined whether in vivo injection of α-GC may modify the clinical course of EAE by stimulating iNKT cells. A study by Singh et al. showed that α-GC is capable of down-modulating EAE, by inducing Th2 bias of iNKT cells (Singh et al. 2001). Furlan et al. also showed an efficacy of α-GC in EAE, but they did not reveal a Th2 bias but rather showed an enhanced IFNγ production by the liver iNKT cells (Furlan et al. 2003). In an independent study by Jahng et al., injection of α-GC with aim to suppress EAE resulted in diverse outcome, which depends on the administration route, timing of injection, and dose of this glycolipid (Jahng et al. 2001). Although the reason for these discrepancies remain unclear, it is possible that source of the mice, quality of the animal facilities, or even gut flora might have influenced the results.

It was subsequently found that CD28-B7 costimulatory signals play a critical role in stimulating iNKT cells with α-GC. When iNKT cells were stimulated with α-GC in the presence of anti-B7 (CD80) antibody in vitro, they selectively produced Th2 cytokines (Pal et al. 2001). In vivo stimulation of iNKT cells along with blocking CD28-B7 interactions was found to suppress the onset of EAE (Pal et al. 2001). These results collectively indicated that proper stimulation of iNKT cells might lead to suppression of pathogenic Th1 responses. We have then explored whether a Th2 polarizing ligand could be identified among α-GC analogs. As discussed briefly in Sect. 3.1.2, we have found that an analog of α-GC, called OCH, bearing a shorter sphingosine chain could selectively induce production of IL-4 but not of IFN-γ and could modulate disease process of EAE when injected in vivo (Miyamoto et al. 2001). This protective effect against the development of EAE was abrogated by a simultaneous injection of anti-IL-4 antibody. Moreover, the protective effect of OCH could not be seen in IL-4 knockout mice, indicating that IL-4 produced from iNKT cells is involved in the disease suppression.

The molecular mechanism for the selective IL-4 production by OCH has been intensively studied in our laboratory. Owing to the truncation of sphingosine chain, OCH binds to CD1d molecule less stably compared to α-GC. We are proposing that the unstable OCH-CD1d interaction, which does not allow continuous TCR stimulation, is a key to understanding the Th2 polarizing character of OCH (Oki et al. 2004). When iNKT cells are stimulated by α-GC, IL-4 is produced within a few hours, which is then followed by production of a large quantity of IFN-γ (Pal et al. 2001). Of note is that de novo protein synthesis is required for the iNKT cell production of IFN-γ but not of IL-4 (Oki et al. 2004). Subsequent analysis has revealed that c-Rel protein is selectively induced, when iNKT cells are simulated by α-GC. Inhibiting c-Rel expression in iNKT cells has led to a selective IL-4 induction as a result of suppressed production of IFN-γ, as seen with OCH stimulation. Taken together, it can be postulated that unstable binding of OCH with CD1d leads to disrupted TCR signaling, which does not induce expression of c-Rel and of its down-stream molecule IFN-γ. Compared with α-GC, which is capable of fully inducing c-Rel and IFN-γ, OCH would exhibit a unique Th2 polarizing effect on iNKT cells in vitro and in vivo. Intriguingly, in vivo injection of OCH induces defective IFN-γ production not only by NKT cells but also by NK cells (Oki et al. 2005). Mechanistic analysis has revealed that an injection of OCH induces an insufficient induction of CD40L in addition to lower primary IFN-γ production by the NKT cells, leading to a marginal IL-12 production by DCs. A combination of these differences between OCH and α-GC stimulation would account for the lower secondary IFN-γ production by NKT and NK cells by OCH. Of note, McCarthy et al. have recently confirmed that shortening of the phytosphingosine chain increased the rate of lipid dissociation from CD1d molecule and induced less sustained TCR signals (McCarthy et al. 2007). In this study, they have also demonstrated the decreased affinity of TCR to OCH bound-CD1d.

Other lipid chain truncated analogs of α-GC have been reported to display a similar skewing of cytokine profile towards Th2 but the mechanism seems to differ from that found in OCH (Goff et al. 2004; Yu et al. 2005). Taken together, altered glycolipid provides attractive means for iNKT cells mediated intervention of inflammatory autoimmune disease such as EAE and human MS.

4   MR1- Restricted Invariant T Cells in MS

Another novel invariant NK cell receptor-positive T cell population besides iNKT cells has been described in mice and humans. They are preferentially located in the gut lamina propria and are generally termed mucosal-associated invariant T (MAIT) cells (Treiner et al. 2003). Of interest, they are absent in germ-free mice, which indicates the role of gut flora for generation and maintenance of this lymphocyte. The discovery of this population is dated back to 1993, when DN T cell population expressing an invariant TCR α-chain was described along with the identification of Vα24 iNKT cells (Porcelli et al. 1993). It is now established that the new invariant T cells are distinct from iNKT cells in the expression of another conserved CDR3α sequence (Vα7.2-Jα33 in humans and Vα19-Jα33 in mice) and restricted use of Vβ2 and Vβ13 in mice and humans. Unlike iNKT cells selected by CD1d, they are selected by another MHC class Ib molecule, MR1, that is also highly conserved among species (Treiner et al. 2003). The mouse MAIT cells were isolated from NK1.1⁺ T cells in the liver of CD1d deficient mice lacking "conventional" iNKT cells, allowing us to call the cells "Vα19-Jα33 NKT cells." As seen with "conventional" NKT cells, human MAIT cells constitutively express memory phenotype and some NK cell markers other than CD57 (Treiner et al. 2005) (Fig. 1). Several lines of evidence suggest that MR1 presents lipid ligands such as α-mannocylceramide (Shimamura et al. 2007). Although the function of MAIT cells is unclear at the moment, their cardinal features such as the semi-invariant repertoire, restriction by monomorphic class I-like molecule and the natural memory phenotype suggest that *i*NKT cells and MAIT cells may exhibit similar and/or complementary functions.

When expression of Vα7.2 invariant TCR for human MAIT cells was investigated in MS patient samples, there was a striking difference between the MAIT and *i*NKT cell invariant TCR in their expression. Expression of the invariant TCR chain for NKT cells was clearly reduced in the peripheral blood of MS patients (Illés et al. 2000), whereas invariant TCR for MAIT cells was clearly detected in the great majority of the patients (Illés et al. 2004). Parallel analysis of CNS lesions from MS patients showed that MAIT cells would infiltrate the majority of the lesions, whereas *i*NKT cells do not (Illés et al. 2000, 2004). The differential expression of the two invariant chains in samples from MS suggests that MAIT cells and NKT cells may complement each other and MAIT cells may substitute deficiency of *i*NKT cells in MS.

The protective role of MAIT cells is further delineated by the study of mouse EAE. We found that overexpression of the invariant Vα19-Jα33 TCR in B6 mice is protective against EAE induction and progression (Croxford et al. 2006). Consistently, EAE was exacerbated in MR1 deficient mice, which lack Vα19-Jα33 invariant T cells. The protective effect was found to accompany a reduced production of inflammatory mediators as well as an increased secretion of IL-10. We have also demonstrated that IL-10 production occurred in part through interactions between B cells and Vα19 MAIT cells involving ICOS costimulatory molecule.

5 Concluding Remarks

NK cells and *i*NKT cells are groups of innate lymphocytes with multi potential qualities. Recent advances in cell biology of these cells have brought our attention to their ability in regulating autoimmune inflammatory responses. Selective induction of their regulatory properties could be an effective means for modification of autoimmune disease affecting the CNS. It is also notable that NK cells and *i*NKT cells change their phenotypes, number, and gene expression profile during disease course of MS. They could be good targets also for those who attempt to identify useful biomarkers for MS.

References

Araki M, Kondo T, Gumperz J, Brenner M, Miyake S, Yamamura T (2003) T$_h$2 bias of CD4$^+$ NKT cells derived from multiple sclerosis in remission. Int Immunol 15:279–288

Araki M, Miyake S, Yamamura T (2004) Continuous oral glucocorticoid therapy restores the NKT cell frequency in multiple sclerosis. Neuroimmunology 12:175–179

Aranami T, Miyake S, Takahashi K (2006) Differential expression of CD11c by peripheral blood NK cells reflects temporal activity of multiple slcerosis. J Immunol 177:5659–5667

Awasthi A, Carrier Y, Peron JP, Bettelli E, Kamanaka M, Flavell RA, Kuchroo VK, Oukka M, Weiner HL (2007) A dominant function for interleukin 27 in generating interleukin 10-producing anti-inflammatory T cells. Nat Immunol 8:1380–1389

Backstrom E, Ljunggren H, Kristensson K (2007) NK cell-mediated destruction of influenza A virus-infected peripheral but not central neurons. Scand J Immunol 65:353–361

Baecher-Allan C, Hafler D (2006) Human regulatory T cells and their role in autoimmune disease. Immunol Rev 212:203–216

Ballas Z, Rasmussen W (1990) NK1.1+ thymocytes, adult murine CD4⁻CD8⁻ thymocytes contain an NK1.1⁺, CD3⁺, CD5⁺, CD44⁺, TCR-Vb 8⁺ subset. J Immunol 145:1039–1045

Benczur M, Petranyi G, Palffy G, Varga M, Talas M, Kotsy B, et al (1980) Dysfunction of natural killer cells in multiple sclerosis: a possible pathogenetic factor. Clin Exp Immunol 39:657–662

Bendelac A, Fearon DT (1997) Innate pathways that control acquired immunity. Curr Opin Immunol 9:1–3

Bendelac A, Savage P, Teyton L (2007) The biology of NKT cells. Annu Rev Immunol 25:297–336

Bielekova B, Goodwin B, Richert J, Cortese I, Kondo T, Afshar G (2000) Encephalitogenic potential of the myelin basic protein peptite (amino acids 83–99) in multiple sclerosis: results of a phase II clinical trial with an altered peptide ligand. Nat Med 6:1167–1175

Bielekova B, Catalfamo M, Reichert-Scrivner S, Packer A, Cerna M, Waldmann T, et al (2006) Regulatory CD56 (bright) natural killer cells mediate immunomodulatory effects of IL-2Rα-targeted therapy (daclizumab) in multiple sclerosis. Proc Natl Acad Sci U S A 103:5941–5946

Bilisland C, Diamond M, Springer T (1994) The leukocyte integrin p150, 95 (CD11c/CD18) as a receptor for ic3b: activation by a heterologous β subunit and localization of a ligand recognition site to the I domain. J Immunol 152:4582–4589

Borg N, Wun K, Kjor-Nielson L, Wilce M, Pellicci D, Koh R, et al (2007) CD1d-lipid-antigen recognition by the semi-invariant NKT T-cell receptor. Nature 448:44–49

Borrego F, Masilamani M, Marusima A, Tang X, Coligan J (2006) The CD94/NKG2 family of receptors from molecules and cells to clinical relevance. Immnol Res 35:263–294

Brigl M, Brenner M (2004) CD1: antigen presentation and T cell function. Annu Rev Immunol 22:817–890

Brossay L, Chioda M, Burdin N, Koezuka Y, Casorati G, Dellabona P, et al (1998) CD1d-mediated recognition of an alpha-galactosylceramide by natural killer T cells is highly conserved through mammalian evolution. J Exp Med 188:1521–1528

Carrol M, Prodeus A (1998) Linkages of innate and adaptive immunity. Curr Opin Immunol 10:36–40

Correale J, McMillan M, McCarthy K, Le T, Weiner L (1995) Isolation and characterization of autoreactive proteolipid protein-peptide specific T cell clones from multiple sclerosis patients. Neurology 45:1370–1378

Croxford J, Miyake S, Huang Y, Shimamura M, Yamamura T (2006) Invariant Vα19i T cells regulate autoimmune inflammation. Nat Immunol 7:987–994

Dellabona P, Padovan E, Casorati G, Brockhaus M, Lanzavecchia A (1994) An invariant Vα24-JαQ/Vβ11 T cell receptor is expressed in all individual by clonally expanded CD4⁻CD8⁻ T cells. J Exp Med 180:1171–1176

Diefenbach A, Jamieson A, Liu S, Shastri N, Raulet D (2000) Ligands for the murine NKG2D receptor: expression by tumor cells and activation of NK cells and macrophages. Nat Immunol 1:119–126

Exley M, Garcia J, Balk S, Porcelli S (1997) Requirements for CD1d recognition by human invariant Vβ24⁺ CD4⁻CD8⁻ T cells. J Exp Med 186:109–120

Exley M, Garcia J, Wilson S, Spada F, Gerdes D, Tahir S, et al (2000) CD1d structure and regulation on human thymocytes, peripheral blood T cells, B cells and monocytes. Immunology 100:37–47

Fearon D, Locksley R (1996) The instructive role of innate immunity in the acquired immune response. Science 272:50–53

Fowlkes B, Kruisbeek A, Ton-That H, Weston M, Coligan J, Schwartz R, et al (1987) A novel population of T-cell receptor αβ-bearing thymocytes which predominantly express a single Vβ gene family. Nature 329:251–254

Furlan R, Bergami A, Cantarella D, Brambilla E, Taniguchi M, Dellabona P, et al (2003) Activation of invariant NKT cells by agalcer administration protects mice from MOG 35–55-induced EAE: critical roles for administration route and IFN-γ. Eur J Immunol 33:1830–1838

Gigli G, Caielli S, Cutuli D, Falcone M (2007) Innate immunity modulates autoimmunity: type 1 interferon-b treatment in multiple sclerosis promotes growth and function of regulatory invariant natural killer T cells through dendritic cell maturation. Immunology 122:409–417

Goff R, Gao Y, Mattner J, Zhou D, Yin N, Cantu C 3rd, et al (2004) Effects of lipid chain lengths in alpha-galactosylceramides on cytokine release by natural killer T cells. J Am Chem Soc 126:13602

Goodnow C, Sprent J, Fazekas de St. Groth B, Vinuesa C (2005) Cellular and genetic mechanism of self-tolerance and immunity. Nature 435:590–597

Gumperz J, Miyake S, Yamamura T, Brenner M (2002) Functionally distinct subsets of CD1d-restricted natural killer T cells revealed by CD1d tetramer staining. J Exp Med 195:625–636

Hafler D (2004) Multiple sclerosis. J Clin Invest 113:788–794

Hauser S, Ault K, Levin M, Garovoy M, Weiner H (1981) Natural killer cell activity in multiple sclerosis. J Immunol 127:1114–1117

Homann D, Jahreis A, Wolfe T, Hughes A, Coon B, van Stipdonk M, et al (2002) CD40L blockade prevents autoimmune diabetes by induction of bitypic NK/DC regulatory cells. Immunity 16:403–415

Hong S, Wilson MT, Serizawa I, Wu L, Singh N, Naidenko OV, et al (2001) The natural killer T-cell ligand α-galactosylceramide prevents autoimmune diabetes in non-obese diabetic mice. Nat Med 7:1052–1056

Hoshino T, WInkler-Pickett RT, Mason AT, Ortaldo JR, Young HA (1999) IL-13 production by NK cells: IL-13-producing NK and T cells are present in vivo in the absence of IFN-γ. J Immunol 162:51–59

Huang D, Shi F, Jung S, Pien G, Wang J, Salazar-Mather T, et al (2006) The neuronal chemokine CX3CR1/fractalkine selectively recruits NK cells that modify experimental autoimmune encephalomyelitis within the central nervous system. FASEB J 20:896–905

Huseby E, Liggitt D, Brabb T, Schnabel B, Ohlen C, Goverman JA (2001) Pathogenic role for myelin-specific CD8$^+$ T cells in a model for multiple sclerosis. J Exp Med 194:669–676

Iglesias A, Bauer J, Litzenburger T, Schubart A, Linington C (2001) T-and B-Cell responses to myelin oligodendrocyte glycoprotein in experimental autoimmune encephalomyelitis and multiple sclerosis. Glia 36:220–234

Illés Z, Kondo T, Yokoyama K, Ohashi T, Tabira T, Yamamura T (1999) Identification of autoimmune T cells among in vivo expanded CD25$^+$ T cells in multiple sclerosis. J Immunol 162:1811–1817

Illés Z, Kondo T, Newcombe J, Oka N, Tabira T, Yamamura T (2000) Differential expression of NK T cell Vα24 JαQ invariant TCR chain in the lesion of multiple sclerosis and chornic inflammatory demyelinating polyneuropathy. J Immunol 164:4375–4381

Illés Z, Shimamura M, Newcombe J, Oka N, Yamamura T (2004) Accumulation of V α7.2- J α33 invariant T cells in human autoimmune inflammatory lesions in the nervous system. Int Immunol 16:223–230

Infante-Duarte C, Weber A, Kratschmar J, Prozorovski T, Pikol S, Hamann I, et al (2005) Frequency of blood CX3CR1-positive natural killer cells correlates with disease activity in multiple sclerosis. FASEB J 19:1902–1904

Jahng A, Maricic I, Pedersen B, Burdin N, Naidenko O, Kronenberg M, et al (2001) Activation of natural killer T cells potentiates or prevents experimental autoimmune encephlomyelitis. J Exp Med 194:1789–1799

Kastrukoff L, Morgan N, Zecchini D, White R, Petkau A, Satoh J, et al (1998) A role for natural killer cells in the immunopathogenesis of multiple sclerosis. J Neuroimmunol 86:123–133

Kastrukoff L, Lau A, Wee R, Zecchini D, White R, Paty D (2003) Clinical relapse of multiple sclerosis are associated with novel valleys in natural killer cell functional activity. J Neuroimmunol 145:103–114

Kawano T, Cui J, Koezuka Y, Toura I, Kaneko Y, Motoki K, et al (1997) CD1d-restricted and TCR-mediated activation of Vα14 NKT cells by glycosylcermides. Science 278:1626–1629

Kim C, Johnston B, Butcher E (2002) Trafficking machinery of NKT cells: shared and differential chemokine receptor expression among Vα24$^+$Vβ11$^+$ NKT cell subsets with distinct cytokine-producing capacity. Blood 100:11–16

Kinjo Y, Wu D, Kim G, Xing G, Poles M, Ho D, et al (2005) Recognition of bacterial glycosphin-golipids by natural killer T cells. Nature 434:520–525

Kirwan S, Burshtyn D (2007) Regulation of natural killer cell activity. Curr Opin Immunol 19:46–54

Koehler N, Genain C, Giesser B, Hauser S (2002) The human T cell response to myelin oligoden-drocyte glycoprotein: a multiple scleorsis family-based study. J Immunol 168:5920–5927

Kondo T, Yamamura T, Inobe J, Ohashi T, Takahashi K, Tabira T (1996) TCR repertoire to proteo-lipid protein (PLP) in multiple sclerosis (MS): homologies between PLP-specific T cells and MS-associated T cells in TCR junctional sequences. Int Immunol 8:123–130

Koseki H, Asano H, Inaba T, Miyashita N, Moriwaki K, Lindahl K, et al (1991) Dominant expression of a distinctive V14+ T-cell antigen receptor a chain in mice. Proc Natl Acad Sci U S A 88:7518–7522

Kronenberg M (2005) Toward understanding of NKT cell biology: progress and paradoxes. Annu Rev Immunol 2005: 877–900

Kyewski B, Derbinski J (2004) Self-representation is the thymus: an extended view. Nat Rev Immunol 4:688–698

Lee P, Benlagha K, Teyton L, Bendelac A (2002) Distinct functional lineages of human Vα24 natural killer T cells. J Exp Med 195:637–641

Lopez-Botet M, Perez-Villar J, Carretero M, Rodriguez A, Melero I, Bellon T (1997) Structure and function of the CD94 C-type lectin receptor complex involved in the recognition of HLA class I molecules. Immunol Rev 155:165–174

Mars L, Laloux V, Goude K, Desbois S, Saoudi A, Van Kaer L, et al (2002) Cutting edge: Vα14-Jα 281 NKT cells naturally regulate experimental autoimmune encephalomyelitis in nonobese diabetic mice. J Immunol 168:6007–6011

Martin R, Howell M, Jaraquemada D, Flerlage M, Richert J, Brostoff S (1991) A myelin basic protein peptide is recognized by cytotoxic T cells in the context of four HLA-DR types associated with multiple sclerosis. J Exp Med 173:19–24

Matsumoto Y, Kohyama K, Aikawa Y, Shin T, Kawazoe Y, Suzuki Y, et al (1998) Role of natural killer cells and TCR γδ T cells in acute autoimmune encephalomyelitis. Eur J Immunol 28:1681–1688

Mattner J, Debord K, Ismail N, Goff R, Cantu 3rd C, Zhou D, et al (2005) Exogenous, and endog-enous glycolipid antigens activate NKT cells during microbial infection. Nature 434:525–529

McCarthy C, Shepherd D, Floire S, Stronge V, Koch M, Illarionov P, et al (2007) The length of lipids bound to human CD1d molecules modulates the affinity of NKT cell TCR and the threshold of NKT cell activation. J Exp Med 204:1131–1144

Medzhitov R, Janeway JC (1997) Innate immunity: impact on the adaptive immune response. Curr Opin Immunol 9:4–9

Mempel M, Ronet C, Suarez F, Gilleron M, Puzo G, Van Kaer L, et al (2002) Natural killer T cells restricted by the monomorphic MHC class 1b CD1d1 molecules behave like inflammatory cells. J Immunol 168:365–371

Mendel I, Kerlero de Rosbo N, Bennun AA (1995) Myeline oligodendrocyte glycoportein peptide induces typical chronic experimental autoimmune encephalomyelitis in H-2β mice: fine specificity and T cell receptor Vβ expression of encephalitogenic T ells. Eur J Immunol 25:1951–1959

Miyamoto K, Miyake S, Yamamura T (2001) A synthetic glycolipid prevents autoimmune enceph-alomyelitis by inducing TH2 bias of natural killer T cells. Nature 413:531–534

Miyara M, Sakaguchi S (2007) Natural regulatory T cells: mechanisms of suppression. Trends Mol Med 13:108–116

Morelli A, Larregina A, Shufesky W, Zahorchak A, Logar A, Papworth G, et al (2003) Internalization of circulating apoptopic cells by splenic marginal zone dendritic cells: depen-dence on complement receptors and effect on cytokine production. Blood 101:611–620

Moretta A, Bottino C, Vitale M, Pende D, Cantoni C, Mingari M (2001) Activating receptors and core-ceptors involved in human natural killer cell-mediated cytolysis. Annu Rev Immunol 19:197–223

Morse RH, Seguin R, McCrea EL, Antel JP (2001) NK cell-mediated lysis of autologous human oligodendrocytes. J Neuroimmunol 116:107–115

Munschauer F, Hartrich L, Stewart C, Jacobs L (1995) Circulating natural killer cells but not cytotoxic T lymphocytes are reduced in patients with active relapsing multiple slcerosis and little clinical disability as compared to controls. J Neuroimmunol 62:177–181

Ohashi T, Yamamura T, J-i Inobe, Kondo T, Kunishita T, Tabira T (1990) Analysis of proteolipid protein (PLP)-specific T cells in multiple sclerosis: identification of PLP 95–116 as an HLA-DR2,w15-associated determinant. Int Immunol 7:1771–1778

Oki S, Chiba A, Yamamura T, Miyake S (2004) The clinical implication and molecular mechanism of preferential IL-4 production by modified glycolipid-stimulated NKT cells. J Clin Invest 113:1631–1640

Oki S, Tomi C, Yamamura T, Miyake S (2005) Preferential T_h 2 polarization by OCH is supported by incompetent NKT cell induction of CD40L and following production of inflammatory cytokine by bystander cells in vivo. Int Immunol 17:1619–1629

Ota K, Matsui M, Milford E, Mackin G, Weiner H, Hafler D (1990) T-cell recognition of an immunodominant myelin basic protein epitope in multiple sclerosis. Nature 346:183–187

Pal E, Tabira T, Kawano T, Taniguchi M, Miyake S, Yamamura T (2001) Costimulation-dependent modulation of experimental autoimmune encephalomyelitis by ligand stimulation of Va14 NK T cells. J Immunol 166:662–668

Pelfrey C, Trotter J, Tranquill L, McFarland H (1993) Identification of a novel T cell epitope of human proteolipid protein (residues 40–60) recognized by proliferative and cytolytic CD4+ T cells from multiple sclerosis patients. J Neuroimmunol 46:33–42

Peritt D, Robertson S, Gri G, Showe L, Aste-Amezaga M, Trinchieri G (1998) Cutting edge. Differentiation of human NK cells into NK1 and NK2 subsets. J Immunol 161:5821–5824

Pette M, Fujita K, Wilkinson D, Altmann D, Trowsdale J, Giegerich G, Wekerle H (1990) Myelin autoreactivity in multiple sclerosis: recognition of myelin basic protein in the context of HLA-DR2 products by T lymphocytes of multiple slcerosis patients and healthy donors. Proc Natl Acad Sci U S A 87:7968–7972

Pillarisetty V, Katz S, Bleier J, Shah A, Dematteo R (2005) Natural killer dendritic cells have both antigen presenting and lytic function and in response to CpG produce IFN-γ via autocrine IL-12. J Immunol 174:2612–2618

Porcelli S, Yockey C, Brenner M, Balk S (1993) Analysis of T cell antigen receptor (TCR) expression by human peripheral blood CD4⁻CD8⁻ α/β T cells demonstrates preferential use of several Vβ genes and an invariant TCRa chain. J Exp Med 178:1–16

Porubsky S, Speak A, Luckow B, Cerundolo V, Platt F, Grone H (2007) Normal development and function of invariant natural killer T cells in mice with isoglobotrihexosylceramide (iGb3) deficiency. Proc Natl Acad Sci U S A 104:5977–5982

Rauch H, Montgomery I, Kaplan J (1985) Natural killer cell activity in multiple sclerosis and myasthenia gravis. Immunol Invest 14:427–434

Rice G, Casali P, Merigan T, Oldstone M (1983) Natural killer cell activity in patients with multiple sclerosis given a interferon. Ann Neurol 1983: 333–338

Richert J, Robinson E, Deibler G, Martenson R, Dragovic L, Kies M (1989) Human cytotoxic T-cell recognition of a synthetic peptide of myelin basic protein. Ann Neurol 26:342–346

Roncarolo M, Gregori S, Battaglia M, Bacchetta R, Fleischhauer K, Levings M (2006) Interleukin-10-secreting type I regulatory T cells in rodents and humans. Immunol Rev 212:28–50

Sakuishi K, Oki S, Araki M, Porcelli S, Miyake S, Yamamura T (2007) Invariant NKT cell biased for IL-5 production act as crucial regulators of inflammation. J Immunol 179:3452–3462

Santoli D, Hall W, Kastrukoff L, Lisak R, Perussia B, Trinchieri G, et al (1981) Cytotoxic activity and interferon production by lymphocytes from patients with multiple sclerosis. J Immunol 126:1274–1278

Saraste M, Irjala H, Airas L (2007) Expansion of CD56[bright] natural killer cells in the peripheral blood of multiple sclerosis patients treated with interferon-b. Neurol Sci 28:121–126

Screpanti V, Wallin R, Grandien A, Ljunggren H (2005) Impact of FASL-induced apoptosis in the elimination of tumor cells. Mol Immunol 42:495–499

Shi F, Van Kaer L (2006) Reciptocal regulation between natural killer cells and autoreactive T cells. Nat Rev Immunol 6:751–760

Shi F, Ljunggren H, Sarventnick N (2001) Innate immunity and autoimmunity: from self-protection to self-destruction. Trends Immunol 22:97–101

Shimamura M, Huang YY, Okamoto N, Suzuki N, Yasuoka J, Morita K, et al (2007) Modulation of Vα19 NKT cell immune responses by α-mannosyl ceramide derivatives consisting of a series of modified sphingosines. Eur J Immunol 37:1836–1844

Singh A, Wilson M, Hong S, Olivares-Villagomez D, Du C, Stanic A, et al (2001) Natural killer T cell activation protects mice against experimental autoimmune encephalomyelitis. J Exp Med 194:1801–1811

Skulina C, Schmidt S, Dornmair K, Babbe H, Roers A, Rajewsky K, Wekerle H, Hohlfeld R, Goebels N (2004) Multiple sclerosis: brain-infiltrating CD8+ T cells persist as clonal expansions in the cerebrospinal fluid and blood. Proc Natl Acad Sci U S A 101:2428–2433

Smeltz R, Wolf N, Swanborg R (1999) Inhibition of autoimmune T cell response in the DA rat by bone marrow-derived NK cells in vitro: implication for autoimmunity. J Immunol 163:1390–1397

Smyth M, Swann J, Cretney E, Zerafa N, Yokoyama W, Hayakawa Y (2005) NKG2D function protects the host from tumor initiation. J Exp Med 202:583–588

Sospedra M, Martin R (2005) Immunology of multiple sclerosis. Annu Rev Immunol 23:683–747

Speak A, Salio M, Nerville D, Fontaine J, Priestman DP, Platt N, Heare T, et al (2007) Implications for invariant natural killer T cell ligands due to the restricted presence of isoglobotrihexosyl-ceramide in mammals. Proc Natl Acad Sci U S A 104:5971–5976

Stacker S, Springer T (1991) Leukocyte intergrin P150, 95 (CD11c/CD18) function as an adhesion molecule binding to a counter-receptor on stimulated endothelium. J Immunol 146:648–655

Steinman L (2001) Multiple sclerosis: a two-stage disease. Nat Immunol 2:762–764

Takahashi K, Miyake S, Kondo T, Terao K, Hatakenaka M, Hashimoto S, et al (2001) Natural killer type 2 bias in remission of multiple sclerosis. J Clin Invest 107:R23–R29

Takahashi K, Aranami T, Endoh M, Miyake S, Yamamura T (2004) The regulatory role of natural killer cells in multiple sclerosis. Brain 127:1917–1927

Takeda K, Hayakawa Y, Smyth M, Kayagaki N, Yamaguchi N, Kakuta S, et al (2001) Involvement of tumor necrosis factor-related apoptosis-inducing ligand in surveillance of tumor metastasis by liver natural killer cells. Nat Med 7:94–100

Taniguchi M, Harada M, Kojo S, Nakayama T, Wakao H (2003) The regulatory role of Vα14 NKT cells in innate and acquired immune response. Annu Rev Immunol 21:483–513

Treiner E, Duban L, Bahram S, Radosavljevic M, Wanner V, Tilloy F, et al (2003) Selection of evolutionary conserved mucosal-associated invariant T cell by MR1. Nature 422:164–169

Treiner E, Duban L, Moura I, Hansen T, Gilfillan S, Lantz O (2005) Mucosal-associated invariant T (MAIT) cells: an evolutionarily conserved T cell subset. Microbes Infect 7:552–559

Trinchieri G (1989) Biology of natural killer cells. Adv Immunol 47:187–193

Vranes Z, Poljakovic Z, Marusic M (1989) Natural killer cell number and activity in multiple sclerosis. J Neurol Sci 94:115–123

Walker L, Abbas A (2002) Keeping self-reactive T cells at bay in the periphery. Nat Rev Immunol 2:11–19

Warren H, Smyth M (1999) NK cells and apoptosis. Immunol Cell Biol 77:64–75

Yamamura T, Sakuishi K, Illés Z, Miyake S (2007) Understanding the behavior of invariant NKT cells in autoimmune disease. J Neuroimmunol 191:8–15

Yoshimoto T, Bendelac A, Hu-Li J, Paul W (1995) Defective IgE production by SJL mice is linked to the absence of CD4+ NK1.1+ T cells that promptly produce interleukin 4. Proc Natl Acad Sci U S A 92:11931–11934

Yu K, Im J, Molano A, Dutronc Y, Illarionov P, Forestier C, et al Modulation of CD1d-restrotced NKT cell responses by using N-acyl variants of α-galactosylceramides. Proc Natl Acad Sci U S A (2005) 102:3383–3388

Zhang B, Yamamura T, Kondo T, Fujiwara M, Tabira T (1997) Regulation of experimental autoimmune encephalomyelitis by natural killer (NK) cells. J Exp Med 186:1677–1687

Zhou D, Mattner J, Cantu 3rd C, Schrantz N, Yin N, Gao Y, et al (2004) Lysosomal glycosphingolipid recognition by NKT cells. Science 306:1786–1789

Antigen Processing and Presentation in Multiple Sclerosis

Christina Stoeckle and Eva Tolosa

Abstract CD4$^+$ T cells play a central role in the pathogenesis of multiple sclerosis (MS). Generation, activation and effector function of these cells crucially depends on their interaction with MHC II-peptide complexes displayed by antigen presenting cells (APC). Processing and presentation of self antigens by different APC therefore influences the disease course at all stages. Selection by thymic APC leads to the generation of autoreactive T cells, which can be activated by peripheral APC. Reactivation by central nervous system APC leads to the initiation of the inflammatory response resulting in demyelination. In this review we will focus on how MHC class II antigenic epitopes are created by different APC from the thymus, the periphery and from the brain, and will discuss the relevance of the balance between creation and destruction of such epitopes in the context of MS. A solid understanding of these processes offers the possibility for designing future therapeutic strategies.

1 Introduction

The common feature of autoimmune diseases is the immune attack on self-components, usually, but not exclusively, self-proteins. Knowing the exact structures that the immune response is directed against and how these structures are generated, is crucial for understanding autoimmune pathogenesis and for designing antigen-specific or antigen-related therapies. In multiples sclerosis the autoimmune attack is directed

C. Stoeckle (✉)
Department of General Neurology, Hertie Institute for Clinical Brain Research,
Otfried-Mueller-Str. 27, 72076, Tuebingen, Germany
e-mail: christina.stoeckle@web.de

E. Tolosa
Center for Molecular Neurobiology Hamburg (ZMNH), Institute for Neuroimmunology
and Clinical MS Research (INiMS), Falkenried 94, 20251, Hamburg, Germany
e-mail: eva.tolosa@zmnh.uni-hamburg.de

Results Probl Cell Differ, DOI 10.1007/400_2009_22
© Springer-Verlag Berlin Heidelberg 2009

149

against central nervous system (CNS) myelin (McFarland and Martin 2007; Sospedra and Martin 2005). Although many cell types are involved in the inflammatory response, especially in later stages of multiple sclerosis (MS), activation of CNS-reactive CD4⁺ T cells is believed to underlie disease initiation. CD4⁺ T cells recognize peptides bound to MHC II (HLA-DR/Q/P) molecules on the surface of antigen presenting cells (APC). These peptides typically derive from proteins that are taken up by the APC via endocytosis and processed in the endolysosomal pathway (Honey and Rudensky 2003; Trombetta and Mellman 2005). Recognition of the cognate antigen in an appropriate context leads to CD4⁺ T cell priming and activation of the immune response. The critical importance of the interaction of CD4⁺ T cells with MHC II-peptide complexes displayed by APC in MS pathogenesis is underlined by the strong genetic association of MS with HLA alleles, especially HLA-DR(2)15, but also HLA-DR4 (Holmes et al. 2005), and by the fact that anti-class II antibodies inhibit or ameliorate disease in the animal model of MS, experimental autoimmune encephalomyelitis (EAE) (McDevitt et al. 1987).

Many autoreactive T cells are already eliminated in the thymus during T cell development by interaction with resident APC (Anderton and Wraith 2002). The deletion of such potentially dangerous cells in the thymus is known as central tolerance. Failure of this safety mechanism leads to escape of autoreactive T cells into the periphery, a prerequisite for the development of autoimmunity. The presence of autoreactive T cells alone, however, is not sufficient to develop an autoimmune disease. In addition to central tolerance, many peripheral tolerance mechanisms prevent these cells from mounting a harmful attack on self-tissues (Goodnow et al. 2005). In MS patients however, both central and peripheral self-tolerance mechanisms fail, leading to activation of CNS-specific T cells. These cells can then migrate into the CNS where recognition of their cognate antigen displayed on MHC II molecules by CNS-resident or -associated APC results in activation of effector function and the ensuing inflammation.

2 Autoantigens

A critical question for understanding MS pathogenesis is the nature of the (auto) antigens recognized by pathogenic CD4⁺ T cells. Since the target of the immune attack is the myelin sheath, and crude myelin extracts can be used to induce EAE in susceptible animals (Laatsch et al. 1962), investigations have mainly focussed on myelin protein components. Several purified proteins and some of their peptides have been found to be able to induce EAE (Baxter 2007). Autoreactive T cells against one or several of these proteins are present in MS patients, and are often clonally expanded and persist for long times in these patients (Sospedra and Martin 2005).

A number of self-proteins have been suggested to play a role in MS pathogenesis. The most prominent and best-described of these are the myelin basic protein (MBP), myelin oligodendrocyte protein (MOG) and the proteolipid protein (PLP). All of them can be used to induce EAE in susceptible animals, and T cells against all of them have been found in MS patients. In addition to these three autoantigens,

T cell responses against myelin oligodendrocytic basic protein (MOBP), myelin-associated glycoprotein (MAG), and oligodendrocyte-specific protein (OSP) have also been described in MS patients. However, their contribution to disease is currently unknown (reviewed in Sospedra and Martin 2005).

2.1 Myelin Basic Protein

MBP is a predominant component of the myelin sheath, accounting for ~35% of total myelin protein. It has been shown to be the target of both T and B cell responses present in MS patients and efficiently induces EAE in susceptible animals (revised in Sospedra and Martin 2005). T cell responses against different regions of MBP are found in patients, but also in healthy donors, the C-terminal region (residues 131–159) and the region spanning amino acids 85–99 being particularly immunodominant in the context of the susceptibility allele HLA-DR2 (Vergelli et al. 1997a). A proportion of mice transgenic for the DR2 allele and a T cell receptor (TCR) from an MS patient recognizing this region spontaneously develop EAE and disease can be accelerated by immunization with MBP85–99 (Madsen et al. 1999). Furthermore, a complex between HLA-DR2b and MBP85–99 is presented by microglia in MS lesions, supporting an involvement in MS pathology (Krogsgaard et al. 2000). MBP111–129 constitutes an immunodominant epitope in the context of HLA-DR4 alleles, and humanized transgenic mice expressing a TCR specific for this epitope together with the restriction element DRB1*0401 develop ascending paralysis typical of EAE, albeit only after adoptive transfer (Quandt et al. 2004).

2.2 Myelin Oligodendrocyte Protein

MOG accounts for less than 0.05% of total myelin protein, but it is located in the outer part of the myelin sheaths, and is thus directly accessible to effector mechanisms. MOG is expressed late in myelogenesis, and is exclusively present in CNS myelin. In susceptible animals, immunization with MOG antigen, MOG-derived peptides or passive transfer of MOG-reactive T cells results in severe disease mimicking many features of MS (Linington et al. 1993). Several MOG-derived peptides are encephalitogenic, MOG35–55 being one of the most prominent in several animal models (Iglesias et al. 2001), and also one of the most immunodominant in MS patients (Kerlero de Rosbo et al. 1997). In addition, a strong response to an intracellular peptide (MOG146–154) can be elicited both in MS patients and healthy donors (Delarasse et al. 2003; Weissert et al. 2002). Recently, a model for spontaneous autoimmune demyelination was developed by crossing transgenic mice with either a T or a B cell receptor specific for MOG (Bettelli et al. 2006; Krishnamoorthy et al. 2006). More than half of the double-transgenic mice spontaneously developed autoimmune demyelination of the spinal cord and optic nerve, exhibiting a pathology reminiscent of human MS, and thus underscoring the importance of this antigen in human disease.

2.3   Proteolipid Protein

Myelin PLP is the most abundant protein in CNS myelin, comprising more than 50% of total protein, and is thought to play a major role in myelin structure and function. PLP is highly encephalitogenic, particularly in SJL/J mice, where the immunodominant PLP139–151 epitope causes a relapsing-remitting type of disease. Interestingly, while the majority of the immune reactivity during the first acute phase of EAE in SJL/J animals immunized with the intact PLP molecule is directed at the above mentioned PLP139–151 epitope, the T cell responses during relapse phases are directed at other epitopes of the same molecule (PLP178–191) or even at epitopes from other myelin antigens (MBP89–101) (McRae et al. 1995). Similarly, if disease is induced with MBP89–101 or with PLP178–191, further waves of the disease involve reactivity to PLP139–151. This phenomenon of reactivity against epitopes distinct from the disease-inducing epitope observed during chronic disease is known as epitope spreading, and supports the notion that T cell responses to epitopes released as a result of acute tissue damage contribute to the pathogenesis of relapsing episodes (Vanderlugt and Miller 2002). The immunogenicity of the PLP139–151 epitope could also be documented in transgenic mice expressing the rearranged TCR genes from either an encephalitogenic or a nonencephalitogenic PLP139–151 specific T cell clone, which interestingly both developed spontaneous EAE (Waldner et al. 2000). A shorter isoform of PLP, DM20, is expressed at lower levels than PLP in CNS myelin, and is also present in peripheral lymphoid organs and in the thymus. Since the lymphoid organs lack the full length PLP, the differential expression of the two isoforms might be relevant for the establishment of peripheral and central tolerance to this antigen (Klein et al. 2000), and thus may explain the immunogenicity of PLP. Immunodominant epitopes that can be processed from whole PLP by human APC lie within the 30–60 and the 180–230 regions of PLP (reviewed in Greer and Pender 2008), and especially the region 184–209 elicits strong T cell responses mostly in MS patients carrying HLA-DR4, -DR7, or -DR13 who tend to develop lesions in the brainstem and cerebellum (Greer et al. 2008). Several other PLP epitopes, particularly within the 40–70 region as well as the 104–117 region were also found immunodominant in healthy donors and MS patients in the context of multiple HLA-DR alleles (Correale et al. 1995; Pelfrey et al. 1993). Thus, the response to PLP seems to be heterogenous concerning both epitopes and MHC restriction.

2.4   Posttranslational Modifications

Posttranslational modifications (PTM) occur during normal cellular events, such as signal transduction, metabolic processes and aging, but also in cases of cellular stress, such as infection, trauma or apoptosis. PTM include phosphorylation, methylation and glycosylation, which regulate the biological function of proteins in many different ways. For other modifications, it is not so clear whether they fulfil specific functions or are simply the result of the chemically reactive cellular environment or oxidative damage. There are several instances where posttranslationally modified self-proteins might contribute to autoimmune pathogenesis. The most obvious one is that the modified self-antigen is not represented in the thymus, and therefore autoreactive T cells escape central tolerance and migrate into the periphery. In addition, since many enzymes involved in antigen processing have some preference/specificity for certain amino acids, PTM may also result in altered processing of the protein, and therefore these modifications can result in the formation of neo-epitopes recognized by pathogenic T cells (Eggleton et al. 2008), or reveal otherwise cryptic epitopes.

Spontaneous or enzymatic deamidation of asparagines residues leads to the formation of iso-aspartate and enzymatic deimination of arginine residues to the formation of citrulline (Doyle and Mamula 2005). In rheumatoid arthritis (RA), antibodies against citrullinated fillagrin are highly predictive and used for diagnosis (Szekanecz et al. 2008) and the RA associated HLA-DRB1*0401 molecule has an increased affinity for citrullinated antigens (Hill et al. 2003). Also, glycosylated or oxidatively modified collagen has been suggested to be involved in disease pathogenesis of RA and its animal model collagen-induced arthritis (CIA) (Corthay et al. 2001; Nissim et al. 2005). Systemic lupus erythematosus (SLE) patients have autoantibodies against the C-terminus of snRNP containing symmetrical dimethyl-arginines (Brahms et al. 2000), and spontaneous iso-aspartate formation in snRNP and cytochrome c peptides renders them immunogenic in an animal model of SLE (Mamula et al. 1999).

A large number of PTM have been demonstrated for MBP, including acylation, deamidation, methylation, phosphorylation, deimination/citrullination, and ADP-ribosylation (Harauz and Musse 2007). Interestingly, MBP isolated from the brain of MS patients shows a number of changes with regard to PTM compared to normal brain (Kim et al. 2003). Deimination of peptide bound arginine to citrulline involves the release of ammonia, and therefore for each conversion one positive charge is lost from the protein. In human MBP, citrullination may occur on six sites, and increased citrullination results in a less cationic protein, thereby diminishing its ability to organize lipid bilayers into compact multilayers, and creating a structure which is more open and susceptible to proteases such as Cathepsin D (CatD) (Cao et al. 1999), which in turn results in myelin instability (Pritzker et al. 2000a, b). The enzymes involved in deimidation, peptidylarginine deiminases (PAD), are upregulated in MS (Mastronardi et al. 2007), and PAD2 overexpression in transgenic mice leads to demyelination in the CNS (Musse et al. 2008). In patients, both the percentage of citrullinated MBP and the number of individual citrulline modifications within each molecule was reported to be increased threefold in MBP isolated from MS brain and six- or sevenfold in fulminating forms such as Marburg disease. Whether such changes are causally related to disease pathogenesis and exacerbation, or are at least partially the result of an inflammatory environment characterised by highly reactive chemical species such as oxygen radicals and other stresses, remains to be investigated. Interestingly though, T cell responses against citrullinated MBP are increased in MS patients (Tranquill et al. 2000).

MBP is not the only myelin antigen undergoing PTM. Thiopalmitoylation of PLP peptides, as it occurs naturally in vivo, increases their immunogenicity and their ability to induce chronic progressive EAE. This increased response is probably due to the improved capacity of such modified peptides to cross cellular membranes, resulting in an enhanced uptake by APC (Pfender et al. 2008). Similarly, DBA/1 mice immunized with malondialdehyde-MOG developed greater proliferative responses and more severe EAE than mice immunized with unmodified MOG. Modified MOG was taken up more effectively by APC, at least partially through scavenger receptors, which led to an increased activation of these APC, suggesting that the posttranslationally modified form of this myelin autoantigen is a more relevant form of the molecule (Wallberg et al. 2007).

3 MHC Molecules

The class II region of the MHC is the major gene locus associated with MS. In the Northern European and the Northern American population, the HLA-DR15 haplotype (containing the two DRB* genes DRB1*1501 and DRB5*0101 and the tightly linked DQ alleles DQA*0102 and DQB1*0602) shows the strongest genetic association for MS (Lincoln et al. 2005). In the DR15 haplotype two beta chains, HLA-DRB1*1501 and -DRB5*0101, are co-expressed resulting in two different surface HLA-DR $\alpha\beta$ heterodimers, DR2b and DR2a, but the individual contribution of these two alleles to MS pathogenesis has remained elusive. Using a transgenic mouse model harbouring these two DR15 alleles individually or in combination, Gregersen et al. elegantly demonstrated a protective epistatic effect of DRB5*0101 over DRB1*1501, the allele eliciting the more severe clinical phenotype, by partially deleting DRB1*1501-restricted autoreactive T cells (Gregersen et al. 2006).

HLA-DM, a non classical HLA molecule, functions as a peptide editor by catalyzing the exchange of peptides from the MHC groove until a peptide with high affinity is finally presented on the cell surface (Denzin and Cresswell 1995). DM affects peptide-MHC class II complex dissociation by a mechanism in which transient interaction between DM and DR causes a conformational change in DR, resulting in destabilization of the interaction of DR with the bound peptide (Narayan et al. 2007). In addition, the activity of HLA-DM is further modulated by another MHC-related molecule, HLA-DO (Denzin et al. 1997). Both DMA and DMB and DOB genes exhibit limited polymorphism, but no associations to MS have been determined (Ristori et al. 1997).

The molecular basis of the contribution of MHC molecules to disease susceptibility still remains elusive. In Caucasian populations, approximately 65% of MS patients carry the HLA-DR15 haplotype compared to 30% of healthy controls. Thus, even though the immunogenic epitope MBP85–99 and several MOG peptides can bind to and be presented by DR2 molecules (Vogt et al. 1994; Wallstrom et al. 1998), presentation of specific autoantigen epitopes by these disease-associated HLA molecules seems unlikely to be sufficient for disease development.

Substantial differences in the expression levels of the two DR2 heterodimers in APC of the CNS or in peripheral APCs between MS patients and healthy donors could also not be demonstrated (Prat et al. 2005). Thus, further studies are required to establish the functional role of MHC molecules in the pathogenesis of MS.

4 Generation of T Cell Epitopes

Professional APC express a wide range of lysosomal proteases involved in MHC II antigen processing, but different APC express distinct protease profiles. The lysosomal proteases comprise the cathepsins and asparagine endopeptidase (AEP) (Honey and Rudensky 2003; Rudensky and Beers 2006). The majority of cathepsins belongs to the papain family of cysteine proteases (CatB, C, F, H, K, L, S, V, W, Z), but a few are serine (CatA and G) or aspartate proteases (CatD and E). Lysosomal proteases are responsible not only for the processing of protein antigens to generate epitopes, but also for the stepwise degradation of the Invariant chain (Ii).

MHC II molecules mainly, but not exclusively, present peptides derived from exogenous proteins that have been taken up by endocytosis. The MHC II molecules themselves are synthesized within the ER where they form trimeric complex with the Ii chaperone. Ii blocks the peptide binding groove and directs the complex via the Golgi into the endosomal/lysosomal compartment where it is degraded in a stepwise fashion by lysosomal proteases such as asparagine endopeptidase (AEP), CatS, L, F or V until only the class II-associated invariant chain peptide (CLIP) fragment remains in the binding groove. Which are the proteases that are involved depends on the type of cell. The CLIP peptide is then exchanged by HLA-DM for peptides derived from the internalized protein and the resulting complex transported to the cell surface where it can be recognized by CD4[+] T cells. As the binding groove is open at both ends, the bound peptides can have different lengths, usually between 12 and 20 amino acids (Trombetta and Mellman 2005).

In addition to the conventional form of antigen presentation involving internalization, lysosomal processing, and loading onto MHC molecules, MBP can be displayed at the cell surface by at least two alternative pathways which are independent of Ii, HLA-DM and antigen processing. In solution and in the absence of lipids, MBP is a relatively small protein with little secondary structure, and in this extended form MBP is flexible enough to bind to MHC heterodimers on the cell surface and elicit a T cell response (Fridkis-Hareli et al. 1994; Vergelli et al. 1997b). In addition, MBP was shown to bind to recycled MHC class II molecules (Pinet et al. 1995). It is known that a fraction of the surface class II molecules is internalized and quickly recycled to the cell membrane. Low affinity peptides binding to these recycled class II molecules can be exchanged (in an HLA-DM-dependent or independent manner) in early endosomes for partially digested peptides deriving from exogenous proteins. These peptides are thus protected from the extensive proteolysis that takes place in the processing and peptide loading compartments (lysosomes) of the classical pathway.

4.1 Destruction Versus Creation of Epitopes by the Processing Machinery

Proteolytic processing of antigens by lysosomal proteases results in the release of peptides which might bind to MHC molecules and be presented as T cell epitopes. Although most lysosomal proteases have a rather broad and often overlapping substrate specificity, accumulating evidence suggests that antigens are not unspecifically degraded by bulk proteolysis but that this process occurs in a stepwise and more regulated fashion. For several antigens it has been shown that processing is dominated by one specific endoprotease, the so-called unlocking protease, which delivers the initial cut(s), opening up the antigen for further degradation by other endoproteases and trimming by exopeptidases. Action of the unlocking protease can both create and destroy epitopes, as has been shown for AEP. AEP is required for the generation of an epitope from tetanus toxin in a human B cell line while its action destroys the MBP85–99 epitope by cleaving it at asparagine 92, limiting its presentation (Manoury et al. 1998, 2002). The nature of the unlocking protease depends on the cell type. While in monocyte-derived DC, AEP and CatS dominate MBP processing, in primary peripheral blood DC and cultured microglia, CatG is the unlocking protease for MBP (Burster et al. 2004, 2005, 2007a) (Table 1). In primary microglia, CatD can also perform this function (Stoeckle et al. 2008). Interestingly, action of any of these proteases can destroy the MBP85–99 epitope. Destruction of a highly dominant epitope, however, can result in overt access for T cells to flanking epitopes, now free of physical constraints or having lost the

Table 1 Proteases involved in the in vitro processing of MBP in different antigen presenting cells

Cell type	Species	Protease dominating MBP processing in this cell type	Reference
Thymus			
Myeloid DC	Human	CatS	Stoeckle et al. (in preparation)
Plasmacytoid DC	Human	CatS	Stoeckle et al. (in preparation)
Cortical TEC (ex vivo)	Human	Candidates: CatV/ CatG	Stoeckle (unpublished data)
Medullary TEC (cultured)	Human	AEP	Stoeckle et al. (2004)
Cortical TEC (cultured)	Human	CatV	Stoeckle (unpublished data)
Myeloid DC	Mouse	AEP	Manoury et al. (2002)
Periphery			
Primary B cells	Human	CatG	Burster et al. (2005)
B cell line	Human	CatS/AEP	Beck et al. (2001), Manoury et al. (2002)
Monocyte-derived DC	Human	CatS/AEP	Burster et al. (2005)
Myeloid DC	Human	CatG	Burster et al. (2005)
Central nervous system			
Microglia (cultured)	Mouse	CatG	Burster et al. (2007a)
Microglia (ex vivo)	Rat	CatD	Stoeckle et al. (2008)

competition posed by the previously more dominant, and now cryptic, epitope (Anderton et al. 2002).

Several factors can in principle affect processing of antigens. These include inflammatory stimuli such as LPS, sex hormones and other steroids as well as cellular stress. For example, LPS treatment of monocyte-derived DC results in an increase in CatS activity (Burster et al. 2005), while oestrogen affects CatD (Westley and May 1987). Processing can also be affected by PTM present in the antigen. As mentioned above, citrullination of MBP was found to be increased in brains of MS patients, and increased citrullination is associated with increased susceptibility to proteolysis by CatD (Cao et al. 1999). Interestingly, the 85–99 region of MBP contains an asparagine, which might be converted into iso-aspartate, for example during normal aging of the protein. This would not only alter the epitope sequence but also remove an important proteolytic cleavage site which is likely to affect epitope processing and presentation. Such altered processing of iso-aspartate containing antigen has been demonstrated for tetanus toxin (Moss et al. 2005). In addition, the epitope contains arginine, which could be converted into citrulline, although so far there is no evidence that this occurs under physiological or pathological conditions. However, citrullination as well as most of the other PTM that can be found in MBP, decrease its overall positive charge, thus affecting not only lipid binding but also its three-dimensional structure (Harauz and Musse 2007). PTM could therefore not only affect processing by affecting cleavage sites but also by altering its three-dimensional structure and, as a consequence, cleavage site accessibility.

The upstream events that lead to the presentation of a specific peptide are not fully understood. One possibility is that peptides generated by proteases bind to class II MHC in the loading compartment. However, the generated peptides could easily be destroyed in the proteolytic environment of this compartment, suggesting that longer precursors may bind first to MHC molecules and then become trimmed. Bound peptides would be – at least partially – protected from destruction. Analysis of peptides eluted from class II molecules supports the notion that processing and trimming can continue after binding of class II molecules to longer precursors (Castellino et al. 1998; Lippolis et al. 2002; Moss et al. 2007).

Which epitope is finally displayed by the MHC heterodimer depends on a number of factors, such as binding affinity and, consequently, competition between different peptides. This implies several potential binders from the same or different proteins for binding to one MHC heterodimer and also that an epitope might be generated but not presented because it is outcompeted. Whether DM is present will also have an influence since it will remove weaker binders. In addition, epitope selection might be altered if the protein is bound to some other molecules, e.g. to the B cell receptor (BCR) after endocytosis, and hence the region bound by the BCR will not be accessible for binding by MHC. On the other hand, epitopes close, but not too close to the BCR binding site might be favoured, since the antigen might be 'handed over' to the MHC (Moss et al. 2007). Hence to summarise, optimal presentation requires limited proteolytic cleavage, followed by binding of long peptide fragments to MHC heterodimers, further trimming by exopeptidases and the final editing of bound peptides until the best fitting ones are presented.

5   Antigen Processing and Presentation in Multiple Sclerosis

The outcome of an interaction between T cells and APC is determined by a number of factors including developmental and activation stage of both partners, the affinity/ avidity of the T cell for the epitope and, of course, whether the APC displays the cognate epitope at levels sufficient for T cell activation, or not. T cell–APC interactions occur at several instances during the life of a T cell. In the case of an autoimmune reaction, there are at least three critical points, where such interactions can influence the balance between tolerance and autoimmune disease. First, the T cell must escape from deletion in the thymus, second, it must become activated in the periphery, and third, it must be reactivated in the target organ, e.g. the CNS, in order to cause the damage underlying disease manifestations.

5.1   T Cell Selection in the Thymus

Tolerance to self is achieved by a number of mechanisms, including regulatory T cells and peripheral antigen presentation by tolerogenic APC (Goodnow et al. 2005). Arguably the most important of these mechanisms preventing autoimmunity is the deletion of autoreactive, and therefore potentially autoaggressive, T cells in the thymus (central tolerance) during T cell development. MHC-peptide complexes displayed by thymic APC provide the matrix on which the TCR repertoire of each individual is selected, thereby shaping the T cell pool available for an (auto) immune response. During T cell development, thymocytes undergo a number of differentiation steps which strongly depend on the interaction with the thymus stroma before they leave the thymus as mature T cells (Anderton and Wraith 2002; Gill et al. 2003). After T cell progenitors enter the thymus, they migrate into the cortical areas of the thymus, expand and start to rearrange their TCR genes. Thymocytes failing to produce a functional TCR (i.e. capable of recognizing self-MHC-peptide complexes) die by apoptosis ('death by neglect'). Only those thymocytes whose TCR can interact with self-peptide-MHC displayed by cortical thymic epithelial cells (cTEC) receive survival signals. This process of positive selection ensures that the T cell repertoire is self-MHC restricted but, due to the nature of the process, self-reactive. In the periphery, low affinity interactions between TCR and self-peptide-MHC do not lead to T cell activation but are required for T cell survival and homeostasis (Kirberg et al. 1997). However, if the strength of the interaction exceeds a certain threshold and appropriate costimulation is present, T cells can become activated and mount an immune response against their target antigen. T cells recognizing self-peptide-MHC with high affinity are therefore potentially dangerous and need to be prevented from escaping into the periphery where they could cause damage. After positive selection, thymocytes therefore migrate into the thymic medulla, where they interact with peptide-MHC complexes on dendritic cells (DC) and medullary thymic epithelial cells (mTEC). Depending on the strength of the interaction, they are allowed to leave the thymus as conventional T cells (low affinity) or regulatory T cells (intermediate affinity) or receive apoptotic signals and die (high affinity) (Hogquist et al. 2005). Deletion of T cells with high affinity to self antigens by thymic APC is known as negative selection and is the basis for central tolerance. Self antigens include those antigens that are ubiquitously expressed or easily reach the thymus, but they also include tissue-specific antigens (TSA), whose expression is very restricted. Tolerance to TSA such as MBP and insulin is achieved through promiscuous expression in the thymus, primarily by mTEC, which is partially dependent on the transcription factor AIRE (autoimmune regulator) (Derbinski et al. 2001; Kyewski and Klein 2006). Although mTEC can mediate negative selection by themselves, the more efficient thymic DC contribute to the process by presenting antigen captured from mTEC (Gallegos and Bevan 2004).

Evidence that alterations in antigen processing or in the degradation of Ii by lysosomal proteases can affect T cell selection comes mostly from CatL knockout mice. CatL is expressed by cTEC and thymic DC in mice, and the lack of this protease results in strongly reduced numbers of CD4+ T cells in the periphery (Nakagawa et al. 1998), increased proportion of CD4+ CD25+ regulatory T cells that could prevent the onset of Type 1 diabetes (Maehr et al. 2005), and in an altered T cell repertoire resulting from the different set of peptides displayed by thymic APC (Honey et al. 2002). In humans, selective overexpression of CatV was found in the thymus of patients with myasthenia gravis, an autoimmune disease with associated thymus pathology. Although the effects of higher levels of CatV in cTEC are unknown, a possible outcome would be an imbalance of the positive/negative selection mechanisms, leading to an increase in potentially autoreactive cells in the periphery (Melms et al. 2006; Tolosa et al. 2003).

The existence of autoreactive T cells against a number of self-epitopes from myelin antigens both in healthy individuals and patients with autoimmune disease (Martin et al. 1990; Meinl et al. 1993) demonstrates that not all potentially autoreactive T cells are deleted in the thymus. Presentation of an epitope in the thymus is a prerequisite for both positive and negative selection and consequently, lack of presentation leads to a failure of negative selection and escape of autoreactive T cells into the periphery. Lack of presentation in the thymus can be due to several factors, including lack of expression of the antigen, expression of splice variants lacking the region in question, destructive processing by the antigen processing machinery or competition with other peptides capable of binding to MHC. MBP and PLP have been shown to be expressed in the thymus, but for MOG, reports are conflicting (Bruno et al. 2002; Derbinski et al. 2001; Sospedra et al. 1998). Since levels of TSA in the thymus differ between individuals (Taubert et al. 2007), it is well conceivable that at least in some, if not all cases, MOG is not expressed in the thymus contributing to the escape of MOG-reactive T cells. The idea that MOG is not or insufficiently expressed in the thymus for induction of tolerance is supported by findings from a MOG knockout mouse. MOG$^{+/+}$ and MOG$^{-/-}$ mice show very similar T cell responses against MOG, including the dominant 35–55 epitope, indicating that deleting the MOG gene does not make a difference for negative selection in the thymus (Delarasse et al. 2003; Fazilleau et al. 2006). For the immunodominant PLP epitope 139–151, the explanation appears to be different. The gene encoding for PLP can undergo alternative splicing to generate the DM20 isoform, which lacks residues 116–150 (and, therefore, the PLP139–151 epitope) (Klein et al. 2000). Both the full length and the DM20 isoforms are expressed in the brain, but the DM20 isoform is preferentially expressed in the thymus. Failure of negative selection against the above epitope due to its absence from the thymus would explain its immunodominance and the unusually high frequency of PLP139–151 reactive T cells (Anderson et al. 2000). However, negative selection against several PLP epitopes may not be very effective in either mice or humans, as numerous studies show that strong autoimmune responses against multiple epitopes of PLP, including epitopes that are also present in DM20, can be induced in experimental animals, and occur naturally in both healthy human subjects and patients with MS (reviewed in Greer and Pender 2008).

Although MBP exists in several splice variants, the longer so-called Golli-MBP variant which is expressed in the thymus and the haematopoietic system, contains the MBP85–99 epitope, so that lack of expression of the epitope cannot explain its immunodominance and the presence of reactive T cells in the periphery (Marty et al. 2002). Destructive processing of this epitope in negatively selecting thymic APC seems a much more likely explanation, since this region is highly susceptible to a number of different prominent lysosomal proteases such as AEP, CatS, CatG and CatD, some of which are also present and active in thymic DC, and can efficiently destroy the epitope (Burster et al. 2007a; Burster et al. 2004; Manoury et al. 2002).

5.2   MBP Processing and Presentation in the Periphery

After having escaped thymic negative selection, T cells encounter APC in peripheral lymph nodes. Depending on the type and activation/maturation status of the APC, presentation of the cognate epitope to a self-reactive T cell will either lead to anergy and tolerance or activation. It is widely accepted that initial activation of encephalitogenic T cells occurs in the periphery, but while in the experimental animal models activation or tolerance induction is routinely achieved, it is not fully understood how and why priming occurs in some individuals and not in others.

Manoury et al. suggested that dysregulation of antigen processing in peripheral APC might contribute to MS pathogenesis (Manoury et al. 2002). Initial findings mainly based on B cell lines suggested that AEP has a central and possibly unique role in limiting MBP85–99 epitope generation. The model they proposed is depicted and explained in more detail in Fig. 1. It suggests that decreased AEP activity in peripheral DC could be responsible for inefficient

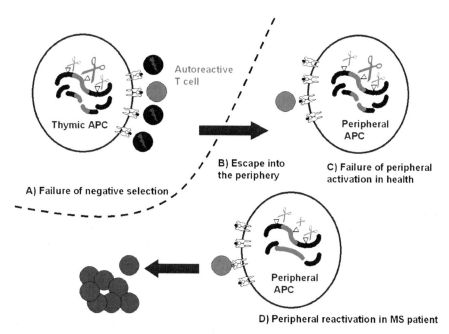

Fig. 1 AEP and multiple sclerosis (MS). (**a**) Autoreactive T cells against the MBP epitope escape negative selection in the thymus due to destruction of the epitope by AEP and can migrate into the periphery. (**b**) Peripheral APC also express active AEP and therefore also destroy the epitope resulting in a failure of activation of autoreactive T cells. (**c**) In MS patients, a reduction in AEP activity in peripheral APC leads to a failure of epitope destruction and, consequently, its presentation to autoreactive T cells (**d**) which then are activated, proliferate and migrate into the CNS where they cause inflammation, ultimately leading to destruction of the myelin sheath and development of MS

destruction and therefore presentation of the epitope in MS patients, which would lead to activation of autoreactive T cells. These T cells could then cross the blood–brain-barrier and become reactivated by resident APC such as the microglia. We have attempted to test this hypothesis using monocyte-derived DC from MS patients and investigated antigen processing in these cells. Although we could confirm the dominant role of AEP (and CatS) in MBP processing, we failed to find evidence for altered AEP activity in MS patients (Stoeckle et al. 2004). Furthermore, inhibition of AEP in B cell lines different from the one used in the initial experiments did not alter MBP85–99 presentation while impairing presentation of tetanus toxin (Costantino et al. 2008), suggesting that AEP may play an important role in epitope destruction in some, but not all cells. More recently, it was shown that in primary peripheral blood DC, B cells and cultured mouse microglia, CatG unlocks MBP at the same time that destroys the epitope (Burster et al. 2004, 2005, 2007a), further supporting the idea that different proteases perform this role in different cell types (Table 1). In accordance with this notion, we have found that CatV and a CatD-like activity can also dominate MBP85–99 processing in human cTEC or primary rat microglia, respectively (unpublished results).

5.2.1 Dendritic Cells

DC play a crucial role in both the initiation of an immune response and in maintenance of tolerance. During infection, immature tissue resident DC take up antigen in the periphery and migrate to the lymph nodes where they display peptide-MHC complexes to T cells (reviewed in detail in Villadangos and Schnorrer 2007). During migration they take on a mature phenotype characterized by upregulation of surface MHC-peptide complexes and costimulatory molecules. If the antigen was originally taken up in a normal, non-inflammatory context, presentation in the lymph node will result in anergy of T cells recognizing the epitope (Wilson et al. 2004). This is an additional tolerance mechanism preventing autoimmunity.

Whether uptake of the autoantigen in an inflammatory context is involved in the misdirected immune response against host tissue is not known, nor are the events leading to priming of autoreactive T cells in patients with MS or other autoimmune diseases. Interestingly, MBP is expressed throughout the lymphoid organs, suggesting that peripheral DC do not need to migrate into the brain in order to take up and present MBP (Marty et al. 2002). However, it has also been shown that DC can take up myelin antigens from brain lesions, migrate to cervical lymph nodes and present the antigen there (de Vos et al. 2002). Furthermore, MBP containing APC are abundant in the lymph nodes of MS patients (de Vos et al. 2002). While CatG can efficiently destroy the MBP85–99 epitope in DC, presentation of other regions of MBP by immature and/or tolerogenic DC could actively contribute to the maintenance of peripheral tolerance to these epitopes (Burster et al. 2005). Breaking of tolerance to the MBP85–99 epitope could occur under inflammatory conditions. We found that the inflammatory stimulus LPS reduces CatG activity in primary peripheral DC (Stoeckle et al. 2009), suggesting that reduced cleavage of the MBP85–99 epitope might occur during inflammation and contribute to activation of autoreactive T cells.

5.2.2 B Cells

The importance of B cells as APC in MS pathology is not well understood. While in some EAE models and also some patients, antibodies appear to contribute to pathogenesis, the APC function of B cells has hardly been addressed at all (Antel and Bar-Or 2006; Cross et al. 2001). Just like DC, presentation of antigen by resting B cells does not result in immunity but rather in tolerance of the corresponding T cell. As the total number of B cells with a given BCR is normally rather low, DC are likely to play a more important role both in tolerance and in induction of immune responses to a particular antigen. Like in DC, CatG activity in B cells might destroy the 85–99 epitope (Burster et al. 2004), preventing its presentation.

However, in B cells processing could be guided by the BCR and potentially affect epitope selection, but this question has not been addressed so far.

The few reports that have investigated the role of B cells in EAE specifically with respect to antigen presentation support the notion that antigen presentation by B cells in CNS inflammation might be beneficial. Coupling of MBP peptide to B cells prevented EAE induction (Day et al. 1992) and presentation of endogenous MBP by B cells resulted in weak T cell proliferation without cytokine production and reduced responsiveness to subsequent stimulation (Seamons et al. 2006).

5.3   Processing and Presentation in the Brain

After being activated in the periphery, myelin-reactive T cells migrate to the CNS where they will be reactivated by local APC. In the CNS, perivascular macrophages/microglia and DC are the only cells constitutively expressing MHC class II molecules, and there is now clear evidence that while CNS-associated DC play a crucial role in target recognition for neuroantigen-reactive T cells, resident microglia are responsible for the maintenance of neuroinflammation.

5.3.1   Macrophages/Microglia

Myeloid cells infiltrating the CNS early in foetal development will become parenchymal microglia, a resident population in the CNS. Peripheral monocytes can also quickly transform into microglia in certain pathologies (Priller et al. 2001). Even in quiescent state, microglia can phagocytose myelin, and under inflammatory conditions, they upregulate MHC class II and costimulatory molecules, suggesting that they can act as local APC. However, most evidence now suggests that microglia are not the initial APC to present antigen to invading myelin-reactive T cells (Greter et al. 2005), but rather sustain and propagate inflammation within the CNS during autoimmune inflammation through antigen presentation and cytokine/chemokine secretion. Indeed, preventing microglial activation resulted in substantial amelioration of the clinical signs in EAE and in strong reduction of CNS inflammation (Heppner et al. 2005). Under non-inflammatory conditions, microglia fulfil a role in immune surveillance (Nimmerjahn et al. 2005).

Microglial cells readily upregulate MHC molecules expression in response to IFNγ, and MHC class II is expressed on microglia in many forms of CNS pathology, including MS. In addition to MHC molecules, inflammatory stimuli also increase CatS activity (Gresser et al. 2001), which – together with other proteases – can generate the immunogenic epitope MBP111–129 (Beck et al. 2001). The same stimulus, however, results in downregulation of CatG activity, a protease that destroys the immunogenic MBP85–99 epitope, suggesting that selective changes in antigen processing in response to inflammatory or danger signals such as IFNγ, IL-1 or LPS could in principle affect the nature of epitopes presented under those conditions (Burster et al. 2007a; Fiebiger et al. 2001). However, when we compared MBP processing in microglia isolated from rats during acute EAE, we did not observe any differences between these and control animals. In both cases, CatD was the dominant protease activity (Stoeckle et al. 2008).

CNS-phagocytes are similarly efficient at antigen processing and presentation as peripheral APC, however, even if apparently CNS APCs are constantly fed with myelin proteins, local autoantigens may not be available at saturating concentrations, as shown in experiments where addition of exogenous antigen rendered these cells fully competent for antigen presentation (Odoardi et al. 2007). Suboptimal antigen concentration in conjunction with less costimulatory molecules may be responsible for the relatively low efficiency of CNS resident macrophages/microglia and of plasmacytoid DC to present ex vivo endogenous myelin protein (Bailey et al. 2007).

5.3.2 Dendritic Cells

The presence of immature and mature myeloid DC in parenchymal lesions and meninges of patients with MS with relapsing and progressive disease courses is well documented (Serafini et al. 2006). Immature DCs, which are highly efficient in antigen uptake and processing, are able to endocytose myelin in the MS brain and at least some of them acquire a mature phenotype. Mature DC, with high expression of MHC class II molecules, stay in the CNS to stimulate autoreactive T cells locally (Greter et al. 2005), or migrate to draining lymph nodes to initiate new waves of autoreactive T cells (Hatterer et al. 2008; Karman et al. 2004). In addition, CNS DC have been shown to play a critical role in initiating epitope spreading, possibly by priming of naive T cells inside the CNS itself (McMahon et al. 2005).

It would be interesting to know if these cells harbour a characteristic set of protease activities, but to our knowledge this issue has not been addressed.

6 Future Trends and Therapies

Since manipulation of autoantigen processing and presentation allows in principle to selectively alter immune responses against self-components without affecting normal immune responses, several such therapeutic approaches are being developed. Inhibitors of CatS have been reported to be beneficial in EAE (Fissolo et al. 2008) and other models of autoimmune diseases, i.e. Sjögren's syndrome (Saegusa 2002), and are currently under investigation for treatment of human disease.

A number of antigen-specific approaches including altered peptide ligands (APL) and tolerization with known encephalitogenic peptides have been reported. By binding directly to MHC in the absence of costimulation, they are expected to induce tolerance to the epitope. Some groups have attempted to increase stability of the tolerizing peptide by circularising it or changing key residues to eliminate proteolytic cleavage sites (Burster et al. 2007b; Tselios et al. 2002).

One peptide, corresponding to amino acids 82–98 of MBP has reached clinical phase II/III (Warren et al. 2006), but other phase II trials with APL based on this region of MBP had to be prematurely terminated due to adverse side effects or exacerbation of the disease (Bielekova et al. 2000; Kappos et al. 2000). Although the advantages of antigen-specific therapies are obvious, they all have several limitations in common. First, only patients with one or two specific HLA types can benefit from a given therapy. Second, since epitope spreading probably already occurs in the very early stage of the disease and may diversify even further as the disease progresses, tolerization to a single epitope might well not be sufficient to ameliorate the disease. Third, even if sharing the same HLA alleles, individual patients have very individual T cell repertoires, recognizing different epitopes on different antigens (Goebels et al. 2000; Greer et al. 2008; Lindert et al. 1999). The diversity and heterogeneity of T cell responses to myelin antigens will therefore likely be an important limiting factor. Also, only limited knowledge is currently available on which epitopes are processed and presented in vivo in human patients and which of these are truly pathogenic is still not known. Solving these problems may allow more patients to benefit from such novel therapeutic approaches.

7 Conclusion

Although some progress has been made in the past few years in understanding especially antigen processing and also presentation in MS, many questions still remain. Especially the events leading to priming of autoreactive T cells still remain poorly understood. Many of the immunodominant myelin epitopes are cryptic epitopes (Anderton et al. 2002; Manoury et al. 2002; Sweenie et al. 2007), implying that changes in processing might be responsible for their display, which is required for activation of pathogenic T cells. Currently, it appears that although a single protease might well be responsible for limiting epitope display by a given cell type, the protease(s) involved differ between cell types and epitopes. Research on how inflammatory and other stimuli as well as apoptosis or PTM affect processing of specific autoantigens and thus epitope display will lead to a better understanding of how normally cryptic epitopes can contribute to initiation of autoimmune disease such as MS.

References

Anderson AC, Nicholson LB, Legge KL, Turchin V, Zaghouani H, Kuchroo VK (2000) High frequency of autoreactive myelin proteolipid protein-specific T cells in the periphery of naive mice: mechanisms of selection of the self-reactive repertoire. J Exp Med 191:761–770. doi:10.1084/jem.191.5.761

Anderton SM, Wraith DC (2002) Selection and fine-tuning of the autoimmune T-cell repertoire. Nat Rev Immunol 2:487–498

Anderton SM, Viner NJ, Matharu P, Lowrey PA, Wraith DC (2002) Influence of a dominant cryptic epitope on autoimmune T cell tolerance. Nat Immunol 3:175–181

Antel J, Bar-Or A (2006) Roles of immunoglobulins and B cells in multiple sclerosis: from pathogenesis to treatment. J Neuroimmunol 180:3–8

Bailey SL, Schreiner B, McMahon EJ, Miller SD (2007) CNS myeloid DCs presenting endogenous myelin peptides 'preferentially' polarize CD4+ T(H)-17 cells in relapsing EAE. Nat Immunol 8:172–180

Baxter AG (2007) The origin and application of experimental autoimmune encephalomyelitis. Nat Rev Immunol 7:904–912

Beck H, Schwarz G, Schroter CJ, Deeg M, Baier D, Stevanovic S, Weber E, Driessen C, Kalbacher H (2001) Cathepsin S and an asparagine-specific endoprotease dominate the proteolytic processing of human myelin basic protein in vitro. Eur J Immunol 31:3726–3736

Bettelli E, Baeten D, Jager A, Sobel RA, Kuchroo VK (2006) Myelin oligodendrocyte glycoprotein-specific T and B cells cooperate to induce a Devic-like disease in mice. J Clin Invest 116:2393–2402

Bielekova B, Goodwin B, Richert N, Cortese I, Kondo T, Afshar G, Gran B, Eaton J, Antel J, Frank JA, McFarland HF, Martin R (2000) Encephalitogenic potential of the myelin basic protein peptide (amino acids 83–99) in multiple sclerosis: results of a phase II clinical trial with an altered peptide ligand. Nat Med 6:1167–1175

Brahms H, Raymackers J, Union A, de Keyser F, Meheus L, Luhrmann R (2000) The C-terminal RG dipeptide repeats of the spliceosomal Sm proteins D1 and D3 contain symmetrical dimethylarginines, which form a major B-cell epitope for anti-Sm autoantibodies. J Biol Chem 275:17122–17129

Bruno R, Sabater L, Sospedra M, Ferrer-Francesch X, Escudero D, Martinez-Caceres E, Pujol-Borrell R (2002) Multiple sclerosis candidate autoantigens except myelin oligodendrocyte glycoprotein are transcribed in human thymus. Eur J Immunol 32:2737–2747

Burster T, Beck A, Tolosa E, Marin-Esteban V, Rotzschke O, Falk K, Lautwein A, Reich M, Brandenburg J, Schwarz G, Wiendl H, Melms A, Lehmann R, Stevanovic S, Kalbacher H, Driessen C (2004) Cathepsin G, and not the asparagine-specific endoprotease, controls the processing of myelin basic protein in lysosomes from human B lymphocytes. J Immunol 172:5495–5503

Burster T, Beck A, Tolosa E, Schnorrer P, Weissert R, Reich M, Kraus M, Kalbacher H, Haring HU, Weber E, Overkleeft H, Driessen C (2005) Differential processing of autoantigens in lysosomes from human monocyte-derived and peripheral blood dendritic cells. J Immunol 175:5940–5949

Burster T, Beck A, Poeschel S, Oren A, Baechle D, Reich M, Roetzschke O, Falk K, Boehm BO, Youssef S, Kalbacher H, Overkleeft H, Tolosa E, Driessen C (2007a) Interferon-gamma regulates cathepsin G activity in microglia-derived lysosomes and controls the proteolytic processing of myelin basic protein in vitro. Immunology 121:82–93

Burster T, Marin-Esteban V, Boehm BO, Dunn S, Rotzschke O, Falk K, Weber E, Verhelst SH, Kalbacher H, Driessen C (2007b) Design of protease-resistant myelin basic protein-derived peptides by cleavage site directed amino acid substitutions. Biochem Pharmacol 74:1514–1523

Cao L, Goodin R, Wood D, Moscarello MA, Whitaker JN (1999) Rapid release and unusual stability of immunodominant peptide 45–89 from citrullinated myelin basic protein. Biochemistry 38:6157–6163

Castellino F, Zappacosta F, Coligan JE, Germain RN (1998) Large protein fragments as substrates for endocytic antigen capture by MHC class II molecules. J Immunol 161:4048–4057

Correale J, McMillan M, McCarthy K, Le T, Weiner LP (1995) Isolation and characterization of autoreactive proteolipid protein-peptide specific T-cell clones from multiple sclerosis patients. Neurology 45:1370–1378

Corthay A, Backlund J, Holmdahl R (2001) Role of glycopeptide-specific T cells in collagen-induced arthritis: an example how post-translational modification of proteins may be involved in autoimmune disease. Ann Med 33:456–465

Costantino CM, Hang HC, Kent SC, Hafler DA, Ploegh HL (2008) Lysosomal cysteine and aspartic proteases are heterogeneously expressed and act redundantly to initiate human invariant chain degradation. J Immunol 180:2876–2885

Cross AH, Trotter JL, Lyons J (2001) B cells and antibodies in CNS demyelinating disease. J Neuroimmunol 112:1–14

Day MJ, Tse AG, Puklavec M, Simmonds SJ, Mason DW (1992) Targeting autoantigen to B cells prevents the induction of a cell-mediated autoimmune disease in rats. J Exp Med 175:655–659

de Vos AF, van Meurs M, Brok HP, Boven LA, Hintzen RQ, van der Valk P, Ravid R, Rensing S, Boon L, t Hart BA, Laman JD (2002) Transfer of central nervous system autoantigens and presentation in secondary lymphoid organs. J Immunol 169:5415–5423

Delarasse C, Daubas P, Mars LT, Vizler C, Litzenburger T, Iglesias A, Bauer J, Della Gaspera B, Schubart A, Decker L, Dimitri D, Roussel G, Dierich A, Amor S, Dautigny A, Liblau R, Pham-Dinh D (2003) Myelin/oligodendrocyte glycoprotein-deficient (MOG-deficient) mice reveal lack of immune tolerance to MOG in wild-type mice. J Clin Invest 112:544–553

Denzin LK, Cresswell P (1995) HLA-DM induces CLIP dissociation from MHC class II alpha beta dimers and facilitates peptide loading. Cell 82:155–165

Denzin LK, Sant'Angelo DB, Hammond C, Surman MJ, Cresswell P (1997) Negative regulation by HLA-DO of MHC class II-restricted antigen processing. Science 278:106–109

Derbinski J, Schulte A, Kyewski B, Klein L (2001) Promiscuous gene expression in medullary thymic epithelial cells mirrors the peripheral self. Nat Immunol 2:1032–1039

Doyle HA, Mamula MJ (2005) Posttranslational modifications of self-antigens. Ann N Y Acad Sci 1050:1–9

Eggleton P, Haigh R, Winyard PG (2008) Consequence of neo-antigenicity of the 'altered self'. Rheumatology (Oxford) 47:567–571

Fazilleau N, Delarasse C, Sweenie CH, Anderton SM, Fillatreau S, Lemonnier FA, Pham-Dinh D, Kanellopoulos JM (2006) Persistence of autoreactive myelin oligodendrocyte glycoprotein (MOG)-specific T cell repertoires in MOG-expressing mice. Eur J Immunol 36:533–543

Fiebiger E, Meraner P, Weber E, Fang IF, Stingl G, Ploegh H, Maurer D (2001) Cytokines regulate proteolysis in major histocompatibility complex class II-dependent antigen presentation by dendritic cells. J Exp Med 193:881–892

Fissolo N, Kraus M, Reich M, Ayturan M, Overkleeft H, Driessen C, Weissert R (2008) Dual inhibition of proteasomal and lysosomal proteolysis ameliorates autoimmune central nervous system inflammation. Eur J Immunol 38:2401–2411

Fridkis-Hareli M, Teitelbaum D, Gurevich E, Pecht I, Brautbar C, Kwon OJ, Brenner T, Arnon R, Sela M (1994) Direct binding of myelin basic protein and synthetic copolymer 1 to class II major histocompatibility complex molecules on living antigen-presenting cells–specificity and promiscuity. Proc Natl Acad Sci USA 91:4872–4876

Gallegos AM, Bevan MJ (2004) Central tolerance to tissue-specific antigens mediated by direct and indirect antigen presentation. J Exp Med 200:1039–1049

Gill J, Malin M, Sutherland J, Gray D, Hollander G, Boyd R (2003) Thymic generation and regeneration. Immunol Rev 195:28–50

Goebels N, Hofstetter H, Schmidt S, Brunner C, Wekerle H, Hohlfeld R (2000) Repertoire dynamics of autoreactive T cells in multiple sclerosis patients and healthy subjects: epitope spreading versus clonal persistence. Brain 123(Pt 3):508–518

Goodnow CC, Sprent J, de St F, Groth B, Vinuesa CG (2005) Cellular and genetic mechanisms of self tolerance and autoimmunity. Nature 435:590–597

Greer JM, Pender MP (2008) Myelin proteolipid protein: an effective autoantigen and target of autoimmunity in multiple sclerosis. J Autoimmun 31:281–287

Greer JM, Csurhes PA, Muller DM, Pender MP (2008) Correlation of blood T cell and antibody reactivity to myelin proteins with HLA type and lesion localization in multiple sclerosis. J Immunol 180:6402–6410

Gregersen JW, Kranc KR, Ke X, Svendsen P, Madsen LS, Thomsen AR, Cardon LR, Bell JI, Fugger L (2006) Functional epistasis on a common MHC haplotype associated with multiple sclerosis. Nature 443:574–577

Gresser O, Weber E, Hellwig A, Riese S, Regnier-Vigouroux A (2001) Immunocompetent astrocytes and microglia display major differences in the processing of the invariant chain and in the expression of active cathepsin L and cathepsin S. Eur J Immunol 31:1813–1824

Greter M, Heppner FL, Lemos MP, Odermatt BM, Goebels N, Laufer T, Noelle RJ, Becher B (2005) Dendritic cells permit immune invasion of the CNS in an animal model of multiple sclerosis. Nat Med 11:328–334

Harauz G, Musse AA (2007) A tale of two citrullines–structural and functional aspects of myelin basic protein deimination in health and disease. Neurochem Res 32:137–158

Hatterer E, Touret M, Belin MF, Honnorat J, Nataf S (2008) Cerebrospinal fluid dendritic cells infiltrate the brain parenchyma and target the cervical lymph nodes under neuroinflammatory conditions. PLoS ONE 3:e3321

Heppner FL, Greter M, Marino D, Falsig J, Raivich G, Hovelmeyer N, Waisman A, Rulicke T, Prinz M, Priller J, Becher B, Aguzzi A (2005) Experimental autoimmune encephalomyelitis repressed by microglial paralysis. Nat Med 11:146–152

Hill JA, Southwood S, Sette A, Jevnikar AM, Bell DA, Cairns E (2003) Cutting edge: the conversion of arginine to citrulline allows for a high-affinity peptide interaction with the rheumatoid arthritis-associated HLA-DRB1*0401 MHC class II molecule. J Immunol 171:538–541

Hogquist KA, Baldwin TA, Jameson SC (2005) Central tolerance: learning self-control in the thymus. Nat Rev Immunol 5:772–782

Holmes S, Siebold C, Jones EY, Friese MA, Fugger L, Bell J (2005) Multiple sclerosis: MHC associations and therapeutic implications. Expert Rev Mol Med 7:1–17

Honey K, Rudensky AY (2003) Lysosomal cysteine proteases regulate antigen presentation. Nat Rev Immunol 3:472–482

Honey K, Nakagawa T, Peters C, Rudensky A (2002) Cathepsin L regulates CD4+ T cell selection independently of its effect on invariant chain: a role in the generation of positively selecting peptide ligands. J Exp Med 195:1349–1358

Iglesias A, Bauer J, Litzenburger T, Schubart A, Linington C (2001) T- and B-cell responses to myelin oligodendrocyte glycoprotein in experimental autoimmune encephalomyelitis and multiple sclerosis. Glia 36:220–234

Kappos L, Comi G, Panitch H, Oger J, Antel J, Conlon P, Steinman L (2000) Induction of a non-encephalitogenic type 2 T helper-cell autoimmune response in multiple sclerosis after administration of an altered peptide ligand in a placebo-controlled, randomized phase II trial. The altered peptide ligand in relapsing MS study group. Nat Med 6:1176–1182

Karman J, Ling C, Sandor M, Fabry Z (2004) Initiation of immune responses in brain is promoted by local dendritic cells. J Immunol 173:2353–2361

Kerlero de Rosbo N, Hoffman M, Mendel I, Yust I, Kaye J, Bakimer R, Flechter S, Abramsky O, Milo R, Karni A, Ben-Nun A (1997) Predominance of the autoimmune response to myelin oligodendrocyte glycoprotein (MOG) in multiple sclerosis: reactivity to the extracellular domain of MOG is directed against three main regions. Eur J Immunol 27:3059–3069

Kim JK, Mastronardi FG, Wood DD, Lubman DM, Zand R, Moscarello MA (2003) Multiple sclerosis: an important role for post-translational modifications of myelin basic protein in pathogenesis. Mol Cell Proteomics 2:453–462

Kirberg J, Berns A, von Boehmer H (1997) Peripheral T cell survival requires continual ligation of the T cell receptor to major histocompatibility complex-encoded molecules. J Exp Med 186:1269–1275

Klein L, Klugmann M, Nave KA, Tuohy VK, Kyewski B (2000) Shaping of the autoreactive T-cell repertoire by a splice variant of self protein expressed in thymic epithelial cells. Nat Med 6:56–61

Krishnamoorthy G, Lassmann H, Wekerle H, Holz A (2006) Spontaneous opticospinal encephalomyelitis in a double-transgenic mouse model of autoimmune T cell/B cell cooperation. J Clin Invest 116:2385–2392

Krogsgaard M, Wucherpfennig KW, Cannella B, Hansen BE, Svejgaard A, Pyrdol J, Ditzel H, Raine C, Engberg J, Fugger L (2000) Visualization of myelin basic protein (MBP) T cell epitopes in multiple sclerosis lesions using a monoclonal antibody specific for the human histocompatibility leukocyte antigen (HLA)-DR2-MBP 85–99 complex. J Exp Med 191:1395–1412

Kyewski B, Klein L (2006) A central role for central tolerance. Annu Rev Immunol 24:571–606

Laatsch RH, Kies MW, Gordon S, Alvord EC Jr (1962) The encephalomyelitic activity of myelin isolated by ultracentrifugation. J Exp Med 115:77–88

Lincoln MR, Montpetit A, Cader MZ, Saarela J, Dyment DA, Tiislar M, Ferretti V, Tienari PJ, Sadovnick AD, Peltonen L, Ebers GC, Hudson TJ (2005) A predominant role for the HLA class II region in the association of the MHC region with multiple sclerosis. Nat Genet 37:1108–1112

Lindert RB, Haase CG, Brehm U, Linington C, Wekerle H, Hohlfeld R (1999) Multiple sclerosis: B- and T-cell responses to the extracellular domain of the myelin oligodendrocyte glycoprotein. Brain 122(Pt 11):2089–2100

Linington C, Berger T, Perry L, Weerth S, Hinze-Selch D, Zhang Y, Lu HC, Lassmann H, Wekerle H (1993) T cells specific for the myelin oligodendrocyte glycoprotein mediate an unusual autoimmune inflammatory response in the central nervous system. Eur J Immunol 23:1364–1372

Lippolis JD, White FM, Marto JA, Luckey CJ, Bullock TN, Shabanowitz J, Hunt DF, Engelhard VH (2002) Analysis of MHC class II antigen processing by quantitation of peptides that constitute nested sets. J Immunol 169:5089–5097

Madsen LS, Andersson EC, Jansson L, Krogsgaard M, Andersen CB, Engberg J, Strominger JL, Svejgaard A, Hjorth JP, Holmdahl R, Wucherpfennig KW, Fugger L (1999) A humanized model for multiple sclerosis using HLA-DR2 and a human T-cell receptor. Nat Genet 23:343–347

Maehr R, Mintern JD, Herman AE, Lennon-Dumenil AM, Mathis D, Benoist C, Ploegh HL (2005) Cathepsin L is essential for onset of autoimmune diabetes in NOD mice. J Clin Invest 115:2934–2943

Mamula MJ, Gee RJ, Elliott JI, Sette A, Southwood S, Jones PJ, Blier PR (1999) Isoaspartyl posttranslational modification triggers autoimmune responses to self-proteins. J Biol Chem 274:22321–22327

Manoury B, Hewitt EW, Morrice N, Dando PM, Barrett AJ, Watts C (1998) An asparaginyl endopeptidase processes a microbial antigen for class II MHC presentation. Nature 396:695–699

Manoury B, Mazzeo D, Fugger L, Viner N, Ponsford M, Streeter H, Mazza G, Wraith DC, Watts C (2002) Destructive processing by asparagine endopeptidase limits presentation of a dominant T cell epitope in MBP. Nat Immunol 3:169–174

Martin R, Jaraquemada D, Flerlage M, Richert J, Whitaker J, Long EO, McFarlin DE, McFarland HF (1990) Fine specificity and HLA restriction of myelin basic protein-specific cytotoxic T cell lines from multiple sclerosis patients and healthy individuals. J Immunol 145:540–548

Marty MC, Alliot F, Rutin J, Fritz R, Trisler D, Pessac B (2002) The myelin basic protein gene is expressed in differentiated blood cell lineages and in hemopoietic progenitors. Proc Natl Acad Sci USA 99:8856–8861

Mastronardi FG, Noor A, Wood DD, Paton T, Moscarello MA (2007) Peptidyl argininedeiminase 2 CpG island in multiple sclerosis white matter is hypomethylated. J Neurosci Res 85:2006–2016

McDevitt HO, Perry R, Steinman LA (1987) Monoclonal anti-Ia antibody therapy in animal models of autoimmune disease. Ciba Found Symp 129:184–193

McFarland HF, Martin R (2007) Multiple sclerosis: a complicated picture of autoimmunity. Nat Immunol 8:913–919

McMahon EJ, Bailey SL, Castenada CV, Waldner H, Miller SD (2005) Epitope spreading initiates in the CNS in two mouse models of multiple sclerosis. Nat Med 11:335–339

McRae BL, Vanderlugt CL, Dal Canto MC, Miller SD (1995) Functional evidence for epitope spreading in the relapsing pathology of experimental autoimmune encephalomyelitis. J Exp Med 182:75–85

Meinl E, Weber F, Drexler K, Morelle C, Ott M, Saruhan-Direskeneli G, Goebels N, Ertl B, Jechart G, Giegerich G et al (1993) Myelin basic protein-specific T lymphocyte repertoire in multiple sclerosis. Complexity of the response and dominance of nested epitopes due to recruitment of multiple T cell clones. J Clin Invest 92:2633–2643

Melms A, Luther C, Stoeckle C, Poschel S, Schroth P, Varga M, Wienhold W, Tolosa E (2006) Thymus and myasthenia gravis: antigen processing in the human thymus and the consequences for the generation of autoreactive T cells. Acta Neurol Scand 183:12–13

Moss CX, Matthews SP, Lamont DJ, Watts C (2005) Asparagine deamidation perturbs antigen presentation on class II major histocompatibility complex molecules. J Biol Chem 280:18498–18503

Moss CX, Tree TI, Watts C (2007) Reconstruction of a pathway of antigen processing and class II MHC peptide capture. EMBO J 26:2137–2147

Musse AA, Li Z, Ackerley CA, Bienzle D, Lei H, Poma R, Harauz G, Moscarello MA, Mastronardi FG (2008) Peptidylarginine deiminase 2 (PAD2) overexpression in transgenic mice leads to myelin loss in the central nervous system. Dis Model Mech 1:229–240

Nakagawa T, Roth W, Wong P, Nelson A, Farr A, Deussing J, Villadangos JA, Ploegh H, Peters C, Rudensky AY (1998) Cathepsin L: critical role in Ii degradation and CD4 T cell selection in the thymus. Science 280:450–453

Narayan K, Chou CL, Kim A, Hartman IZ, Dalai S, Khoruzhenko S, Sadegh-Nasseri S (2007) HLA-DM targets the hydrogen bond between the histidine at position beta81 and peptide to dissociate HLA-DR-peptide complexes. Nat Immunol 8:92–100

Nimmerjahn A, Kirchhoff F, Helmchen F (2005) Resting microglial cells are highly dynamic surveillants of brain parenchyma in vivo. Science 308:1314–1318

Nissim A, Winyard PG, Corrigall V, Fatah R, Perrett D, Panayi G, Chernajovsky Y (2005) Generation of neoantigenic epitopes after posttranslational modification of type II collagen by factors present within the inflamed joint. Arthritis Rheum 52:3829–3838

Odoardi F, Kawakami N, Klinkert WE, Wekerle H, Flugel A (2007) Blood-borne soluble protein antigen intensifies T cell activation in autoimmune CNS lesions and exacerbates clinical disease. Proc Natl Acad Sci USA 104:18625–18630

Pelfrey CM, Trotter JL, Tranquill LR, McFarland HF (1993) Identification of a novel T cell epitope of human proteolipid protein (residues 40–60) recognized by proliferative and cytolytic CD4+ T cells from multiple sclerosis patients. J Neuroimmunol 46:33–42

Pfender NA, Grosch S, Roussel G, Koch M, Trifilieff E, Greer JM (2008) Route of uptake of palmitoylated encephalitogenic peptides of myelin proteolipid protein by antigen-presenting cells: importance of the type of bond between lipid chain and peptide and relevance to autoimmunity. J Immunol 180:1398–1404

Pinet V, Vergelli M, Martin R, Bakke O, Long EO (1995) Antigen presentation mediated by recycling of surface HLA-DR molecules. Nature 375:603–606

Prat E, Tomaru U, Sabater L, Park DM, Granger R, Kruse N, Ohayon JM, Bettinotti MP, Martin R (2005) HLA-DRB5*0101 and -DRB1*1501 expression in the multiple sclerosis-associated HLA-DR15 haplotype. J Neuroimmunol 167:108–119

Priller J, Flugel A, Wehner T, Boentert M, Haas CA, Prinz M, Fernandez-Klett F, Prass K, Bechmann I, de Boer BA, Frotscher M, Kreutzberg GW, Persons DA, Dirnagl U (2001) Targeting gene-modified hematopoietic cells to the central nervous system: use of green fluorescent protein uncovers microglial engraftment. Nat Med 7:1356–1361

Pritzker LB, Joshi S, Gowan JJ, Harauz G, Moscarello MA (2000a) Deimination of myelin basic protein. 1. Effect of deimination of arginyl residues of myelin basic protein on its structure and susceptibility to digestion by cathepsin D. Biochemistry 39:5374–5381

Pritzker LB, Joshi S, Harauz G, Moscarello MA (2000b) Deimination of myelin basic protein. 2. Effect of methylation of MBP on its deimination by peptidylarginine deiminase. Biochemistry 39:5382–5388

Quandt JA, Baig M, Yao K, Kawamura K, Huh J, Ludwin SK, Bian HJ, Bryant M, Quigley L, Nagy ZA, McFarland HF, Muraro PA, Martin R, Ito K (2004) Unique clinical and pathological features in HLA-DRB1*0401-restricted MBP 111–129-specific humanized TCR transgenic mice. J Exp Med 200:223–234

Ristori G, Carcassi C, Lai S, Fiori P, Cacciani A, Floris L, Montesperelli C, Di Giovanni S, Buttinelli C, Contu L, Pozzilli C, Salvetti M (1997) HLA-DM polymorphisms do not associate with multiple sclerosis: an association study with analysis of myelin basic protein T cell specificity. J Neuroimmunol 77:181–184

Rudensky A, Beers C (2006) Lysosomal cysteine proteases and antigen presentation. Ernst Schering Res Found Workshop. 56:81–95

Saegusa K (2002) Cathepsin S inhibitor prevents autoantigen processing and autoimmunity. J Clin Invest 110:361–369

Seamons A, Perchellet A, Goverman J (2006) Endogenous myelin basic protein is presented in the periphery by both dendritic cells and resting B cells with different functional consequences. J Immunol 177:2097–2106

Serafini B, Rosicarelli B, Magliozzi R, Stigliano E, Capello E, Mancardi GL, Aloisi F (2006) Dendritic cells in multiple sclerosis lesions: maturation stage, myelin uptake, and interaction with proliferating T cells. J Neuropathol Exp Neurol 65:124–141

Sospedra M, Martin R (2005) Immunology of multiple sclerosis. Annu Rev Immunol 23:683–747

Sospedra M, Ferrer-Francesch X, Dominguez O, Juan M, Foz-Sala M, Pujol-Borrell R (1998) Transcription of a broad range of self-antigens in human thymus suggests a role for central mechanisms in tolerance toward peripheral antigens. J Immunol 161:5918–5929

Stoeckle C, Burster T, Gnau V, Driessen C, Kalbacher H, Melms A, Tolosa E (2004) Is processing regulated in patients with multiple sclerosis? Immunology 3:285–291

Stoeckle C, Herrmann M, Burster T, Beck A, Weissert R, Melms A, Tolosa E (2008) Autoantigen processing in the CNS during autoimmune inflammation. J Neuroimmunol 203:226

Stoeckle C, Sommandas V, Adamopoulou E, Belisle K, Schiekofer S, Melms A, Weber E, Driessen C, Boehm BO, Tolosa E, Burster T (2009) Cathepsin G is differentially expressed in primary human antigen-presenting cells. Cell Immunol 255:41–45

Sweenie CH, Mackenzie KJ, Rone-Orugboh A, Liu M, Anderton SM (2007) Distinct T cell recognition of naturally processed and cryptic epitopes within the immunodominant 35–55 region of myelin oligodendrocyte glycoprotein. J Neuroimmunol 183:7–16

Szekanecz Z, Soos L, Szabo Z, Fekete A, Kapitany A, Vegvari A, Sipka S, Szucs G, Szanto S, Lakos G (2008) Anti-citrullinated protein antibodies in rheumatoid arthritis: as good as it gets? Clin Rev Allergy Immunol 34:26–31

Taubert R, Schwendemann J, Kyewski B (2007) Highly variable expression of tissue-restricted self-antigens in human thymus: implications for self-tolerance and autoimmunity. Eur J Immunol 37:838–848

Tolosa E, Li W, Yasuda Y, Wienhold W, Denzin LK, Lautwein A, Driessen C, Schnorrer P, Weber E, Stevanovic S, Kurek R, Melms A, Bromme D (2003) Cathepsin V is involved in the degradation of invariant chain in human thymus and is overexpressed in myasthenia gravis. J Clin Invest 112:517–526

Tranquill LR, Cao L, Ling NC, Kalbacher H, Martin RM, Whitaker JN (2000) Enhanced T cell responsiveness to citrulline-containing myelin basic protein in multiple sclerosis patients. Mult Scler 6:220–225

Trombetta ES, Mellman I (2005) Cell biology of antigen processing in vitro and in vivo. Annu Rev Immunol 23:975–1028

Tselios T, Apostolopoulos V, Daliani I, Deraos S, Grdadolnik S, Mavromoustakos T, Melachrinou M, Thymianou S, Probert L, Mouzaki A, Matsoukas J (2002) Antagonistic effects of human cyclic MBP(87–99) altered peptide ligands in experimental allergic encephalomyelitis and human T-cell proliferation. J Med Chem 45:275–283

Vanderlugt CL, Miller SD (2002) Epitope spreading in immune-mediated diseases: implications for immunotherapy. Nat Rev Immunol 2:85–95

Vergelli M, Kalbus M, Rojo SC, Hemmer B, Kalbacher H, Tranquill L, Beck H, McFarland HF, De Mars R, Long EO, Martin R (1997a) T cell response to myelin basic protein in the context of the multiple sclerosis-associated HLA-DR15 haplotype: peptide binding, immunodominance and effector functions of T cells. J Neuroimmunol 77:195–203

Vergelli M, Pinet V, Vogt AB, Kalbus M, Malnati M, Riccio P, Long EO, Martin R (1997b) HLA-DR-restricted presentation of purified myelin basic protein is independent of intracellular processing. Eur J Immunol 27:941–951

Villadangos JA, Schnorrer P (2007) Intrinsic and cooperative antigen-presenting functions of dendritic-cell subsets in vivo. Nat Rev Immunol 7:543–555

Vogt AB, Kropshofer H, Kalbacher H, Kalbus M, Rammensee HG, Coligan JE, Martin R (1994) Ligand motifs of HLA-DRB5*0101 and DRB1*1501 molecules delineated from self-peptides. J Immunol 153:1665–1673

Waldner H, Whitters MJ, Sobel RA, Collins M, Kuchroo VK (2000) Fulminant spontaneous autoimmunity of the central nervous system in mice transgenic for the myelin proteolipid protein-specific T cell receptor. Proc Natl Acad Sci USA 97:3412–3417

Wallberg M, Bergquist J, Achour A, Breij E, Harris RA (2007) Malondialdehyde modification of myelin oligodendrocyte glycoprotein leads to increased immunogenicity and encephalitogenicity. Eur J Immunol 37:1986–1995

Wallstrom E, Khademi M, Andersson M, Weissert R, Linington C, Olsson T (1998) Increased reactivity to myelin oligodendrocyte glycoprotein peptides and epitope mapping in HLA DR2(15)+multiple sclerosis. Eur J Immunol 28:3329–3335

Warren KG, Catz I, Ferenczi LZ, Krantz MJ (2006) Intravenous synthetic peptide MBP8298 delayed disease progression in an HLA Class II-defined cohort of patients with progressive multiple sclerosis: results of a 24-month double-blind placebo-controlled clinical trial and 5 years of follow-. Eur J Neurol 13:887–895

Weissert R, Kuhle J, de Graaf KL, Wienhold W, Herrmann MM, Muller C, Forsthuber TG, Wiesmuller KH, Melms A (2002) High immunogenicity of intracellular myelin oligodendrocyte glycoprotein epitopes. J Immunol 169:548–556

Westley BR, May FE (1987) Oestrogen regulates cathepsin D mRNA levels in oestrogen responsive human breast cancer cells. Nucleic Acids Res 15:3773–3786

Wilson NS, El-Sukkari D, Villadangos JA (2004) Dendritic cells constitutively present self antigens in their immature state in vivo and regulate antigen presentation by controlling the rates of MHC class II synthesis and endocytosis. Blood 103:2187–2195

Immune-Mediated CNS Damage

Katrin Kierdorf, Yiner Wang, and Harald Neumann

Abstract Multiple sclerosis (MS) is a demyelinating autoimmune disease. However, the persisting neurological deficits in MS patients result from acute axonal injury and chronic neurodegeneration, which are both triggered by the autoreactive immune response. Innate immunity, mainly mediated by activated microglial cells and invading macrophages, appears to contribute to chronic neurodegeneration. Activated microglia produce several reactive oxygen species and proinflammatory cytokines which affect neuronal function, integrity and survival. Adaptive immunity, particularly in cytotoxic CD8+ T cells, participates in acute demyelination and axonal injury by directly attacking oligodendrocytes and possibly neurons as well. Understanding the mechanisms of immune-mediated neuronal damage might help to design novel therapy strategies for MS.

1 Introduction

Under physiological conditions, the central nervous system (CNS) is often regarded as an immune privileged organ. A coordinated immune response needs time to be established in the normal CNS parenchyma since most of the relevant molecules are not expressed and have to be induced (Lowenstein 2002). To maintain this immune compromised status, the CNS displays many structural modifications and adaptations. Firstly, a blood–brain barrier locked by endothelial tight junctions shields the CNS from invading immune cells like lymphocytes and neutrophils, and only a few transport systems allow the entrance of necessary macromolecules into the parenchyma at a minimum level (Huber et al. 2001). Secondly, there are no activated resident immune cells in the healthy CNS tissue to produce cytokines and chemokines, which can attract other immune cells to invade into the tissue parenchyma. Furthermore,

H. Neumann(✉), K. Kierdorf and Y. Wang
Institute of Reconstructive Neurobiology, University Bonn LIFE and BRAIN Center,
University Bonn and Hertie-Foundation, Sigmund-Freud-Str. 25, 53127, Bonn, Germany
e-mail: hneuman1@uni-bonn.de

Results Probl Cell Differ, doi:10.1007/400_2008_15

adequate numbers of cell adhesion molecules such as the intercellular adhesion molecule (ICAM) are not expressed in the brain endothelium under healthy conditions, which would have served to reduce the invasion or attraction of activated immune cells into the CNS. Circulating nonactivated lymphocytes are incapable of reaching the CNS due to the blood–brain barrier. Finally, the expression of major histocompatibility complex (MHC) molecules is low or absent on all the cell types in the CNS, preventing effective antigen presentation to specific T lymphocytes and the propagation of an adaptive immune response (Fig. 1).

However, the immune privileged status is restricted to the normal CNS. During microbial infections or after acute injury, this status is completely changed and the CNS is then able to coordinate an innate immune response, which finally leads to the mobilization of a specific adaptive immune response against any invading pathogen. The same breakdown is also seen, although at lower levels, in most primary neurodegenerative diseases such as Alzheimer's disease, Parkinson's disease, Huntington's disease and amyotrophic lateral sclerosis (ALS). Under these pathological conditions, an induction of molecules of the innate as well as the adaptive immune response like the upregulation of MHC class I and II molecules is observed.

The rising inflammatory processes in the CNS tissue change the gene expression pattern of the resident cells. More and more immunologically relevant molecules are expressed on brain cells, particularly on the microglia, and allow specific communication between the immune cells and the neural cells. However, all proinflammatory responses are counter-regulated by local immunosuppressive mechanisms. It was shown that electrically active and functionally intact neurons are key players in maintaining the immune privileged status of the CNS and in inhibiting the expression of immune molecules such as MHC class II on astrocytes and microglia (Neumann 2001). Neurons appear to keep the immune competent microglial cell in a quiescent state with a defined cytokine and tissue growth factor milieu. The electrically active neurons are producers of several neurotrophins including brain-derived neurotrophic factor (BDNF) and neurotrophin-3 (NT3), which can inhibit the inducibility of MHC class II molecules on the microglial cells (Fig. 1). Moreover, astrocytes help to maintain the microglia in a resting state. They are the main source of transforming growth factor-β (TGFβ), which is known to have an immune suppressive function in the CNS.

Fig. 1 (See the next page) The immune privileged status of the CNS is maintained by healthy neurons and an immunosuppressive environment. Under physiological conditions, sporadic invading T cells are undergoing apoptosis in the CNS parenchyma without recognizing specific antigens or receiving activation signals. After CNS damage, the injured neurons secrete chemotactic and cytotoxic factors such as fractalkine (CX3CL1) and glutamate that can further contribute to the chemoattraction and activation of the surrounding microglial cells. Microglia as well as astrocytes start to secrete additional chemokines that attract monocytes and lymphocytes from the blood circulation. Attraction of T-lymphocytes is supported by the chemokines CCL21 and CCL19, released from endothelium cells during inflammatory CNS lesions. Invading monocytes receive activation signals from the impaired neurons and activated microglia. The activated macrophages secrete proinflammatory cytokines like IL1β and TNFα that contribute to the proinflammatory milieu

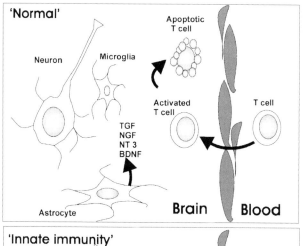
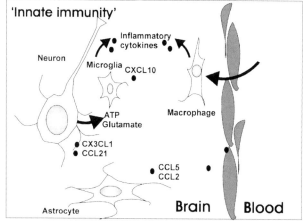
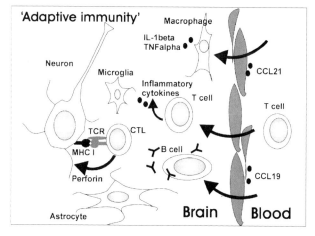

Fig. 1

Thus, once the neurons are impaired in their function, proinflammatory processes will be allowed. Even very weak inflammatory signals are then able to stimulate a whole subset of immune reactions like the release of different chemokines, the expression of immune relevant molecules and the secretion of more inflammatory mediators, all of which are necessary to initiate organized innate and adaptive immune response within the CNS.

In the immune privileged CNS, cells of the adaptive immune system are almost absent. Sometimes sporadic lymphocytes are found in the healthy nervous tissue. These are activated T cells from the lymph nodes or other secondary immune organs, which can migrate through the intact blood–brain barrier. However, they have been shown to undergo apoptosis in the normal brain tissue that lacks expression of MHC molecules and costimulatory molecules (Gold et al. 1996).

Multiple sclerosis (MS) is an autoimmune disease hallmarked by demyelination in the CNS white matter associated with invading T cells and proinflammatory cytokine production at the lesion sites. In MS, immune reactions of the adaptive immune response are mainly confined to the inflammatory brain foci, while the innate immune system is activated outside the inflammatory foci. Most of the immune reactions of the adaptive immune response are highly destructive and contribute to the process of brain injury. Inflamed glial cells produce proinflammatory cytokines like Interleukin-1β (IL1β) and tumor necrosis factor-α (TNFα), and chemoattractant molecules like CXCL10, CX3CL1, CCL2 and CCL5. These secreted molecules attract microglial cells to the lesion sites and promote invasion of blood-born immune cells such as monocytes and lymphocytes into the CNS (Fig. 1).

The inflammatory reaction also activates the microglial cells next to the injured tissue to become antigen presenting cells (APCs) and to present brain derived antigens to the invading T lymphocytes, which might become competent effector cells as cytotoxic T lymphocytes (CTLs) or cytokine secreting cells under the stimulation of the antigen. These cells can also coordinate the B cell directed antibody production. Within the inflammatory foci, the endothelial cells produce the chemokines CCL19 and CCL21, which guide the T lymphocytes to enter the injured tissue and facilitate the transmigration through the endothelia (Alt et al. 2002; Campbell et al. 1998).

2 Innate Immunity in Multiple Sclerosis

2.1 Microglia Activation

Microglia are the only resident immune cells of the CNS and among the major players in establishing the innate immune reaction within the CNS. It is widely accepted that microglia derive from myeloid precursors of the hematopoietic system and invade into the CNS from the yolk sac in early development. The density and derivation of microglial cells differs among different brain regions and approximately 12% of brain cells are microglial cells. Specifically, the density can differ between

0.5 and 16.5%, with the highest concentration of microglia within the hippocampus, olfactory telencephalon, basal ganglia and substantia nigra (Block et al. 2007).

In the adult normal brain, microglia keep a so called 'resting' state which is characterized by a ramified structure and low expression of immunological molecules. Under these normal conditions, the microglial cell is actually not resting at all, but is continuously moving its multitude of processes (Davalos et al. 2005; Nimmerjahn et al. 2005). Each microglial cell is monitoring a defined area in the nervous tissue (Davalos et al. 2005; Nimmerjahn et al. 2005). Upon receiving immunological stimuli, microglia can be activated and maturate into active immune competent cells (Fetler and Amigorena 2005; Nimmerjahn et al. 2005). During the activation process, the cell morphology changes from a ramified structure to a hyper ramified and finally to an amoeboid morphology (Raivich 2005). The amoeboid structure facilitates the migration of the microglial cell through the parenchyma to lesion sites or invading pathogens. In addition, microglia proliferate in response to several stimuli and the cell size increases. Another significant parts of the transformation is the change in the cell surface receptor expression pattern. Many important immunological molecules such as MHC molecules and chemokine receptors are upregulated during microglial activation (Cho et al. 2006; Oehmichen and Gencic 1975). Other receptors including the pattern recognition receptors, which are essential to initiate innate immune responses, are constitutively expressed on microglia. The activated microglia also express several phagocytic receptors and become professional phagocytes.

During brain development, microglia also exhibit an amoeboid morphology and play important roles in tissue homeostasis. They remove excess neural cells by phagocytosis (Marin-Teva et al. 2004) and promote neuronal survival by release of neurotrophic factors (Liao et al. 2005; Morgan et al. 2004; Muller et al. 2006).

The activated microglial cell is often regarded as a double-edged sword. On the one hand, the activation enables microglia to maintain and support neuronal survival (Harry et al. 2004; Streit 2002) by release of neurotrophic and antiinflammatory molecules, the clearance of toxic products or invading pathogens, as well as the guidance of stem cells to inflammatory lesion sites to promote neurogenesis (Liu et al. 2002; Walton et al. 2006; Ziv et al. 2006). On the other hand, over-activated microglia can be neurotoxic by releasing several cytotoxic substances like nitric oxide (NO) or superoxide (Colton and Gilbert 1987), the proinflammatory factors IL1β and TNFα (Lee et al. 1993; Sawada et al. 1989), which can be harmful to neurons. Investigations of Jin et al. showed that the neurotoxicity mediated by activated microglial is inhibited by the administration of interferon-β (IFNβ) to cortical neuronal and microglial cocultures (Jin et al. 2007). However, this prevention of microglial induced neuronal death seems to be due to the inhibition of a direct interaction between the microglia and the neurons, while the release of proinflammatory cytokines is still present (Jin et al. 2007). However, one has to consider different types of microglia activation, prompt stimulation of microglia after acute injury and the chronic overactivation of microglia after constant proinflammatory stimuli. Overactivation can be induced by environmental toxins and products released from neuronal death or damage as occurring in many neurodegenerative diseases. A pathological phenomenon called microgliosis, which describes the accumulation of activated microglia next to lesion sites, is seen in MS and several other neurodegenerative diseases, indicating that these cells might be involved in the disease process. It is not clear what conditions will lead to microglia activation or over-activation. However, there is more and more evidence suggesting that microglia–neuronal crosstalk is a key point in guiding microglia activation. Therefore, it is important to identify characteristics of microglia activation and over-activation in order to better understand the pathological mechanism of MS and other neurodegenerative diseases.

2.2 Microglial Pattern Recognition Receptors and Neurotoxicity

Pattern recognition receptors (PRR) are expressed on 'resting' microglia (Fig. 2). They are genetically conserved and bind to pathogen-associated molecular patterns (PAMPs), small molecular motifs consistently found on pathogens

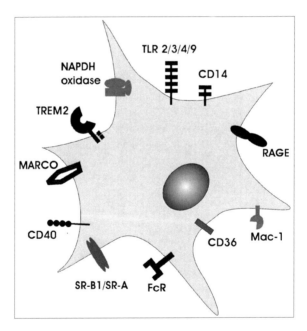

Fig. 2 Microglia express different classes of innate immune receptors which mediate either pro- or anti- inflammatory signalings and contribute to the activation state of microglia in MS. Toll-like receptors (TLR2/3/4/9), NOD-like receptors, RNA helicases, triggering receptors expressed on myeloid cells-2 (TREM2) and scavenger receptors B1 and A (SR-B1/SR-A) belong to the class of pattern recognition receptors, which mediate the immune response directed against invading pathogens, but also damage neurons by triggering microglial release of cytotoxic cytokines and reactive oxygen species (ROS). These receptors, although constitutively expressed on microglia, are up-regulated during EAE and/or MS and seem to be involved in disease onset and progression

(e.g., lipopolysaccharide LPS). The recognition of PAMPs is crucial in host defense to establish an innate immune response (Akira et al. 2006). Although the PRRs are receptors for microbes, they can also play a harmful role in many chronic autoimmune diseases as shown for Crohn's disease (Karin et al. 2006) and lupus-prone mice (Christensen et al. 2006). Three major types of PRRs exist in the human immune system: the class of membrane bound PRRs, which includes the toll-like receptors (TLRs) and the mannose receptor, the class of cytoplasmic PRRs with NOD-like receptors and RNA helicases, and the class of secreted PRRs including complement receptors and others.

In addition, there are also classes of PRRs that recognize endogenous molecules. These receptors have either an activatory or an inhibitory function and play a key role in self tolerance versus activation of immune responses. Members of these expanding families are the T cell immunoglobulin mucin (TIM) receptors and, possibly, the triggering receptors expressed on myeloid cells (TREM).

2.2.1 Toll-Like Receptors

The best studied class of PRRs is the membrane-bound TLR which belongs to the interleukin-1 receptor superfamily. So far, 11 TLRs have been identified in humans and 13 in mice, which recognize different molecules derived from bacteria, fungi and viruses. On microglia, TLRs 1–9 appear to be expressed, but the expression levels vary highly in different development stages and pathogen invasion circumstances. TLRs are also widely expressed on dendritic cells and macrophages. It is not clear whether TLRs are involved in MS pathology or not, although they have been shown to be upregulated in MS patients.

TLR4 is the best analyzed TLR which contributes to microglial activation. After binding of its ligand LPS, it activates nuclear factor-κB (NFκB) and induces massive transcription of proinflammatory genes via a MyD88-dependent signaling pathway (Walter et al. 2007). In addition to its major ligand LPS derived from bacteria, TLR4 appears to have several other endogenous ligands including gangliosides or sialic acid-containing glycosphingolipids of the neuronal membrane which are released from apoptotic neurons (Jou et al. 2006). After binding of these molecules, microglia are rapidly activated through a defined TLR4 signaling pathway (Jou et al. 2006). Interestingly, the activation of TLR4 on microglia can have both neurodestructive and neuroprotective effects. On one side, it induces neurodegeneration by release of proinflammatory molecules. On the other side, it can recruit oligodendrocyte progenitor cells (OPC) to lesion sites in MS and promote remyelination (Glezer et al. 2006).

By contrast, TLR2 on microglial cells has been less investigated, although its involvement in MS is suggested by a few studies (Zekki et al. 2002). TLR2 was first reported to be important in the immune response against herpes simplex virus (Aravalli et al. 2005). In addition, peptidoglycan (PGN), a major cell wall component of gram-positive bacteria, can also signal through TLR2. It was shown that PGN can be used as an adjuvant to induce experimental autoimmune encephalomyelitis

(EAE), the animal model of multiple sclerosis, by its capability of priming autoreactive T cells and triggering dendritic cell maturation (Visser et al. 2005).

TLR3 is a member of the second class of PRRs. It is located in endosomes in dendritic cells and microglia (Town et al. 2006). By binding to viral double stranded RNA or a synthetic analog of dsRNA, polyinosinic acid–cytidylic acid (Poly I:C), it can activate microglia (Kawai and Akira 2007). The TLR3 expression on microglia is upregulated in MS cases but no further involvement was described (Bsibsi et al. 2002).

A more important role in MS is possibly attributed to TLR9. This receptor recognizes single-stranded unmethylated CpG-motif rich DNA (CpG–DNA) mainly reflecting bacterial DNA. It was demonstrated in organotypic cultures that CpG–DNA can activate microglia in a neurotoxic manner, leading to strong microglia activation with the release of proinflammatory molecules including TNFα and NO which are toxic to the surrounding neurons (Iliev et al. 2004). TLR9 is upregulated in MS patients and EAE mice. Mice deficient in TLR9 or the adaptor protein MyD88 develop only a very mild form of EAE. Indeed, TLR9 activation seems to be crucial to the development of clinical symptoms in EAE, but the ligand activating this pathway is still under question since typical CpG motifs are missing in any destructed tissue released during EAE (Prinz et al. 2006).

In conclusion, the Toll-like receptors are essential components to induce innate immune response in the CNS. Multiple TLRs lead to microglia activation followed by neuronal damage, others are suggested to be involved in the disease progress of EAE.

2.2.2 NOD like Receptors and RNA Helicases

Besides the membrane associated TLRs, the nucleotide binding and oligomerization domain (NOD)-like receptors (NLRs) and the retinoid acid-inducible gene I (RIG-I)-like receptors (RLRs) are two newly identified classes of PRRs that are located in the cell cytoplasm. The NLRs consist of three domains: an N- terminal effector binding domain, an NOD domain and an array of C-terminal leucine rich repeat (LRR) motifs. Both of the two NLR proteins (NOD1 and NOD2) bind to peptidoglycan and it is suggested that the NLRs recognize peptidoglycan through the LRR motifs. Upon binding, they get oligomerized via the NOD domain and consequently activate the transcription factor NF-κB by interacting with serine–threonine kinase RICK through the effector domain (Inohara et al. 1999). The transcription of proinflammatory genes and genes involved in the innate and adaptive immune response are then enhanced (Kanneganti et al. 2007). It was demonstrated in demyelinating diseased primates that peptidoglycan can bind to extracellular TLR2 and intracellular NOD receptors of phagocytes (Visser et al. 2006). Activation of the NOD receptors in microglia might lead to release of proinflammatory cytokines and neurotoxic substances.

RIG-I and MDA5 are cytoplasmic RNA helicases that recognize double stranded viral RNA. The expression of RIG-I is induced by invading pathogens (e.g., viruses), retinoic acid and interferons. In addition to the TLRs, RIG-I shows an alternative antiviral function by inducing a strong immune response against viruses (Melchjorsen et al. 2005). However, it is still unclear whether RIG-I is involved in EAE or MS.

2.2.3   Triggering Receptors Expressed on Myeloid Cells

In addition to all the above discussed receptors that are necessary for the activation of immune cells, microglia and other immune cells also express inhibitory receptors which are essential for self tolerance in the immune system (Fig. 2). The inhibitory receptors are characterized by one or several immunoglobulin-like domains in their extracellular part and an immunoreceptor tyrosine-based inhibitory motif (ITIM) in their cytoplasmic domain which can recruit protein phosphatases and mediate inhibitory signals in the immune cells. The genes of the inhibitory receptors are often located in clusters where activating isoforms are also found (Trowsdale et al. 2001). In vertebrates the activating receptors differ from the inhibitory receptors by lacking the signaling domain in the cytoplasmic tail. Instead they have a positively charged transmembrane domain that binds to adaptor proteins carrying immunoreceptor tyrosine-based activating motifs (ITAM).

Triggering receptors expressed on myeloid cells-1 (TREM1) and TREM2 are two receptors of the immunoglobulin superfamily in the TREM cluster which signal through the ITAM-containing adaptor protein DNAX activating protein 12 (DAP12). Once associated with TREM, DAP12 can be phosphorylated in its ITAM region and dock the tyrosine kinase, spleen tyrosine kinase (SYK). This promotes the tyrosine phosphorylation of an adaptor complex containing Cbl and GRB2. Finally, the intracellular release of Ca^{2+} is enhanced, the actin cytoskeleton is rearranged, and the activation of other transcription complexes is also enhanced (McVicar et al. 1998). TREM1 is involved in the immune response against bacteria and fungi by inducing the release of proinflammatory factors from the phagocytes. Expression of TREM2 is detected on myeloid precursors, dendritic cells, osteoclasts and also on microglial cells in the CNS (Schmid et al. 2002). Recently, it was shown that activation of TREM2 stimulates phagocytosis in microglia and counter-regulates proinflammatory signals (Takahashi et al. 2005). TREM2 seems to be a key receptor for the phagocytosis of apoptotic neurons in the brain with immunosuppressive effect. The lack of TREM2 in microglia in vitro impairs phagocytosis of apoptotic neurons and enhances inflammation, which could lead to neuronal damage. Investigations also suggest that microglia deficient in TREM2 could play a major role in the pathogenesis of Nasu-Hakola disease, an autosomal recessive disease also called polycystic lipomembranous osteodysplasia with sclerosing leukoencephalopathy (PLOSL), which is associated with loss-of-function mutations in the DAP12 gene or the TREM2 gene. The phenotype of PLOSL is very heterogeneous with bone cysts and presenile dementia, but is only found in two organ systems: bone and brain. In the CNS, the disease is characterized by sclerosing leukoencephalopathy with demyelination, axonal loss and gliosis. In a recent study, TREM2 was detected on infiltrating macrophages in peripheral tissues and was shown to attenuate the immune response of macrophages against several TLR ligands (Turnbull et al. 2006).

Furthermore, bone-marrow derived myeloid precursors lentivirally transduced with TREM2 (TREM2 + BM-MC) were shown to be able to remove apoptotic brain cells and resolve inflammation in the CNS (Takahashi et al. 2007). Takahashi et al. showed that TREM2 + BM-MC might be a potential novel therapy approach to the animal model of MS. They demonstrated that after intravenous injection, the TREM2 + BM-MC migrated into the CNS, improved the disease symptoms and triggered prompt recovery from EAE (Takahashi et al. 2007). The cells migrated directly to the lesion sites showed an enhanced lysosomal activity and phagocytosed surrounding myelin debris. The reduction of TNFα together with the increase of interleukin-10 expression in the spinal cord lesion indicated that TREM2 promotes anti-inflammatory processes in EAE.

2.2.4 Scavenger Receptors

Scavenger receptors (SRs) are a class of structurally distinct receptors expressed on macrophages, endothelial cells and smooth muscle cells, which bind to acetylated low-density lipoprotein (acLDL) or other modified lipoproteins (Goldstein et al. 1979). Some of these receptors also appear to recognize other ligands like fibrillary amyloid-β or lipidic membranes of apoptotic cells which are negatively charged similar to the modified LDL. According to their structural similarities, the receptors are categorized into eight subclasses. In the CNS, SR-A and SR-B (including SR-BI and CD36) were detected on glial cells, where they seemed to be involved in phagocytosis of apoptotic neurons, uptake of denatured or modified proteins and lipoproteins, and homeostasis of the lipid metabolism (Husemann et al. 2002). On microglial cells, the expression of SRs is regulated developmentally. For example, high expression levels of CD36, SR-A and SR-BI are detected on cultured fetal microglia (Fig. 2), but in adult mice only low levels of CD36 and SR-A are expressed while SR-BI is undetectable. It was shown in Alzheimer Disease (AD) that SR-A is upregulated, but SR-A and SR-BI are downregulated after stimulation with interferon-γ (IFNγ) and TNFα (Buechler et al. 1999; Yamada et al. 1998). Fadok et al. observed that apoptotic cells losing membrane asymmetry and expressing membrane phosphatidylserine, were phagocytosed by macrophages (Fadok et al. 1992). On neonatal microglia SR-A seems to be a possible binding partner of phosphatidylserine and SR-BI might mediate the clearance of apoptotic cells by astrocytes in vivo (Husemann and Silverstein 2001). SR-A is upregulated in microglial cells that accumulated around lesion sites in AD and ischemia (Christie et al. 1996), strongly indicating that SR-A might be involved in the disease progression. SR-A was one of the first receptors shown to be involved in the binding and uptake of fibrillary amyloid-β (Aβ) protein by cultured microglia (Paresce et al. 1996).

Another important role in the CNS is assigned to the macrophage receptor with collagenous structure (MARCO) which belongs to the SR-A family and is expressed both in mice and humans. It can bind to gram-positive and gram-negative bacteria, but not to yeast (Kraal et al. 2000). It is constitutively expressed on macrophages in the marginal zone of the spleen and in the lymph node medullary cord (Fig. 2). A correlation between actin cytoskeleton rearrangement and MARCO expression is shown in LPS treated dendritic cells and GM–CSF treated microglia followed by a reduced phagocytic activity of these cells (Granucci et al. 2003). Alarćon et al claimed that MARCO is responsible for the uptake of Aβ by microglia and astrocytes according to the finding that the clearance of Aβ by neonatal rat microglia and astrocytes is mediated via fucoidan- and poly(I)-sensitive mechanisms (Alarcon et al. 2005).

Another multiligand receptor family is the receptors for advanced glycation end products (RAGE), whose ligands are mainly S100 and other β-sheet fibrils including the amyloid-β protein. RAGEs are widely expressed during embryonic development. In adults the expression is downregulated, but can be enhanced at inflammatory sites when the transcription factors AP-2 and NF-κB become activated. RAGE is detectable in microglia in the inflamed nervous tissue and in neurodegenerative diseases. It was shown that RAGE is involved in the activation of microglia through amyloid-β protein which binds to RAGE and leads to increased secretion of M–CSF (Lue et al. 2001).

2.3   Macrophage Antigen Complex-1 Receptor and Other Microglial Integrin-Associated Receptor Complexes

The macrophage antigen complex-1 (Mac-1) receptor, also known as complement receptor-3, is composed of CD11b and noncovalently associated CD18. It belongs to the leukocyte β2-integrin family that is constitutively expressed on monocytes and is constituted by the same CD18 β-chain and one out of three α-chains. For example, leukocyte function antigen (LFA1) is composed of CD11a and CD18, and complement receptor-4 is composed of CD11c and CD18. All these receptor complexes are intercellular adhesion molecules. Mac-1 is widely expressed on both phagocytic active cells and 'resting' microglial cells in the central nervous tissue (Fig. 2). It is also known as a pattern recognition receptor that recognizes the C3b on opsonized bacteria and takes part in the complement mediated phagocytosis of pathogens.

In MS, adhesion molecules are crucial for trafficking of invading effector cells into lesion sites (Archelos et al. 1999). Members of the β2-integrin family are involved in the development of EAE. During EAE, microglia and invading macrophages phagocytose the hydrolyzed myelin proteins. Microglial cells phagocytose myelin via Mac-1 accompanied by the production of inflammatory mediators and respiratory burst of the cells (van der Laan et al. 1996). Many reports indicate that Mac-1 also contributes to the trafficking of macrophages to lesion sites in the EAE model (Liu et al. 2002). However, new findings of Bullard et al reported that Mac-1 is required for the trafficking of T cells and macrophages in the chronic phase of EAE rather than in early EAE progression. Mac-1 seems to be essential for the leukocyte infiltration. In Mac-1 knock-out mice, fewer infiltrates were detected in the chronic phase of EAE than in control mice (Bullard et al. 2005). Reduced phagocytosis of myelin by microglia and macrophages was also observed in Mac-1 deficient mice.

CD11a/CD18, also called leukocyte common antigen 1 (LFA-1), is also found on microglial cells, and seems to bind to the ICAM-1 (Martin et al. 1987). Although the role of CD11a is not completely clear, there is evidence that it is involved in the microglial migration to lesion sites in the inflamed nervous tissue. In an ischemic model of organotypic slice cultures from rat, a CD11a-deficient microglial cell line, namely BV-2, was no longer able to migrate into inflamed lesion sites. This leads to the enhanced loss of neurons and no further neuroprotection in these regions as seen on control slices (Neumann et al. 2006).

All b2-integrin receptor complexes seem to be important in the migration and activation of microglia in the inflamed nervous tissue. CD11c is a marker specifically expressed on dendritic cells; the major antigen presenting cells of the immune system. After induction of primary demyelination in the corpus callosum of mice treated with cuprizone, microglia are activated and migrate to the lesion sites. Analysis of this microglial population showed that it is a heterogeneous population with an emerging subpopulation of CD11c positive microglia (Remington et al. 2007). This could indicate that a microglial subpopulation activated by a demyelinating lesion might present antigens to invading T cells.

2.4 Reactive Oxygen Species

In MS, microglial cells are often over-activated upon tissue damage and the presence of proinflammatory mediators. The production of reactive oxygen radicals and especially the extracellular superoxide as a result of microglia activation can further lead to neuronal damage and reactive microgliosis at lesion sites. Several cellular enzymes, including the mitochondrial oxidases, cytochrom P450, cyclooxygenases (COXs), nitric oxide synthase (NOS) and NADPH oxidase are involved in the production of reactive oxygen species (ROS) such as superoxide anion or hydrogen peroxide. In the CNS, a small amount of some ROS, like NO plays a special role as signaling intermediates to neurons. However, overproduction of NO by activated immune cells can lead to neuronal damage. The inducible nitric oxide synthase (iNOS, or NOS2) and NADPH oxidase are two very important enzymes contributing to the production of ROS in the CNS. The NADPH oxidase is involved in pathogen killing by phagocytic cells (El-Benna et al. 2005). It is detected in neurons, astrocytes and microglia in the CNS (Fig. 2). This membrane-located enzyme catalyzes oxygen into superoxide anion. It is inactive in the normal brain, but becomes activated under certain inflammatory stimuli (Babior 2000) and results in a massive efflux of superoxide. The NADPH oxidase contains three cytoplasmic subunits (p47, p67, and p40) and three membrane-bound subunits (p91, p22, and Rac-2). Upon activation, the cytoplasmic subunits translocate to the plasma membrane and form the functional enzyme together with the membrane-bound subunits. NADPH oxidase plays an important role in the production of ROS modulating neuronal excitation (Kishida et al. 2005) and is supposed to be the major source of neuronal superoxide, which is involved in triggering synaptic plasticity and memory formation (Kishida and Klann 2007). Superoxide-anions (O_2^-) generated by NADPH oxidase upon inflammation have a major role in mediating oligodendrocyte damage in MS. The LPS-induced oligodendrocyte death is mediated by microglial-derived peroxynitrite (ONOO⁻), which is a product of superoxide-anion and nitric oxide derived from iNOS. It was shown that NADPH oxidase is induced after LPS treatment to microglia and the activated microglial cells can mediate toxicity to oligodendrocytes in vitro (Li et al. 2005).

It is known that inducible NO synthase (iNOS; NOS2) is not expressed on astrocytes and microglia in the normal brain, but can be induced by inflammation and leads to enhanced release of NO. The iNOS is one of the possible key players of microglia that induces the inflammatory immune response in EAE. It is also indirectly involved in modulating the tonus of the brain's vasculature. In EAE and MS, ROS production is not always harmful; it could also play a beneficial role. In vitro, iNOS is induced in primary microglia by inflammatory cytokines like IFNγ, TNFα, and IL1β, which are detected at relevant levels in the plasma, the brain tissue and blood derived leukocyte cultures in MS patients (Hofman et al. 1989; Merrill 1992). In contrast, TGFβ inhibits the transcription and translation of iNOS (Vodovotz and Bogdan 1994). In vitro studies indicate that the phagocytosis of myelin by primary microglia leads to an early proinflammatory response, which is accompanied by the release of TNFα and IL1β. However, in the late phase the inflammatory reaction of the microglial cell is suppressed and antiinflammatory cytokines like interleukin-10 (IL10) are upregulated. Release of ROS by the microglial cells was identified as a key mechanism of switching the cell from proinflammatory to anti inflammatory reaction (Liu et al. 2006). The p47-phagocyte oxidase (PHOX), which is an adaptor molecule of the NADPH oxidase, is a key molecule involved in this process. It has been demonstrated that PHOX deficiency leads to less production of ROS in cells that express NADPH oxidase (Olofsson et al. 2003). Liu et al. further showed that knock-down of p47-PHOX during the in vitro myelin phagocytosis suppresses the anti-inflammatory response. In conclusion, ROS generation could also mediate a negative feedback-regulation to the inflammatory reaction of microglial cells during neuroinflammation.

Interleukin-4 plays an ambivalent role in regulation of NO production. In activated human macrophages, it was shown to upregulate NO production. While in spontaneously activated monocytes from allergic patients, NO release is decreased by interleukin-4 (Mautino et al. 1994). In further studies trying to verify the role of NO production in EAE, both isolated monocytes from the CNS and the periphery of EAE animals expressed high levels of iNOS and released high amounts of reactive oxygen radicals (MacMicking et al. 1992). A relatively high mRNA level of iNOS was also detected in EAE mice and other EAE rodents (Koprowski et al. 1993). All these findings indicate a central role of NO production in the disease progression of MS. Particularly, NO derived from activated microglia or other glial cells could be involved in the oligodendrocyte damage as seen in MS and might further enhance the neuronal death in the injured tissue. As already discussed above NO can efficiently react with the superoxide anion to form peroxynitrite, which is known as a reactive oxidant that can damage cell membranes. Recent studies showed that the released peroxynitrite and NO are toxic to oligodendrocytes and neurons by forming iron–NO complexes of iron-containing enzymes (Drapier and Hibbs 1988), oxidating sulphhydryl groups in enzymes (Radi et al. 1991), nitrating enzymes and inducing DNA strand breaks (Wink et al. 1991). Selective inhibition of ROS on the neural target tissue might be an appropriate therapy approach to MS. Intensive research is now focusing on small molecule inhibitors. To summarize, NADPH oxidase as well as iNOS play an important role in mediating neural toxicity in MS and are potential targets for therapy strategies.

3 Adaptive Immunity in Multiple Sclerosis

3.1 Antigen Presentation and Immune Cell Invasion

MS is an autoimmune disease characterized by demyelination and axonal loss maintained by auto-reactive T cells directed against myelin antigens. To prime the myelin-specific T cell response, naive CD4 + T cells have to interact antigen-specifically with APC, which present myelin antigens via MHC class II in conjunction with costimulatory molecules on their cell surface. In the CNS, myelin antigens can be principally presented via the MHC class II molecules by activated parenchymal microglia or perivascular macrophages, but not by the myelin containing oligodendrocytes themselves. 'Resting' microglia and perivascular macrophages appear to be relatively weak antigen presenting cells with almost no capacity to activate naive T lymphocytes. In vitro studies using fetal or neonatal primary microglia showed that these cultured microglial cells are capable of phagocytosing myelin and activating CD4 + myelin-specific T cells (Williams et al. 1994). However, studies on microglia isolated from adult rodents failed to demonstrate the priming capacity of CD4 + T cells (Havenith et al. 1998) except when they were stimulated with interferon-γ (IFNγ) (Carson et al. 1999). IFNγ treatment promotes differentiation of the naive T cell from the Th1 helper cells, but does not enhance T cell proliferation.

In secondary immune organs the major APC is the dendritic cell (DC). Upon activation through PAMPs, which bind to TLRs, the DC upregulates costimulatory molecules and secretes many cytokines and chemokines to further recruit CD4 + and CD8 + T cells. Typical DCs are not detected within the CNS parenchyma, but cells expressing dendritic specific molecules are located in the Virchow–Robin space together with perivascular macrophages, where myelin antigens can be presented to invading CD4 + T cells (McMahon et al. 2005). These 'CNS-associated' dendritic-like cells seem to fully prime and further activate CD4 + T cells in a manner similar to lymphoid or splenic APCs. Interestingly, it was shown that myelin specific CD4 + T cells transplanted intravenously were found in secondary lymphoid organs before CNS infiltration indicating that the priming phase takes place in the secondary immune organs instead of in the CNS. Myelin antigen-containing APCs were also observed in the cervical lymph nodes in EAE and MS (de Vos et al. 2002). This indicates that the myelin antigens stay within the CNS and do not spread to the periphery, whereas myelin-containing APCs migrate to the lymph nodes after uptake of myelin (de Vos et al. 2002). Greter et al. suggested that only CNS-associated dendritic-like cells, but not resident APC of the brain parenchyma are necessary for EAE development and immune attack to the CNS (Greter et al. 2005). In their investigations, no CNS infiltration and disease development were observed after transfer of myelin oligodendrocyte glycoprotein (MOG)-specific T cells in bone-marrow chimeric mice with CNS- restricted MHC class II expression. But bone-marrow chimeric mice, whose MHC class II expression was restricted to the systemic immune system, were permissive for disease development (Greter et al. 2005). They also showed that DCs alone were able to reactivate already primed T cells in vivo and that no other MHC class II expressing cell types were essential for this reactivation. However, all studies with bone marrow chimeras are difficult to interpret due to irradiation damage leading to recruitment of immune cells (Mildner et al. 2007; Ransohoff 2007).

Thus the question of which role the microglial cell plays versus that played by the invaded marcrophages or 'CNS-associated' dendrite-like cells in the pathogenesis of EAE, is still open. As already discussed above, there are several different pathways through which the microglial cell can be activated and change from resting into an activated phenotype that is prone to induce neuronal damage. In MS the activated microglia upregulate costimulatory molecules which are essential for the interaction with invading CD4 + T cells including B7.2 (CD86), B7.1 (CD80), MHC class II and CD40 (van Kooten and Banchereau 2000). Early in 1996, scientists found that in lesion sites of MS patients and EAE animals, activated T cells expressing CD40 ligand are in close contact with CD40 expressing cells. The CD40 expressing cells are further characterized by CD11b expression as monocytes or microglia rather than dendritic cells. To verify the role of CD40–CD40L interaction in EAE, animals were treated with a monoclonal CD40 antibody during disease induction and later on during the disease course (Gerritse et al. 1996). Treatment of the animals with the antibody during the induction phase resulted in no or only slight development of clinical symptoms. Administration of the antibody after the onset of the disease led to blockade of the disease in 67–80% of all the animals (Gerritse et al. 1996). Further studies showed that CD40 –/– and CD40L –/– mice did not develop EAE (Becher et al. 2001). Mice lacking CD40 expression specifically on microglial cells displayed an ameliorated disease course. Analysis of the activation stage of microglia at the onset of the disease and during the course revealed a two step mechanism of microglial cell activation. Microglia can be activated without CD40 interaction which is characterized by upregulation of activation markers like CD45, but this activation is incomplete and insufficient to induce T cell infiltration and proliferation as seen in wildtype EAE mice (Becher et al. 2006). Under the stimulation of CD40, microglia released interleukin-12 (IL12), which stimulated T cell proliferation. The importance of CD40–CD40L interaction in microglia activation and T cell function indicates that CD40/CD40L might be a new target as therapy approach for MS. There is evidence that the CD40 stimulation induced by LPS in primary microglia can be inhibited by the administration of interleukin-10 (IL10) (Qin et al. 2006). The invading CD4 + T cells in the CNS did not further differentiate into effector T cell except they were reactivated by recognizing their cognate antigen (Qin et al. 2006). Microglia can present myelin antigens to the already primed invading CD4 + T cells via the MHC class II that are recognized by T cell receptor (Qin et al. 2006).

In summary, the microglial cell is not the initial APC for priming and full activation of invading T cells, but seems to be crucial in the establishment of an inflammatory milieu and the maintenance of the auto-reactive T cells in the CNS. Microglia guide primed CD4 + T cells having entered the CNS to the inflamed parts by releasing a set of chemokines and establish an inflammatory milieu through the production of proinflammatory cytokines.

3.2 Cytotoxic CD8 + T Cells

Most data on CD8 + T cells in CNS autoimmunity are derived from the EAE model which is primarily mediated by CD4 + encephalitogenic T cells and characterized by demyelination and axonal loss at different stages, as well as blood–brain barrier permeability and inflamed lesion sites in the CNS. EAE can be induced in animals by immunization using myelin antigens like myelin-glycoprotein (MOG) and a suitable adjuvant. The immunization leads to a strong CD4 + T cell response in the CNS, mainly at the inflammatory lesion sites (Zamvil and Steinman 1990). For many years, CD8 + T cells were thought to play a minor role in the EAE model. Cytotoxic CD8 + T cells develop into cytotoxic effector T cells (CTL) after recognizing their specific target antigens presented on the MHC class I complex in conjunction with a costimulatory signal. Then the CTL is ready to attack a target cell after secondary antigen recognition (Fig. 3). CTLs can act on the target cell through three different effector pathways. Firstly, they can secrete cytotoxic granules containing perforin and serine proteases by exocytosis. The perforins can insert into the plasma membrane of the target cell by forming pores which enable the serine proteases to enter the cell. In the target cell cytoplasm the serine proteases cleave precursor caspases which then initiate an apoptosis pathway. The second pathway of effector function is mediated through the CD95–CD95L interaction with the target cell. CD95 is expressed on CTLs which can bind to CD95L of the target cell after recognizing target cells antigen-specifically via MHC class I (Siegel et al. 2000). CD95L signaling is then leading to the activation of apoptosis pathways (Siegel et al. 2000). The third effector pathway is the secretion of cytokines belonging to the TNF family. Binding of TNFα to its receptor TNF receptor I activates the receptor's death domain and initiates an apoptosis pathway.

Both CD4 + and CD8 + T cells appear to be involved in MS disease pathogenesis since depletion of CD4 + T cells in MS patients by a monoclonal anti CD4 antibody failed to ameliorate the disease progress (Lindsey et al. 1994). Furthermore, CD8 + T cell populations were found in the inflammatory lesion sites of MS patients (Traugott 1983). In contrast to the CD4 + T cells, the CD8 + CTLs invaded deeply into the parenchyma of the inflamed CNS, whereas the CD4 + T cells preferentially stayed in the perivascular cuffs and close to meninges (Kawakami et al. 2005). A direct comparison of the number of CD4 + T cells and CD8 + T cells

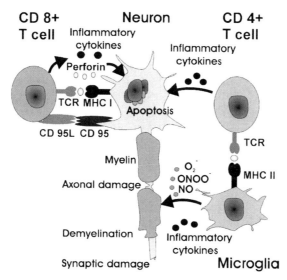

Fig. 3 Indirect and direct neurotoxicity mediated by invading T cells in the inflamed CNS. Invading CD8 + T cells are recognizing myelin antigens presented on MHC class I molecules on oligodendrocytes, but possibly also other auto-antigens presented by neurons. Principally, CD8 + cytotoxic T lymphocyte (CTL) could attack the targeted neuron directly through three different effector pathways, first, secretion of cytotoxic cytokines such as tumor necrosis factor-α (TNFα), second, CD95/CD95L interaction and third, release of cytotoxic granules such as perforin and serine proteases. All three cytotoxic pathways can induce apoptosis in the targeted neuron. CD4 + T cells are unable to directly interact with neurons via their T cell receptor (TCR) since neurons do not express MHC class II molecules. Myelin specific CD4 + T cells are recognizing myelin antigens presented by microglial cells via their MHC class II molecules. Through this interaction microglia become activated and start to secrete reactive oxygen species (ROS) including NO, ONOO⁻ and O₂⁻ as well as pro-inflammatory cytokines. ROS can impair axonal and synaptic function and integrity

in MS lesion sites revealed that CD8 + T cells invaded the lesions tenfold more as compared to CD4 + T cells (Booss et al. 1983). Through the characterization of CD4 + and CD8 + T cell clones with single cell PCR in lesion sites of MS patients, Babbe et al. demonstrated that preferentially CD8 + T cells appear to be clonally expanded within the lesion (Babbe et al. 2000).

In vitro, CD8 + T cells showed the capacity to differentiate to myelin-specific cytotoxic T cells (Huseby et al. 2001). Early in 1989, Grenier et al demonstrated that oligodendrocytes express only MHC class I but not MHC class II on their surface even after stimulation with IFNγ (Grenier et al. 1989), indicating that myelin antigens presented via MHC class I on oligodendrocytes can be recognized by myelin specific CD8 + T cells. Thus, MHC class I expressing oligodendrocytes can be directly attacked by CD8 + cytotoxic T cells (Fig. 3).

The toxicity mediated by CD4 + T cells is most probably not a direct attack against the oligodendrocytes. The CD4 + T cells recruit macrophages to the inflamed CNS which then release proinflammatory cytokines and toxic molecules such as NO (Fig. 3). All neural cell types and microglia are able to express MHC class I after stimulation. In the healthy CNS parenchyma, where CD8 + T cells are very low in number or undetectable, MHC class I expression is also low or undetectable. In vitro, glial cells express low levels of MHC class I antigens. Cultures of primary microglial cells are lysed by CD8 + CTLs, when they present myelin peptides on MHC class I. However, the antigen presentation capacity appears to be low even after IFNγ stimulation, indicating that microglia are not able to efficiently present endogenous antigens on MHC class I (Bergmann et al. 1999). Oligodendrocytes induced by inflammatory cytokines to express MHC class I molecules could present myelin antigens which are then potential targets of myelin-specific CD8 + CTLs. Astrocytes are also lysed by CTLs in vitro and can be targeted through both pathways, either cytotoxic granules or Fas (CD95)–FasL (CD95L) interaction (Medana et al. 2001).

Neurons do not express classical MHC class I molecules under normal conditions. However, isolated hippocampal neurons can be induced to de novo transcribe MHC class I genes, if they were electrically silenced (Neumann et al. 1995). MHC class I cell surface expression was induced in cultured hippocampal neurons under stimulation by IFNγ, when they were electrically paralyzed with tetrodotoxin (Neumann et al. 1995, 1997). Experiments on the effector mechanism of CD8 + CTLs directed against neurons lead to the conclusion that neurons, particularly their processes, are susceptible to cytotoxic granules derived from CTL. It was observed that neurons are susceptible to the direct application of cytotoxic granules and the neuritic processes were transected by CTLs in an antigen-restricted fashion, but their cell bodies were resistant to perforin-mediated lysis by CTLs due to the expression of FasL (CD95L) (Medana et al. 2001, 2000). However, the expression of FasL (CD95L) failed to protect the neurons against the Fas (CD95)- mediated apoptosis (Medana et al. 2001, 2000). Investigation on the relevance of CTLs in EAE by Huseby et al. identified myelin antigen specific CD8 + T cells in the secondary immune organs of wild-type C3H mice after immunization. They isolated and activated these cells before injecting them intravenously to recipient mice to show that myelin-specific CD8 + T cells have an effector function in CNS autoimmunity. Mice that received the adoptively transferred cytotoxic CD8 + T cells, developed severe neurological disease with symptoms quite different from classical EAE, while transfer of CD4 + T cells led to classical EAE in these mice (Huseby et al. 2001).

4 Conclusion

Under physiological conditions the CNS is immune privileged. This status is lost after tissue injury or microbial infections, which result in activation of microglia and invasion of immune cells from the periphery into the CNS. The transmigration of immune cells through the blood–brain barrier is essential to fight against CNS-invading pathogens. However in EAE, the animal model of MS, auto-immune myelin-specific T-lymphocytes pass the blood–brain barrier and lead to the disease onset and progression. During EAE and MS, a full blown adaptive immune response becomes involved. The neuronal damage seen in MS is not uniformly mediated by one immune cell type but seems to be a multifactor event. Activated autoreactive CD8 + T cells as well as CD4 + T cells pass the blood–brain barrier and enter the CNS. In MS, the key players of the disease progression appear to be the myelin specific autoreactive CD4 + T cells which have developed in the periphery and escaped tolerance induction. In contrast, the executor of acute demyelination and axonal injury might be the cytotoxic CD8 + T cells. Cytotoxic T cells could directly attack oligodendrocytes, as well as possibly neurons. Neuronal damage is also caused indirectly by CD4 + T cells and their proinflammatory cytokines. Activated microglia and macrophages are involved in mediating neurotoxicity by guiding and activating T cells in the CNS and establishing a proinflammatory milieu in the CNS. Furthermore, activated microglia and macrophages contribute to chronic neurodegeneration. Activated microglia and invaded macrophages produce several reactive oxygen species including NO, which impair axonal and synaptic function and integrity. Identification of the key molecular pathways of immune-mediated neuronal damage could help to find novel targets for therapies of MS.

Acknowledgments The group of H.N. is supported by the Hertie Foundation, the Rose Foundation, the Deutsche Forschungsgemeinschaft and the European Union (LSHM-CT-2005-018637). The Neural Regeneration Group at the University Hospital Bonn is supported by the Hertie-Foundation, the Walter-und-Ilse-Rose-Foundation, the DFG (KFO177, SFB704) and the EU (LSHM-CT-2005-018637).

References

Akira S, Uematsu S, Takeuchi O (2006) Pathogen recognition and innate immunity. Cell 124:783–801

Alarcon R, Fuenzalida C, Santibanez M, von Bernhardi R (2005) Expression of scavenger receptors in glial cells. Comparing the adhesion of astrocytes and microglia from neonatal rats to surface-bound beta-amyloid. J Biol Chem 280:30406–30415

Alt C, Laschinger M, Engelhardt B (2002) Functional expression of the lymphoid chemokines CCL19 (ELC) and CCL 21 (SLC) at the blood-brain barrier suggests their involvement in G-protein-dependent lymphocyte recruitment into the central nervous system during experimental autoimmune encephalomyelitis. Eur J Immunol 32:2133–2144

Aravalli RN, Hu S, Rowen TN, Palmquist JM, Lokensgard JR (2005) Cutting edge: TLR2-mediated proinflammatory cytokine and chemokine production by microglial cells in response to herpes simplex virus. J Immunol 175:4189–4193

Archelos JJ, Previtali SC, Hartung HP (1999) The role of integrins in immune-mediated diseases of the nervous system. Trends Neurosci 22:30–38

Babbe H, Roers A, Waisman A, Lassmann H, Goebels N, Hohlfeld R, Friese M, Schroder R, Deckert M, Schmidt S et al. (2000) Clonal expansions of CD8(+) T cells dominate the T cell infiltrate in active multiple sclerosis lesions as shown by micromanipulation and single cell polymerase chain reaction. J Exp Med 192:393–404

Babior BM (2000) Phagocytes and oxidative stress. Am J Med 109:33–44

Becher B, Durell BG, Miga AV, Hickey WF, Noelle RJ (2001) The clinical course of experimental autoimmune encephalomyelitis and inflammation is controlled by the expression of CD40 within the central nervous system. J Exp Med 193:967–974

Becher B, Bechmann I, Greter M (2006) Antigen presentation in autoimmunity and CNS inflammation: how T lymphocytes recognize the brain. J Mol Med 84:532–543

Bergmann CC, Yao Q, Stohlman SA (1999) Microglia exhibit clonal variability in eliciting cytotoxic T lymphocyte responses independent of class I expression. Cell Immunol 198:44–53

Block ML, Zecca L, Hong JS (2007) Microglia-mediated neurotoxicity: uncovering the molecular mechanisms. Nat Rev Neurosci 8:57–69

Booss J, Esiri MM, Tourtellotte WW, Mason DY (1983) Immunohistological analysis of T lymphocyte subsets in the central nervous system in chronic progressive multiple sclerosis. J Neurol Sci 62:219–232

Bsibsi M, Ravid R, Gveric D, van Noort JM (2002) Broad expression of Toll-like receptors in the human central nervous system. J Neuropathol Exp Neurol 61:1013–1021

Buechler C, Ritter M, Quoc CD, Agildere A, Schmitz G (1999) Lipopolysaccharide inhibits the expression of the scavenger receptor Cla-1 in human monocytes and macrophages. Biochem Biophys Res Commun 262:251–254

Bullard DC, Hu X, Schoeb TR, Axtell RC, Raman C, Barnum SR (2005) Critical requirement of CD11b (Mac-1) on T cells and accessory cells for development of experimental autoimmune encephalomyelitis. J Immunol 175:6327–6333

Campbell JJ, Bowman EP, Murphy K, Youngman KR, Siani MA, Thompson DA, Wu L, Zlotnik A, Butcher EC (1998) 6-C-kine (SLC), a lymphocyte adhesion-triggering chemokine expressed by high endothelium, is an agonist for the MIP-3beta receptor CCR7. J Cell Biol 141:1053–1059

Carson MJ, Sutcliffe JG, Campbell IL (1999) Microglia stimulate naive T-cell differentiation without stimulating T-cell proliferation. J Neurosci Res 55:127–134

Cho BP, Song DY, Sugama S, Shin DH, Shimizu Y, Kim SS, Kim YS, Joh TH (2006) Pathological dynamics of activated microglia following medial forebrain bundle transection. Glia 53:92–102

Christensen SR, Shupe J, Nickerson K, Kashgarian M, Flavell RA, Shlomchik MJ (2006) Toll-like receptor 7 and TLR9 dictate autoantibody specificity and have opposing inflammatory and regulatory roles in a murine model of lupus. Immunity 25:417–428

Christie RH, Freeman M, Hyman BT (1996) Expression of the macrophage scavenger receptor, a multifunctional lipoprotein receptor, in microglia associated with senile plaques in Alzheimer's disease. Am J Pathol 148:399–403

Colton CA, Gilbert DL (1987) Production of superoxide anions by a CNS macrophage, the microglia. FEBS Lett 223:284–288

Davalos D, Grutzendler J, Yang G, Kim JV, Zuo Y, Jung S, Littman DR, Dustin ML, Gan WB (2005) ATP mediates rapid microglial response to local brain injury in vivo. Nat Neurosci 8:752–758

de Vos AF, van Meurs M, Brok HP, Boven LA, Hintzen RQ, van der Valk P, Ravid R, Rensing S, Boon L, t Hart BA, Laman JD (2002) Transfer of central nervous system autoantigens and presentation in secondary lymphoid organs. J Immunol 169:5415–5423

Drapier JC, Hibbs JB Jr (1988) Differentiation of murine macrophages to express nonspecific cytotoxicity for tumor cells results in L-arginine-dependent inhibition of mitochondrial iron-sulfur enzymes in the macrophage effector cells. J Immunol 140:2829–2838

El-Benna J, Dang PM, Gougerot-Pocidalo MA, Elbim C (2005) Phagocyte NADPH oxidase: a multicomponent enzyme essential for host defenses. Arch Immunol Ther Exp (Warsz) 53:199–206

Fadok VA, Voelker DR, Campbell PA, Cohen JJ, Bratton DL, Henson PM (1992) Exposure of phosphatidylserine on the surface of apoptotic lymphocytes triggers specific recognition and removal by macrophages. J Immunol 148:2207–2216

Fetler L, Amigorena S (2005) Neuroscience. Brain under surveillance: the microglia patrol. Science 309:392–393

Gerritse K, Laman JD, Noelle RJ, Aruffo A, Ledbetter JA, Boersma WJ, Claassen E (1996) CD40-CD40 ligand interactions in experimental allergic encephalomyelitis and multiple sclerosis. Proc Natl Acad Sci U S A 93:2499–2504

Glezer I, Lapointe A, Rivest S (2006) Innate immunity triggers oligodendrocyte progenitor reactivity and confines damages to brain injuries. FASEB J 20:750–752

Gold R, Schmied M, Tontsch U, Hartung HP, Wekerle H, Toyka KV, Lassmann H (1996) Antigen presentation by astrocytes primes rat T lymphocytes for apoptotic cell death. A model for T-cell apoptosis in vivo. Brain 119(Pt 2):651–659

Goldstein JL, Brown MS, Krieger M, Anderson RG, Mintz B (1979) Demonstration of low density lipoprotein receptors in mouse teratocarcinoma stem cells and description of a method for producing receptor-deficient mutant mice. Proc Natl Acad Sci U S A 76:2843–2847

Granucci F, Petralia F, Urbano M, Citterio S, Di Tota F, Santambrogio L, Ricciardi-Castagnoli P (2003) The scavenger receptor MARCO mediates cytoskeleton rearrangements in dendritic cells and microglia. Blood 102:2940–2947

Grenier Y, Ruijs TC, Robitaille Y, Olivier A, Antel JP (1989) Immunohistochemical studies of adult human glial cells. J Neuroimmunol 21:103–115

Greter M, Heppner FL, Lemos MP, Odermatt BM, Goebels N, Laufer T, Noelle RJ, Becher B (2005) Dendritic cells permit immune invasion of the CNS in an animal model of multiple sclerosis. Nat Med 11:328–334

Harry GJ, McPherson CA, Wine RN, Atkinson K, Lefebvre d'Hellencourt C (2004) Trimethyltin-induced neurogenesis in the murine hippocampus. Neurotox Res 5:623–627

Havenith CE, Askew D, Walker WS (1998) Mouse resident microglia: isolation and characterization of immunoregulatory properties with naive CD4 + and CD8 + T-cells. Glia 22:348–359

Hofman FM, Hinton DR, Johnson K, Merrill JE (1989) Tumor necrosis factor identified in multiple sclerosis brain. J Exp Med 170:607–612

Huber JD, Egleton, RD, Davis, TP (2001) Molecular physiology and pathophysiology of tight junctions in the blood-brain barrier. Trends Neurosci 24:719–725

Huseby ES, Liggitt D, Brabb T, Schnabel B, Ohlen C, Goverman J (2001) A pathogenic role for myelin-specific CD8(+) T cells in a model for multiple sclerosis. J Exp Med 194:669–676

Husemann J, Silverstein SC (2001) Expression of scavenger receptor class B, type I, by astrocytes and vascular smooth muscle cells in normal adult mouse and human brain and in Alzheimer's disease brain. Am J Pathol 158:825–832

Husemann J, Loike JD, Anankov R, Febbraio M, Silverstein SC (2002) Scavenger receptors in neurobiology and neuropathology: their role on microglia and other cells of the nervous system. Glia 40:195–205

Iliev AI, Stringaris AK, Nau R, Neumann H (2004) Neuronal injury mediated via stimulation of microglial toll-like receptor-9 (TLR9). FASEB J 18:412–414

Inohara N, Koseki T, del Peso L, Hu Y, Yee C, Chen S, Carrio R, Merino J, Liu D, Ni J, Nunez G (1999) Nod1, an Apaf-1-like activator of caspase-9 and nuclear factor-kappaB. J Biol Chem 274:14560–14567

Jin S, Kawanokuchi J, Mizuno T, Wang J, Sonobe Y, Takeuchi H, Suzumura A (2007) Interferon-beta is neuroprotective against the toxicity induced by activated microglia. Brain Res 1179:140–146

Jou I, Lee JH, Park SY, Yoon HJ, Joe EH, Park EJ (2006) Gangliosides trigger inflammatory responses via TLR4 in brain glia. Am J Pathol 168:1619–1630

Kanneganti TD, Lamkanfi M, Nunez G (2007) Intracellular NOD-like receptors in host defense and disease. Immunity 27:549–559

Karin M, Lawrence T, Nizet V (2006) Innate immunity gone awry: linking microbial infections to chronic inflammation and cancer. Cell 124:823–835

Kawai T, Akira S (2007) Antiviral signaling through pattern recognition receptors. J Biochem 141:137–145

Kawakami N, Nagerl UV, Odoardi F, Bonhoeffer T, Wekerle H, Flugel A (2005) Live imaging of effector cell trafficking and autoantigen recognition within the unfolding autoimmune encephalomyelitis lesion. J Exp Med 201:1805–1814

Kishida KT, Klann E (2007) Sources and targets of reactive oxygen species in synaptic plasticity and memory. Antioxid Redox Signal 9:233–244

Kishida KT, Pao M, Holland SM, Klann E (2005) NADPH oxidase is required for NMDA receptor-dependent activation of ERK in hippocampal area CA1. J Neurochem 94:299–306

Koprowski H, Zheng YM, Heber-Katz E, Fraser N, Rorke L, Fu ZF, Hanlon C, Dietzschold B (1993) In vivo expression of inducible nitric oxide synthase in experimentally induced neurologic diseases. Proc Natl Acad Sci U S A 90:3024–3027

Kraal G, van der Laan LJ, Elomaa O, Tryggvason K (2000) The macrophage receptor MARCO. Microbes Infect 2:313–316

Lee SC, Liu W, Dickson DW, Brosnan CF, Berman JW (1993) Cytokine production by human fetal microglia and astrocytes. Differential induction by lipopolysaccharide and IL-1 beta. J Immunol 150:2659–2667

Li J, Baud O, Vartanian T, Volpe JJ, Rosenberg PA (2005) Peroxynitrite generated by inducible nitric oxide synthase and NADPH oxidase mediates microglial toxicity to oligodendrocytes. Proc Natl Acad Sci U S A 102:9936–9941

Liao H, Bu WY, Wang TH, Ahmed S, Xiao ZC (2005) Tenascin-R plays a role in neuroprotection via its distinct domains that coordinate to modulate the microglia function. J Biol Chem 280:8316–8323

Lindsey JW, Hodgkinson S, Mehta R, Siegel RC, Mitchell DJ, Lim M, Piercy C, Tram T, Dorfman L, Enzmann D et-al. (1994) Phase 1 clinical trial of chimeric monoclonal anti-CD4 antibody in multiple sclerosis. Neurology 44:413–419

Liu B, Gao HM, Wang JY, Jeohn GH, Cooper CL, Hong JS (2002) Role of nitric oxide in inflammation-mediated neurodegeneration. Ann N Y Acad Sci 962:318–331

Liu Y, Hao W, Letiembre M, Walter S, Kulanga M, Neumann H, Fassbender K (2006) Suppression of microglial inflammatory activity by myelin phagocytosis: role of p47-PHOX-mediated generation of reactive oxygen species. J Neurosci 26:12904–12913

Lowenstein PR (2002) Immunology of viral-vector-mediated gene transfer into the brain: an evolutionary and developmental perspective. Trends Immunol 23:23–30

Lue LF, Walker DG, Brachova L, Beach TG, Rogers J, Schmidt AM, Stern DM, Yan SD (2001) Involvement of microglial receptor for advanced glycation endproducts (RAGE) in Alzheimer's disease: identification of a cellular activation mechanism. Exp Neurol 171:29–45

MacMicking JD, Willenborg DO, Weidemann MJ, Rockett KA, Cowden WB (1992) Elevated secretion of reactive nitrogen and oxygen intermediates by inflammatory leukocytes in hyperacute experimental autoimmune encephalomyelitis: enhancement by the soluble products of encephalitogenic T cells. J Exp Med 176:303–307

Marin-Teva JL, Dusart I, Colin C, Gervais A, van Rooijen N, Mallat M (2004) Microglia promote the death of developing Purkinje cells. Neuron 41:535–547

Martin DE, Chiu FJ, Gigli I, Muller-Eberhard HJ (1987) Killing of human melanoma cells by the membrane attack complex of human complement as a function of its molecular composition. J Clin Invest 80:226–233

Mautino G, Paul-Eugene N, Chanez P, Vignola AM, Kolb JP, Bousquet J, Dugas B (1994) Heterogeneous spontaneous and interleukin-4-induced nitric oxide production by human monocytes. J Leukoc Biol 56:15–20

McMahon EJ, Bailey SL, Castenada CV, Waldner H, Miller SD (2005) Epitope spreading initiates in the CNS in two mouse models of multiple sclerosis. Nat Med 11:335–339

McVicar DW, Taylor LS, Gosselin P, Willette-Brown J, Mikhael AI, Geahlen RL, Nakamura MC, Linnemeyer P, Seaman WE, Anderson SK et al. (1998) DAP12-mediated signal transduction in natural killer cells. A dominant role for the Syk protein-tyrosine kinase. J Biol Chem 273:32934–32942

Medana IM, Gallimore A, Oxenius A, Martinic MM, Wekerle H, Neumann H (2000) MHC class I-restricted killing of neurons by virus-specific CD8 + T lymphocytes is effected through the Fas/FasL, but not the perforin pathway. Eur J Immunol 30:3623–3633

Medana I, Li Z, Flugel A, Tschopp J, Wekerle H, Neumann H (2001) Fas ligand (CD95L) protects neurons against perforin-mediated T lymphocyte cytotoxicity. J Immunol 167:674–681

Melchjorsen J, Jensen SB, Malmgaard L, Rasmussen SB, Weber F, Bowie AG, Matikainen S, Paludan SR (2005) Activation of innate defense against a paramyxovirus is mediated by RIG-I and TLR7 and TLR8 in a cell-type-specific manner. J Virol 79:12944–12951

Merrill JE (1991) Proinflammatory and antiinflammatory cytokines in multiple sclerosis and central nervous system acquired immunodeficiency syndrome. J Immunother 12:167–170

Mildner A, Schmidt H, Nitsche M, Merkler D, Hanisch UK, Mack M, Heikenwalder M, Bruck W, Priller J, Prinz M (2007) Microglia in the adult brain arise from Ly-6ChiCCR2 + monocytes only under defined host conditions. Nat Neurosci 10:1544–1553

Morgan SC, Taylor DL, Pocock JM (2004) Microglia release activators of neuronal proliferation mediated by activation of mitogen-activated protein kinase, phosphatidylinositol-3-kinase/Akt and delta-Notch signalling cascades. J Neurochem 90:89–101

Muller FJ, Snyder EY, Loring JF (2006) Gene therapy: can neural stem cells deliver? Nat Rev Neurosci 7:75–84

Neumann H (2001) Control of glial immune function by neurons. Glia 36:191–199

Neumann H, Cavalie A, Jenne DE, Wekerle H (1995) Induction of MHC class I genes in neurons. Science 269:549–552

Neumann H, Schmidt H, Cavalie A, Jenne D, Wekerle H (1997) Major histocompatibility complex (MHC) class I gene expression in single neurons of the central nervous system: differential regulation by interferon (IFN)-γ and tumor necrosis factor (TNF)-α. J Exp Med 185:305–316

Neumann J, Gunzer M, Gutzeit HO, Ullrich O, Reymann KG, Dinkel K (2006) Microglia provide neuroprotection after ischemia. FASEB J 20:714–716

Nimmerjahn A, Kirchhoff F, Helmchen F (2005) Resting microglial cells are highly dynamic surveillants of brain parenchyma in vivo. Science 308:1314–1318

Oehmichen W, Gencic M (1975) Experimental studies on kinetics and functions of monuclear phagozytes of the central nervous system. Acta Neuropathol Suppl (Suppl 6):285–290

Olofsson P, Holmberg J, Tordsson J, Lu S, Akerstrom B, Holmdahl R (2003) Positional identification of Ncf1 as a gene that regulates arthritis severity in rats. Nat Genet 33:25–32

Paresce DM, Ghosh RN, Maxfield FR (1996) Microglial cells internalize aggregates of the Alzheimer's disease amyloid beta-protein via a scavenger receptor. Neuron 17:553–565

Prinz M, Garbe F, Schmidt H, Mildner A, Gutcher I, Wolter K, Piesche M, Schroers R, Weiss E, Kirschning CJ et al. (2006) Innate immunity mediated by TLR9 modulates pathogenicity in an animal model of multiple sclerosis. J Clin Invest 116:456–464

Qin H, Wilson CA, Roberts KL, Baker BJ, Zhao X, Benveniste EN (2006) IL-10 inhibits lipopolysaccharide-induced CD40 gene expression through induction of suppressor of cytokine signaling-3. J Immunol 177:7761–7771

Radi R, Beckman JS, Bush KM, Freeman BA (1991) Peroxynitrite oxidation of sulfhydryls. The cytotoxic potential of superoxide and nitric oxide. J Biol Chem 266:4244–4250

Raivich G (2005) Like cops on the beat: the active role of resting microglia. Trends Neurosci 28:571–573

Ransohoff RM (2007) Microgliosis: the questions shape the answers. Nat Neurosci 10:1507–1509

Remington LT, Babcock AA, Zehntner SP, Owens T (2007) Microglial recruitment, activation, and proliferation in response to primary demyelination. Am J Pathol 170:1713–1724

Sawada M, Kondo N, Suzumura A, Marunouchi T (1989) Production of tumor necrosis factor-alpha by microglia and astrocytes in culture. Brain Res 491:394–397

Schmid CD, Sautkulis LN, Danielson PE, Cooper J, Hasel KW, Hilbush BS, Sutcliffe JG, Carson MJ (2002) Heterogeneous expression of the triggering receptor expressed on myeloid cells-2 on adult murine microglia. J Neurochem 83:1309–1320

Siegel RM, Chan FK, Chun HJ, Lenardo MJ (2000) The multifaceted role of Fas signaling in immune cell homeostasis and autoimmunity. Nat Immunol 1:469–474

Streit WJ (2002) Microglia as neuroprotective, immunocompetent cells of the CNS. Glia 40:133–139

Takahashi K, Rochford CD, Neumann H (2005) Clearance of apoptotic neurons without inflammation by microglial triggering receptor expressed on myeloid cells-2. J Exp Med 201:647–657

Takahashi K, Prinz M, Stagi M, Chechneva O, Neumann H (2007) TREM2-transduced myeloid precursors mediate nervous tissue debris clearance and facilitate recovery in an animal model of multiple sclerosis. PLoS Med 4:e124

Town T, Jeng D, Alexopoulou L, Tan J, Flavell RA (2006) Microglia recognize double-stranded RNA via TLR3. J Immunol 176:3804–3812

Traugott U (1983) Acute experimental autoimmune encephalomyelitis. Differences between T cell subsets in the blood and meningeal infiltrates in susceptible and resistant strains of guinea pigs. J Neurol Sci 61:81–91

Trowsdale J, Barten R, Haude A, Stewart CA, Beck S, Wilson MJ (2001) The genomic context of natural killer receptor extended gene families. Immunol Rev 181:20–38

Turnbull IR, Gilfillan S, Cella M, Aoshi T, Miller M, Piccio L, Hernandez M, Colonna M (2006) Cutting edge: TREM-2 attenuates macrophage activation. J Immunol 177:3520–3524

van der Laan LJ, Ruuls SR, Weber KS, Lodder IJ, Dopp EA, Dijkstra CD (1996) Macrophage phagocytosis of myelin in vitro determined by flow cytometry: phagocytosis is mediated by CR3 and induces production of tumor necrosis factor-alpha and nitric oxide. J Neuroimmunol 70:145–152

van Kooten C, Banchereau J (2000) CD40-CD40 ligand. J Leukoc Biol 67:2–17

Visser L, Jan de Heer H, Boven LA, van Riel D, van Meurs M, Melief MJ, Zahringer U, van Strijp J, Lambrecht BN, Nieuwenhuis EE, Laman JD (2005) Proinflammatory bacterial peptidoglycan as a cofactor for the development of central nervous system autoimmune disease. J Immunol 174:808–816

Visser L, Melief MJ, van Riel D, van Meurs M, Sick EA, Inamura S, Bajramovic JJ, Amor S, Hintzen RQ, Boven LA et al. (2006) Phagocytes containing a disease-promoting Toll-like receptor/Nod ligand are present in the brain during demyelinating disease in primates. Am J Pathol 169:1671–1685

Vodovotz Y, Bogdan C (1994) Control of nitric oxide synthase expression by transforming growth factor-beta: implications for homeostasis. Prog Growth Factor Res 5:341–351

Walter S, Letiembre M, Liu Y, Heine H, Penke B, Hao W, Bode B, Manietta N, Walter J, Schulz-Schuffer W, Fassbender K (2007) Role of the toll-like receptor 4 in neuroinflammation in Alzheimer's disease. Cell Physiol Biochem 20:947–956

Walton NM, Sutter BM, Laywell ED, Levkoff LH, Kearns SM, Marshall GP II, Scheffler B, Steindler DA (2006) Microglia instruct subventricular zone neurogenesis. Glia 54:815–825

Williams K, Ulvestad E, Waage A, Antel JP, McLaurin J (1994) Activation of adult human derived microglia by myelin phagocytosis in vitro. J Neurosci Res 38:433–443

Wink DA, Kasprzak KS, Maragos CM, Elespuru RK, Misra M, Dunams TM, Cebula TA, Koch WH, Andrews AW, Allen JS et al. (1991) DNA deaminating ability and genotoxicity of nitric oxide and its progenitors. Science 254:1001–1003

Yamada Y, Doi T, Hamakubo T, Kodama T (1998) Scavenger receptor family proteins: roles for atherosclerosis, host defence and disorders of the central nervous system. Cell Mol Life Sci 54:628–640

Zamvil SS, Steinman L (1990) The T lymphocyte in experimental allergic encephalomyelitis. Annu Rev Immunol 8:579–621

Zekki H, Feinstein DL, Rivest S (2002) The clinical course of experimental autoimmune encephalomyelitis is associated with a profound and sustained transcriptional activation of the genes encoding toll-like receptor 2 and CD14 in the mouse CNS. Brain Pathol 12:308–319

Ziv Y, Ron N, Butovsky O, Landa G, Sudai E, Greenberg N, Cohen H, Kipnis J, Schwartz M (2006) Immune cells contribute to the maintenance of neurogenesis and spatial learning abilities in adulthood. Nat Neurosci 9:268–275

Neuro-Immune Crosstalk in CNS Diseases

Martin Kerschensteiner, Edgar Meinl, and Reinhard Hohlfeld

Abstract Immune cells infiltrate the central nervous system (CNS) in many neurological diseases, with a primary or secondary inflammatory component. In the CNS, immune cells employ shared mediators to promote crosstalk with neuronal cells. The net effect of this neuro-immune crosstalk critically depends on the context of the interaction. It has long been established that inflammatory reactions in the CNS can cause or augment tissue injury in many experimental paradigms. However, emerging evidence suggests that in other paradigms inflammatory cells can contribute to neuroprotection and repair. This dual role of CNS inflammation is also reflected on the molecular level as it is becoming increasingly clear that immune cells can release both neurodestructive and neuroprotective molecules into CNS lesions. It is thus the balance between destructive and protective factors that ultimately determines the net result of the neuro-immune interaction.

1 Introduction

Immune cells enter the nervous systems not only during autoimmune diseases like multiple sclerosis but also as a part of the physiological response to tissue damage after stroke or trauma, and during chronic neurodegenerative diseases. The interplay between the infiltrating immune cells and the resident cells of the central nervous system (CNS) is, however, complex, because a growing body of evidence suggests that inflammation in the CNS can, in addition to well-documented neurotoxic effects, also convey neuroprotection (Rapalino et al. 1998; Moalem et al. 1999; Serpe et al. 1999; Hammarberg et al. 2000). A possible molecular explanation for these findings is that the nervous and the immune systems

M. Kerschensteiner (✉), E. Meinl, and R. Hohlfeld
Institute of Clinical Neuroimmunology, Ludwig-Maximilians University Munich,
Marchioninistr, 17, 81377 Munich, Germany
e-mail: Martin.Kerschensteiner@med.uni-muenchen.de

Results Probl Cell Differ, DOI 10.1007/400_2009_6
© Springer-Verlag Berlin Heidelberg 2009

engage in an intense crosstalk (Kerschensteiner et al. 2003). Both systems produce a range of factors (e.g., cytokines or chemokines for the immune system and neurotrophic factors for the CNS) that modulate cell growth and differentiation. Interestingly, the factors expressed by immune and nervous system cells overlap, and neither cytokines nor neurotrophic factors are completely exclusive to either systems (Kerschensteiner et al. 2003).

In this review, we analyze the role of neuro-immune crosstalk in neurological diseases with presumed "primary" or "secondary" inflammation of the CNS. We first describe the molecules that can mediate the communication between immune cells and neurons and then assess experimental evidence for functional neurodestructive and neuroprotective effects of immune cells. Finally, we explore the implications of neuro-immune crosstalk for the treatment of neurological diseases.

2 Molecular Aspects of Neuro-Immune Crosstalk

The communication between infiltrating immune cells and their neuronal counterparts is likely based on shared molecules with overlapping expression pattern in immune and nervous systems (see Fig. 1). Indeed, many of these molecules also have shared effects in both the immune and the nervous systems (examples are listed in Table 1). In the following paragraphs, we want to highlight the examples of these shared mediators that can convey signals from immune cells to neurons or vice versa.

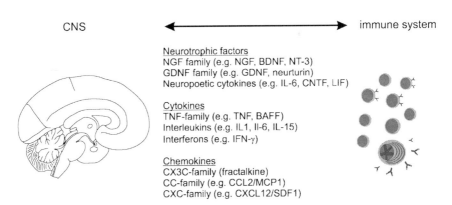

Fig. 1 Molecular mediators of neuro-immune crosstalk. Molecular mediators that are shared between the immune and nervous system likely mediate neuro-immune crosstalk. Neurotrophic factors, cytokines, and chemokines are among the molecules that can mediate bidirectional signaling between immune cells and the resident cells of the nervous system

Table 1 Molecular interactions between the immune and nervous system

Neurotrophic factors	Effects on the immune system	References
NGF	Acts as a survival factor for memory B cells	Torcia et al. (1996)
	Influences the production of Th1 and Th2 cytokines	Villoslada et al. (2000)
	Promotes the maturation of dendritic cells	Jiang et al. (2007)
BDNF	Increases survival of thymocyte precursors	Maroder et al. (1996)
	Influences cytokine production by T cells	Bayas et al. (2003)
NT3	Downregulates MHC II expression on microglia	Neumann et al. (1998)
GDNF	Reduces TNF-α release by activated peripheral blood mononuclear cells	Vargas-Leal et al. (2005)
Neurturin	Reduces TNF-α release by activated peripheral blood mononuclear cells	Vargas-Leal et al. (2005)
IL 6	Multiple immune effects in acute and chronic inflammation	Kishimoto (2005)
CNTF	Reduces CNS inflammation in animal model of multiple sclerosis	Kuhlmann et al. (2006)
LIF	Required for IFN-γ production of T cells	Linker et al. (2008)
Cytokines	**Effects on the nervous system**	**References**
TNF-α	Inhibits neurite outgrowth and branching	Neumann et al. (2002)
	Inhibits hippocampal long-term potentiation	Butler et al. (2004)
IL 1beta	Can depolarize and excite neurons	Xia et al. (1999)
	Enhances neuronal NMDA receptor phosphorylation	Zhang et al. (2008)
IL 18	Stimulates synaptic release of glutamate	Kallo et al. (2004)
	Inhibits hippocampal long-term potentiation	Cumiskey et al. (2007)
IFN-γ	Induces MHC I expression in neurons	Neumann et al. (1997)
	Promotes neuronal differentiation of adult neuronal stem cells	Wong et al. (2004)
Chemokines	**Effects on the nervous system**	**References**
Fractalkine	Regulates neuronal migration and adhesion	Lauro et al. (2006)
	Modulates glutamatergic neurotransmission	Ragozzino et al. (2006)
CCL 2/MCP 1	Enhances excitability of nociceptive neurons	Sun et al. (2006)
	Activates the p38 MAPK pathway in neurons	Cho and Gruol (2008)
CXCL 12/SDF 1	Acts as chemoattractant for embryonic neurons	Zhu et al. (2002)
	Regulates GABAergic inputs to neuronal progenitors	Bhattacharyya et al. (2008)
	Promotes axonal elongation and pathfinding	Ohshima et al. (2008)

2.1 Neurotrophic Factors

Neurotrophic factors comprise a family of proteins that are essential for the development of the CNS in vertebrates. Neurotrophic action is pleiotropic; in addition to mediating neuronal survival, neurotrophic factors regulate a host of neuronal and glial cell activities, including axonal and dendritic growth, synaptic structure and plasticity, neurotransmitter expression, and long-term potentiation (Lewin and Barde 1996; Thoenen and Sendtner 2002; Nagappan et al. 2008). Furthermore, many neurotrophic factors can influence immune cell functions, such as migration, activation, differentiation, and local antigen presentation (Torcia et al. 1996;1998 Neumann et al. ; Villoslada et al. 2000; Flügel et al. 2001). Neurotrophic factors have always held high promise for the treatment of neurodegenerative conditions. Although the initial enthusiasm has been replaced with more a realistic perspective, these molecules still hold enormous potential for various therapeutic applications (Thoenen and Sendtner 2002). In addition, there has been much interest in the protective role of endogenously produced neurotrophic factors, and particularly, in the hypothetical neuroprotective potential of immune-cell-derived neurotrophic factors (see below).

Three families of neurotrophic factors have been characterized and they include (1) the NGF-related neurotrophic factors, called "neurotrophins," namely NGF, BDNF, neurotrophin-3 (NT-3), and neurotrophin-4/5 (NT-4/5); (2) the glial cell line-derived neurotrophic factor (GDNF) family ligands, namely GDNF, neurturin, artemin, and persephin; (3) the neuropoietic cytokines, such as ciliary neurotrophic factor and leukemia inhibitory factor; and (4) miscellaneous other factors that can also exert neurotrophic effects.

BDNF. BDNF is a member of the NGF-related neurotrophin family. It plays an essential role in neuronal plasticity and survival and also regulates neurotransmitter release and dendritic growth (Lewin and Barde 1996; Nagappan et al. 2008). Further, BDNF can prevent neuroaxonal damage in animal models following various pathologic insults and injuries (Thoenen and Sendtner 2002). These actions affect key cell populations, including sensory, cerebellar, and spinal neurons. BDNF binds preferentially to TrkB receptor (gp145trkB), which is expressed on neuronal cells. In addition, BDNF may also regulate astrocyte functions via a truncated form of TrkB (Rose et al. 2003). Like other neurotrophins, BDNF also binds to the p75 neurotrophin receptor. Although neurons were considered to be the main cellular source of BDNF, work from our group and others have demonstrated that various immune cells secrete BDNF in vitro (Besser and Wank 1999; Braun et al. 1999; Kerschensteiner et al. 1999; Moalem et al. 2000). Specifically, BDNF expression is increased following antigen stimulation in T helper (Th)1 and Th2 CD4+ cell lines specific for myelin autoantigens such as myelin basic protein (MBP) and myelin oligodendrocyte glycoprotein (Kerschensteiner et al. 1999; Ziemssen et al. 2002). The neurotrophin was bioactive, as it supported neuronal survival in vitro. Moreover, inflammatory cells in brain lesions and in perivascular locations in patients with acute disseminated encephalomyelitis and multiple sclerosis also expressed BDNF (Kerschensteiner et al. 1999; Stadelmann et al. 2002). Notably, lesion areas with high numbers of demyelinating macrophages showed enhanced BDNF immunoreactivity. Subsequently, Stadelmann and colleagues found that the BDNF receptor gp145trkB is also found in neurons adjacent to multiple sclerosis lesions and in reactive astrocytes within the plaques (Stadelmann et al. 2002). While it is well established that BDNF can protect CNS neurons, it is important to note that BDNF can also affect neuronal function, in particular synaptic plasticity (McAllister et al. 1999; Gartner et al. 2006; Arancio and Chao 2007; Naggapan et al. 2008). It is thus possible that immune-cell-derived BDNF might, on the one hand, improve neuronal survival, while, on the other hand, it alters the functional connectivity of neurons and thereby might even contribute to cognitive disturbances. The complexity of neurotrophin signaling is further increased by the existence of neurotrophin precursor proteins. These precursor proteins, known as pro-neurotrophins, are cleaved proteolytically to produce mature BDNF, NGF, NT3, and NT4 (Lu et al. 2005). Although mature neurotrophins bind with high affinity to cognate Trk receptors to foster neuronal cell survival, pro-neurotrophins bind preferentially to the p75 neurotrophin receptor to promote cell death. The predisposition to produce both pro- and anti-apoptotic responses has been described as the "yin and yang" of neurotrophin action (Lu et al. 2005). In this concept, the particular effect depends on the form of the activated neurotrophin (pro- or mature) and the class of targeted receptor (Trk or p75). In addition to BDNF, other members of the neurotrophin family such as NGF, Neurotrophin 3, and Neurotrophin4/5 are also widely expressed in the immune system (Ehrhard et al. 1993; Torcia et al. 1996; Besser and Wank 1999; Moalem et al. 2000). Members of the neurotrophin family are thus likely to contribute to neuro-immune communication, and due to their dual properties may contribute to both neuroprotective and neurodestructive effects of inflammation. The complex effects of neurotrophins are highlighted by recent studies that have started to address the contribution of individual immune-cell-derived neurotrophic factors in autoimmune CNS disease using conditional gene knock-out mice. Emerging insights suggest, for example, that protective effects of BDNF in the CNS may be counter-balanced by immunomodulatory effects of the same neurotrophin in the periphery (Linker et al. 2007).

GDNF. Ligands of the GDNF family (GFL), GDNF, neurturin, artemin, and persephin, promote central and peripheral neuronal growth and differentiation (Ariaksinen and Saarma 2002). GDNF protects dopaminergic neurons in animal models of Parkinson's disease and promotes motor neuron survival in vivo (Tomac et al. 1995; Wang et al. 2002). This factor also has a number of important nonneural functions, which range from regulating kidney development to mediating spermatogonial differentiation (Ariaksinen and Saarma 2002). Neuroprotective effects in animal models of ischemia have also been shown for the GFL Persephin (Tomac et al. 2002). GFLs signal through the RET ("rearranged during transfection" proto-oncogene) receptor tyrosine kinase. To activate this pathway, GFLs must first link with a second protein class, the GDNF family receptor-α (GFRα) receptors, which in turn bind to the RET plasma membrane. Four such GFRα receptors that determine the ligand specificity have been characterized: GDNF binds to GFRα1, neurturin to GFRα2, artemin to GFRα3, and persephin to GFRα4. In addition to these high-affinity interactions, more promiscuous bindings between GFLs and GFRαs have been observed (Ariaksinen and Saarma 2002). Like the NGF family ligands, GFLs also function in the immune system and are expressed by immune cells. Vargas-Leal et al. reported that human CD4$^+$ and CD8$^+$ T cells, B cells, and monocytes express neurturin transcript and protein (Vargas-Leal et al. 2005). These immune cells also express RET and GFRα2, allowing the formation of the GFRα2–RET complex. The addition of GDNF or neurturin to activated peripheral blood mononuclear cells reduced the amount of detectable tumor necrosis factor (TNF) protein without altering its transcription (Vargas-Leal et al. 2005). These findings suggest that intercellular communication between immune cell populations may be mediated by neurturin.

2.2 Cytokines

Over the last years, a large number of studies have shown that cytokines that were primarily known to mediate communication between different cell types within the immune system are expressed by neurons and glial cells as well. Expression in the nervous system has now been described for cytokines of many families, including interleukins, interferons, and members of the tumor necrosis (TNF) family (Neumann et al. 1997; Knoblach et al. 1999; Krumbholz et al. 2005; Liu et al. 2005).

BAFF. TNF family members regulate many aspects of immune cell survival, differentiation, and effector function. Their secretion in the CNS could thus provide important signals to infiltrating immune cells. One particularly interesting example is the TNF family member BAFF (B cell-activating factor of the tumor necrosis factor family), which is required for peripheral B-cell survival and homeostasis (Kalled 2006). It was long believed that immune cells such as monocytes, macrophages, and neutrophils were the only source of BAFF. However, Krumbholz and colleagues found that BAFF is also expressed in the normal human brain and its production by reactive astrocytes could foster B-cell survival in multiple sclerosis (Krumbholz et al. 2005). Specifically, the level of BAFF expression in normal human brain was approximately 10% of the level observed in lymphatic tissues, for example, tonsils and adenoids. Immunohistochemical analysis of CNS tissue revealed astrocytes as a major source of CNS-derived BAFF. In multiple sclerosis plaques, BAFF expression was strongly upregulated to levels observed in lymphatic tissues (Krumbholz et al. 2005). BAFF was localized in astrocytes close to BAFF-R-expressing immune cells, suggesting that astrocyte-derived BAFF can directly act on BAFF-R expressing B cells. Furthermore, in vitro studies showed that stimulation of cultured human astrocytes with interferon-gamma and TNF-alpha induced the secretion of functionally active BAFF via a furin-like protease-dependent pathway. In these experiments, BAFF secretion per cell was substantially higher in activated astrocytes than in monocytes and macrophages. Taken together, these observations identified astrocytes as a major "nonimmune" source of BAFF. It is tempting to speculate that CNS-derived BAFF supports B-cell survival in and around inflammatory lesions. This mechanism would offer one plausible explanation for the notorious persistence of the intrathecal oligoclonal immunoglobulin response in MS patients (Meinl et al. 2006).

2.3 Chemokines

Chemokines guide the migration of immune cells throughout the body and are central to the immigration of immune cells into the CNS. The expression of chemokines in the nervous system would thus allow resident CNS cells to directly influence the immune infiltration of their environment. In addition to regulating immune cell trafficking and activation, chemokines play an essential role in CNS development and during demyelination and remyelination: The CXCR2/CXCL1 system controls positioning of oligodendrocyte precursor cells (Tsai et al. 2002), and CXC chemokine receptors have been detected on human oligodendrocytes in situ (Omari et al. 2005). CXCR4 and its ligand CXCL12 regulate neuronal migration and axonal pathfinding during development, and in the mature CNS, this system helps to maintain homeostasis (Li et al. 2008). In fact, on the basis of evolutionary studies, it has been hypothesized that CXC chemokines originate from the CNS (Huising et al. 2003).

Fractalkine. Fractalkine or CX3CL1 is the only member of the δ-family of chemokines and the exclusive ligand for the chemokine receptor CX3CR1 (Bazan et al. 1997; Imai et al. 1997). Fractalkine and its receptor show a peculiar expression pattern: while fractalkine is predominantly expressed in CNS neurons, its receptor CX3CR1 is primarily expressed by monocytes and dendritic cells in the immune system and in the microglial cells in the CNS (Harrison et al. 1998). This expression pattern strongly suggests that fractalkine can relay neuronal signals to the immune system. Indeed a recent study by Cardona et al. demonstrates that fractalkine is a crucial regulator of inflammatory neurotoxicity in the brain (Cardona et al. 2006). The authors use CX3CR1-deficient mice to investigate how the lack of fractalkine receptor expression affects microglial function. The induction of systemic inflammation by peripheral injections of lipopolysaccharide in mice lacking the CX3CR1 receptor leads to a massive activation of microglial cells and subsequently to neuronal cell death. Interestingly, the increased activation of microglial cells in these mice appears to be a cell-autonomous process and can be transferred by the injection of CX3CR1$^{-/-}$ microglia into the cortex of the wild-type recipients. In line with these finding, CX3CR-deficient mice also show increased microglial neurotoxcity in animal models of Parkinson disease and amyotrophic lateral sclerosis. In summary, these results suggest that neurons can regulate the activation of microglial cells through the release of fractalkine. The constitutive expression of neuronal fractalkine in the CNS indicates that it provides a tonic signal that limits microglial activation in the healthy nervous system. To what extend fractalkine signaling can also contribute to the control of autoimmune inflammation is not yet fully understood. It seems, however, that the neuronal expression of fractalkine is unaltered both in multiple sclerosis and in its animal model EAE (Hulshof et al. 2003; Sunnemark et al. 2005). The functional role of fractalkine signaling in EAE was recently studied using CX3CR1-deficient mice (Huang et al. 2006). Interestingly CXCR1-deficiency did affect the recruitment of NK cells into the CNS, but did not affect the recruitment of macrophages. The impaired infiltration of NK cells was associated with increased disease severity, suggesting a regulatory role for NK cells in this scenario.

Further studies will be necessary to fully understand how fractalkine signaling regulates CNS inflammation. With regard to the dual nature of immune responses in the brain, it is interesting to note that fractalkine-deficient mice which show an increased activation of microglial cells are partially protected from focal ischemic stroke (Soriano et al. 2002).

CXCL12. CXCL12 has essential functions in both the immune system and the CNS. Its secretion by bone marrow stroma cells is important during embryogenesis for the colonization of bone marrow by fetal liver-derived hematopoietic stem cells. In adult life, it plays an essential role in retention and homing of these cells into the marrow microenvironment (Kucia et al. 2004). In secondary lymphatic organs, CXCL12 synergizes with CXCL13 in the organization of the dark and light zone of germinal centers (Allen et al. 2004). In the adult brain, CXCL12 is constitutively expressed (Krumbholz et al. 2006) and there it modulates neurotoxicity, neurotransmission, and neuroglial interactions (Li and Ransohoff 2008). The expression of this chemokine is up-regulated in both active and inactive MS lesions and could be localized to astrocytes (Krumbholz et al. 2006). Since CXCL12 is a strong survival factor for plasma cells (Manz et al. 2005), the high expression of this chemokine in MS lesions has been linked to the long-term survival of plasma cells in MS brains (Meinl et al. 2006). This chemokine may also be involved in neurodegeneration in different disease: CXCL12 is converted to a neurotoxic mediator by different metalloproteinases such as MMP2 and has been implicated in AIDS related dementia (Zhang et al. 2003; Vergote et al. 2006). As both CXCL12 and metalloproteinases that can transfer it to its neurotoxic form are induced in MS lesions, it is possible that cleaved CXCL12 contributes to tissue destruction in MS lesions (Krumbholz et al. 2006).

CCL19. This chemokine shares its receptor CCR7 with CCL21 and this system is crucial for immune cell migration to lymphatic tissue. Remarkably, CCL19 is constitutively expressed in the adult CNS, as seen by in-situ hybridization in rats (Alt et al. 2002) and PCR, ELISA, and CSF analysis in humans (Krumbholz et al. 2007). This chemokine is further upregulated in both active and inactive MS lesions and its CSF level correlates with intrathecal Ig production (Krumbholz et al. 2007). In the adult CNS, CCL19 might play a role in immunosurveillance, as it was localized to endothelial cells (Alt et al. 2002) and basically all T cells in the CSF express the receptor for this chemokine (which is CCR7, Kivisäkk et al. 2003).

In summary, the growing list of shared molecules between the immune and nervous system illustrates the complexity of neuro-immune crosstalk and suggests that the result of this cross-talk is not determined by the single molecules or even single classes of molecules, but rather by the integration of multiple signals that individually may favor either destruction or repair.

3 Functional Aspects of Neuro-Immune Crosstalk

The production of both neurotoxic and neuroprotective mediators by immune cells supports the notion that neuro-immune crosstalk can in principle have both beneficial as well as destructive consequences. While the destructive capability of neuroinflammation has long been documented, evidence for its beneficial properties has emerged more recently. Experimental evidence for "protective autoimmunity" has been provided by the work of Schwartz and colleagues, who showed that T cells specific to a CNS self-antigen, such as myelin basic protein (MBP), can protect damaged neurons from secondary degeneration (Moalem et al. 1999). Although anti-MBP T cells are encephalitogenic in rodent animals models, these cells are also found in the immune systems of healthy subjects (Hohlfeld and Wekerle 2004). To determine whether accumulating T cells exert a beneficial or deleterious effect following axonal injury, Moalem and colleagues injected anti-MBP T cells into rats that experienced injury to the optic nerve (Moalem et al. 1999). Compared with control rats, anti-MBP-injected rats maintained more than twice as many retinal ganglion cells with functional axons. Electrophysiological analysis suggested that the neuroprotective effect of the anti-MBP cell clones was due to a transient reduction in energy requirements, which caused a temporary state of inactivity in the damaged nerve. However, the investigators also considered other explanations, such as the expression of growth factors by anti-MBP cells, which has been shown, for example, for members of the NGF and GDNF neurotrophin families (Besser and Wank 1999; Kerschensteiner et al. 1999; Moalem et al. 2000; Vargas-Leal et al. 2005). Subsequent studies, mainly in Lewis rats with spinal cord injury (SCI), provided further support for the neuroprotective role of CNS autoreactive T lymphocytes (Hauben et al. 2000; Yoles et al. 2001). In addition, Hammarberg et al. reported that T and natural killer (NK) cells in the spinal cord of rats with EAE produced BDNF, NT-3, and GDNF, and can reduce the extent of neuronal injury after ventral root avulsion (Hammarberg et al. 2000).They also found that bystander recruited NK and T cells displayed comparable or increased neurotrophic factor levels, compared with the anti-MBP T cell populations. Recently, Ziv et al. showed that hippocampal neurogenesis could be restored in mice with SCI by the transfer of CNS-reactive T cells (Ziv et al. 2006). These regulatory T-lymphocytes were also necessary for the completion of spatial learning and memory tasks and, notably, for BDNF expression in the dentate gyrus.

It is important to note, however, that a series of studies by Jones and colleagues challenged the notion that autoreactive T lymphocytes can minimize neuronal and glial cell death following CNS injury (Jones et al. 2002; Jones et al. 2004). These authors reported that TCR transgenic mice in which more than 95% of CD4[+] T cells are reactive to MBP experienced impaired recovery of locomotor and reflex function after spinal cord injury compared with non-transgenic mice. This impairment correlated with aggravated demyelination and axonal loss, along with the increased expression of pro-inflammatory cytokines (Jones et al. 2002). Likewise, the immunization of rats with MBP enhanced tissue damage and functional disability after spinal cord injury (Jones et al. 2004). The investigators were unable to observe neuroprotection or functional improvement in MBP-immunized rats and thus concluded that myelin-reactive T cells are pathologic effector cells that impair recovery and exacerbate tissue injury at and beyond trauma sites.

The discrepancies between theses studies and previous work illustrate an important point, namely that the net effect of inflammation crucially depends on the respective experimental setting. Immunostimulation in one setting may lead to net

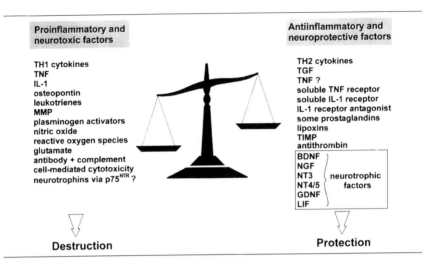

Fig. 2 The neuro-immune balance. Immune cells produce both mediators that can mediate neurodestruction as well as neuroprotection. The net effect of inflammation in a particular pathologic situation is determined by which process outweighs the other (Adapted, with permission, from Kerschensteiner et al. (2003; p 299)

neuroprotection while immunostimulation in a slightly different setting or genetic background that favors an encephalitogenic immune response leads to net neurodestruction. This notion should also caution us about the use of immunostimulatory treatment strategies for human CNS disease, which are often heterogenous and affect an outbred population. A concept that can reconcile these discrepancies is illustrated by the "neuro-immune balance" (Kerschensteiner et al. 2003, see Fig. 2). This concept emphasizes that immune cells can in principle release both neuroprotective and neurodestructive mediators. The net effect of the immune reaction – the position of the balance – can vary with the nature as well the particular stage of the underlying disease process.

4 Implications for Inflammatory CNS Disease

4.1 CNS Disease with Primary Inflammation

We have argued that CNS inflammation is, in essence, an ambivalent process that can have both protective and destructive effects. In multiple sclerosis, several lines of evidence indicate that the net effect of CNS inflammation is destructive at least in the acute stage of the disease. Animal models of multiple sclerosis like experimental autoimmune encephalomyelitis (EAE) illustrate that an inflammatory reaction against myelin targets can lead to the formation of destructive CNS lesions, which reflect the pathological hallmarks of multiple sclerosis (Storch et al. 1998; Kornek et al. 2000). As immune reactivity against myelin proteins – albeit certainly more diverse than in the EAE model – can also be detected in MS patients (Hohlfeld and Wekerle 2004) it is likely that anti-myelin autoimmunity contributes to tissue destruction in multiple sclerosis. Experimental studies further suggest that antigen-independent activation of immune cells, in particular macrophages, can lead to CNS damage (Newman et al. 2001). In line with these findings, the histopathological analysis of MS lesions reveals close contacts between infiltrating immune cells and damaged axons (Trapp et al. 1998). Furthermore, the extent of axonal damage in a given lesion correlates with the number of both macrophages and CD8+ T cells present in these lesions (Bitsch et al. 2000). One of the strongest arguments supporting a damaging role of the immune response in MS is based on the undisputable beneficial effects of various immunomodulatory and immunosuppressive therapies (Kieseier et al. 2007). The most direct evidence is probably that the inhibition of immune cell infiltration in the CNS using the anti-integrin antibody natalizumab dramatically reduces the number of relapses and limits disease progression (Polman et al. 2006). This clinical benefit is paralleled by a reduction of apparent tissue damage as assessed by MRI measurements.

Although it thus seems established that CNS inflammation is destructive in the acute stages of multiple sclerosis, the net effect of the inflammation in the chronic disease stages is more difficult to assess. In particular, as recent studies suggest, that neurodegenerative, and not the neuroinflammatory, mechanisms drive disease progression in the advanced stages of multiple sclerosis (Bjartmar and Trapp 2001; Hauser and Oksenberg 2006). Histopathological and neuroimaging studies support the view that axon damage in the chronic disease stages evolves with a slow constant progression compatible with a degenerative disease mechanism (Lossefff et al. 1996; Bjartmar et al. 2000; Kornek et al. 2000). Likewise the clinical course of multiple sclerosis follows a more uniform progression pattern in the chronic compared with the acute stages of the disease (Confavreux et al. 2000). A predominance of a degenerative disease mechanism would indeed explain why many of the established immunomodulatory or immunosuppressive therapies fail in the chronic stages of the disease (Killestein and Polman 2005; Miller and Leary, 2007). At present, however, it is worth noting that no definitive evidence for a predominant neurodegenerative disease mechanism in multiple sclerosis has been provided. Recently, an alternative hypothesis for the disease progression in chronic multiple sclerosis has been provided by the group of Hans Lassmann (Kutzelnigg et al. 2005). The concept of "smoldering inflammation" implies that CNS inflammation is driving the disease process in chronic as well as in acute stages of multiple sclerosis. However, the type of CNS inflammation changes as the disease progresses. Although focal demyelinated lesions densely infiltrated by macrophages and T cells dominate the histopathological presentation of acute multiple sclerosis, chronic MS is primarily characterized by a disseminated activation of microglial cells, which lead to widespread progression of axon damage. It is thought that the immune reaction in the CNS is increasingly secluded from the peripheral immune system ("compartmentalization" of the inflammatory process). This model might explain why many immunomodulatory and immunosuppressive therapies fail during chronic stages of the disease, despite the fact that the inflammatory process is still active. It is of obvious importance for the evolution of MS therapy to clarify to which extent neurodegeneration or neuroinflammation are responsible for the devastating progression of axon loss and clinical disability in chronic MS. If indeed neurodegeneration dominates chronic MS, one may argue that it is worthwhile to explore whether endogenous neuroprotective components of CNS inflammation can be exploited therapeutically.

4.2 CNS Disease with Secondary Inflammation

Damage to the CNS caused, for example, by trauma or ischemia is generally followed by a pronounced inflammatory response that is dominated by granulocytes in the first days after injury and by macrophages and microglial cells in the later stages (Wang et al. 2007). Over the last years, it has become increasingly clear that immune responses both inside and outside the CNS can critically affect the clinical outcome of CNS damage. Work pioneered by the group of Ulrich Dirnagl has shown that CNS insults like ischemic stroke lead to massive systemic immunodepression (Prass et al. 2003; Dirnagl et al. 2007). Evidence from experimental and clinical work suggests that this systemic immunosupression is induced by the activation of the hypothalamic pituitary axis (and the subsequent release of glucocorticoids) and stimulation of the sympathetic nervous system (and subsequent release of norephinrine). Resulting death and deactivation of lymphocytes and macrophages then weaken the host defense. The systemic immunosuppression has immediate therapeutic consequences, as it suggests that immunocompromised stroke patients need to be protected against subsequent infections. Ongoing clinical trials are currently evaluating the effects of prophylactic antibiotic treatment in stroke patients (Dirnagl et al. 2007).

While few would argue that the peripheral immunosuppression is detrimental, the role of the stroke-induced immune response within the CNS is much less clear. There is ample evidence that a number of destructive mediators including pro-inflammatory cytokines, matrix metalloproteinases, and reactive oxygen species are released by invading immune cells after stroke (Wang et al. 2007). In line with these findings, a number of studies have reported beneficial effects of immunosuppressive manipulations targeting adhesion molecules like ICAM-1 (Chopp et al. 1996; Connolly et al. 1996) or pro-inflammatory cytokines like IL-1 (Yang et al. 1997; Mulcahy et al. 2003) in experimental stroke models. However, beneficial effects of immunosuppression are not found in all studies in experimental stroke models and more importantly could at least so far not be reproduced in clinical trials with human stroke patients (Enlimomab Acute Stroke Trial Investigators 2001; Krams et al. 2003). The divergence of these findings is in line with the conflicting results obtained with immunomodulation protocols in traumatic models of CNS injury as introduced earlier. Taken together, this divergence again highlights the critical influence of the experimental paradigm on the neuro-immune balance and the net effect of inflammation in many CNS conditions. In the future, it will thus be crucial to first identify the key destructive and protective mediators supplied by infiltrating immune cells and use this knowledge for the development of more specific immunomodulatory interventions. Ideally these therapies should be targeted selectively to the CNS to avoid "collateral damage" as, for example, worsening of the stroke-induced immunosuppression in the peripheral immune system.

5 Perspectives

To better understand how the immune cells interact with the nervous system, we have to elucidate the factors that lead to the differential activation of immune cells in different experimental and clinical paradigms. In principle, differential activation patterns have already been described for many immune cell populations, including macrophages and T cells. T cells, for example, can be classified into different T helper subsets based on the cytokine pattern they secrete (Liew 2002). It was recently suggested that the neuro-immune balance critically depends on T helper cell polarization (Hendrix and Nitsch 2007). This concept suggests that Th1 and Th17 cells, which mainly produce pro-inflammatory cytokines like IFN-γ, TNF-α and IL-17, are primarily responsible for the destructive effects of CNS inflammation. Th2 cells, on the other hand, which produce anti-inflammatory cytokine like IL-4 and IL-10, would then be primarily responsible for CNS protection and repair. This idea is based on recent findings that therapeutic compounds thought to promote Th2 polarization like glatiramer acetate or statins as well as vaccinations using Th2-inducing adjuvants improve neuroprotection and repair (Kipnis et al. 2000; Sicotte et al. 2003; Lu et al. 2004).

Similar to T cells, macrophages display a spectrum of differential activation patterns that result in clearly distinct cellular phenotypes (Gordon and Taylor 2005; Mantovani et al. 2005). Classically activated macrophages or M1 macrophages, for example, produce high levels of destructive mediators like pro-inflammatory cytokines as well as nitric oxide. Alternatively, activated macrophages, or M2, on the other hand, express predominantly anti-inflammatory cytokines and can contribute to debris phagocytosis and tissue repair. It is thus tempting to speculate that distinct macrophage subpopulations contribute differentially to CNS destruction and repair. In this context, it is interesting to note that glatiramer acetate also modulated monocyte activation in vivo in treated MS patients (Weber et al. 2004) and induces a shift towards a M2 phenotype for macrophages in addition to the Th2 shift in an animal model of MS (Weber et al. 2007). This example may thus provide proof-of-principle that the neuro-immune balance can be targeted therapeutically.

To develop more specific therapeutic interventions in the future, we will first have to identify the regulatory switches that govern the balance between protective and destructive responses in different immune cell populations. Selective targeting of the neuro-immune balance would then provide exciting new therapeutic options for many neurological disease characterized by primary or secondary CNS inflammation.

Acknowledgements The authors have been supported by the Deutsche Forschungsgemeinschaft (Emmy Noether Programm and SFB 571), Hermann and Lilly Schilling Foundation; Max Planck Society and Verein "Therapieforschung für Multiple Sklerose Kranke e.V". The authors thank Markus Krumbholz for the help with figures. This chapter is reprinted, with permission from the publisher, from our recent review article (Kerschensteiner, M., Meinl E., Hohlfeld R. (in press) Neuro-immune crosstalk in CNS disease. *Neuroscience*).

References

Airaksinen MS, Saarma M (2002) The GDNF family: signalling, biological functions and therapeutic value. *Nat Rev Neurosci* 3:383–394

Allen CD, Ansel KM, Low C, Lesley R, Tamamura H, Fujii N, Cyster JG (2004) Germinal center dark and light zone organization is mediated by CXCR4 and CXCR5. *Nat Immunol* 5:943–952

Alt C, Laschinger M, Engelhardt B (2002) Functional expression of the lymphoid chemokines CCL19 (ELC) and CCL 21 (SLC) at the blood-brain barrier suggests their involvement in G-protein-dependent lymphocyte recruitment into the central nervous system during experimental autoimmune encephalomyelitis. *Eur J Immunol* 32:2133–2144

Arancio O, Chao MV (2007) Neurotrophins, synaptic plasticity and dementia. *Curr Opin Neurobiol* 17:325–330

Bayas A, Kruse N, Moriabadi NF, Weber F, Hummel V, Wohleben G, Gold R, Toyka KV, Rieckmann P (2003) Modulation of cytokine mRNA expression by brain-derived neurotrophic factor and nerve-growth factor in human immune cells. *Neurosci Lett* 335:155–158

Bazan JF, Bacon KB, Hardiman G, Wang W, Soo K, Rossi D, Greaves DR, Zlotnik A, Schall TJ (1997) A new class of membrane-bound chemokine with a CX3C motif. *Nature* 385:640–644

Besser M, Wank R (1999) Cutting edge: clonally restricted production of the neurotrophins brain-derived neurotrophic factor and neurotrophin-3 mRNA by human immune cells and Th1/Th2-polarized expression of their receptors. *J Immunol* 162:6303–6306

Bhattacharyya BJ, Banisadr G, Jung H, Ren D, Cronshaw DG, Zou Y, Miller RJ (2008) The chemokines stromal cell-derived factor-1 regulates GABAergic inputs to neuronal progenitors in the postnatal dentate gyrus. *J Neurosci* 28:6720–6730

Bitsch A, Schuchardt J, Bunkowski S, Kuhlmann T, Brück W (2000) Acute axonal injury in multiple sclerosis. Correlation with demyelination and inflammation. *Brain* 123:1174–1183

Bjartmar C, Kidd G, Mörk S, Rudick R, Trapp BD (2000) Neurological disability correlates with spinal cord axonal loss and reduced N-acetyl aspartate in chronic multiple sclerosis patients. *Ann Neurol* 48:893–901

Bjartmar C, Trapp BD (2001) Axonal and neuronal degeneration in multiple sclerosis: mechanisms and functional consequences. *Curr Opin Neurol* 14:271–278

Braun A, Lommatzsch M, Mannsfeldt A, Neuhaus-Steinmetz U, Fischer A, Schnoy N, Lewin GR, Renz H (1999) Cellular sources of enhanced brain-derived neurotrophic factor production in a mouse model of allergic inflammation. *Am J Respir Cell Mol Biol* 21:537–546

Butler MP, O'Connor JJ, Moynagh PN (2004) Dissection of tumor-necrosis factor-alpha inhibition of long-term potentiation (LTP) reveals a p38 mitogen-activated protein kinase-dependent mechanism which maps to early- but not late-phase LTP. *Neuroscience* 124:319–326

Cardona AE, Pioro EP, Sasse ME, Kostenko V, Cardona SM, Dijkstra IM, Huang D, Kidd G, Dombrowski S, Dutta R, Lee JC, Cook DN, Jung S, Lira SA, Littman DR, Ransohoff RM (2006) Control of microglial neurotoxicity by the fractalkine receptor. *Nat Neurosci* 9:917–924

Cho J, Gruol DL (2008) The chemokines CCL2 activates p38 mitogen-activated protein kinase pathway in cultured rat hippocampal cells. *J Neuroimmunol* 199:94–103

Chopp M, Li Y, Jiang N, Zhang RL, Prostak J (1996) Antibodies against adhesion molecules reduce apoptosis after transient middle cerebral artery occlusion in rat brain. *J Cereb Blood Flow Metab* 16:578–584

Confavreux C, Vukusic S, Moreau T, Adeleine P (2000) Relapses and progression of disability in multiple sclerosis. *N Engl J Med* 343:1430–1438

Connolly ES Jr, Winfree CJ, Springer TA, Naka Y, Liao H, Yan SD, Stern DM, Solomon RA, Gutierrez-Ramos JC, Pinsky DJ (1996) Cerebral protection in homozygous null ICAM-1 mice after middle cerebral artery occlusion. Role of neutrophil adhesion in the pathogenesis of stroke. *J Clin Invest* 97:209–216

Cumiskey D, Pickering M, O'Connor JJ (2007) Interleukin-18 mediated inhibition of LTP in the rat dentate gyrus is attenuated in the presence of mGluR antagonists. *Neurosci Lett* 412:206–210

Dirnagl U, Klehmet J, Braun JS, Harms H, Meisel C, Ziemssen T, Prass K, Meisel A (2007) Stroke-induced immunodepression. Experimental evidence and clinical relevance. *Stroke* 38:770–773

Ehrhard PB, Erb P, Graumann U, Otten U (1993) Expression of nerve growth factor and nerve growth factor receptor tyrosine kinase Trk in activated CD4-positive T-cell clones. *Proc Natl Acad Sci USA* 90:10984–10988

Enlimomab Acute Stroke Trial Investigators (2001) Use of anti-ICAM-1 therapy in ischemic stroke: results of the Enlimomab Acute Stroke Trial. *Neurology* 57:1428–1434

Flügel A, Matsumuro K, Neumann H, Klinkert WE, Birnbacher R, Lassmann H, Otten U, Wekerle H (2001) Anti-inflammatory activity of nerve growth factor in experimental autoimmune encephalomyelitis: inhibition of monocyte transendothelial migration. *Eur J Immunol* 31:11–22

Gartner A, Polnau D, Staiger V, Sciarretta C, Minichiello L, Thoenen H, Bonhoeffer T, Korte M (2006) Hippocampal long-term potentiation is supported by presynaptic and postsynaptic tyrosine receptor kinase B-mediated phospholipase Cγ signaling. *J Neurosci* 26:3496–3504

Gordon S, Taylor PR (2005) Monocyte and macrophage heterogeneity. *Nat Rev Immunol* 5:953–964

Hammarberg H, Lidman O, Lundberg C, Eltayeb SY, Gielen AW, Muhallab S, Svenningsson A, Lindå H, van Der Meide PH, Cullheim S, Olsson T, Piehl F (2000) Neuroprotection by encephalomyelitis: rescue of mechanically injured neurons and neurotrophin production by CNS-infiltrating T and natural killer cells. *J Neurosci* 20:5283–5291

Harrison JK, Jiang Y, Chen S, Xia Y, Maciejewski D, McNamara RK, Streit WJ, Salafranca MN, Adhikari S, Thompson DA, Botti P, Bacon KB, Feng L (1998) Role for neuronally derived fractalkine in mediating interactions between neurons and CX3CR1-expressing microglial. *Proc Natl Acad Sci USA* 95:10896–10901

Hauben E, Butovsky O, Nevo U, Yoles E, Moalem G, Agranov E, Mor F, Leibowitz-Amit R, Pevsner E, Akselrod S, Neeman M, Cohen IR, Schwartz M (2000) Passive or active immunization with myelin basic protein promotes recovery from spinal cord contusion. *J Neurosci* 20:6421–6430

Hauser SL, Oksenberg JR (2006) The neurobiology of multiple sclerosis: genes, inflammation, and neurodegeneration. *Neuron* 52:61–76

Hendrix S, Nitsch R (2007) The role of T helper cells in neuroprotection and regeneration. *J Neuroimmunol* 184:100–112

Hohlfeld R, Wekerle H (2004) Autoimmune concepts of multiple sclerosis as a basis for selective immunotherapy: from pipe dreams to (therapeutic) pipelines. *Proc Natl Acad Sci USA* 101:14599–14606

Huang D, Shi FD, Jung S, Pien GC, Wang J, Salazar-Mather TP, He TT, Weaver JT, Ljunggren HG, Biron CA, Littman DR, Ransohoff RM (2006) The neuronal chemokine CX3CL1/frac-talkine selectively recruits NK cells that modify experimental autoimmune encephalomyelitis within the central nervous system. *FASEB J* 20:896–905

Huising MO, Stet RJ, Kruiswijk CP, Savelkoul HF, Lidy Verburg-van Kemenade BM (2003) Molecular evolution of CXC chemokines: extant CXC chemokines originate from the CNS. *Trends Immunol* 24:307–313

Hulshof S, van Haastert ES, Kuipers HF, van den Elsen PJ, De Groot CJ, van der Valk P, Ravid R, Biber K (2003) CX3CL1 and CX3CR1 expression in human brain tissue: noninflammatory control versus multiple sclerosis." *J Neuropathol Exp Neurol* 62:899–907

Imai T, Hieshima K, Haskell C, Baba M, Nagira M, Nishimura M, Kakizaki M, Takagi S, Nomiyama H, Schall TJ, Yoshie O (1997) Identification and molecular characterization of fractalkine recep-tor CX3CR1, which mediates both leukocyte migration and adhesion. *Cell* 91:521–530

Jiang Y, Chen G, Zhang Y, Lu L, Liu S, Cao X (2007) Nerve growth factor promotes the TLR4 signaling-induced maturation of human dendritic cells in vitro through inducible p75NTR 1. *J Immunol* 179:6297–6304

Jones TB, Basso DM, Sodhi A, Pan JZ, Hart RP, MacCallum RC, Lee S, Whitacre CC, Popovich PG (2002) Pathological CNS autoimmune disease triggered by traumatic spinal cord injury: implications for autoimmune vaccine therapy. *J Neurosci* 22:2690–2700

Jones TB, Ankeny DP, Guan Z, McGaughy V, Fisher LC, Basso DM, Popovich PG (2004) Passive or active immunization with myelin basic protein impairs neurological function and exacer-bates neuropathology after spinal cord injury in rats. *J Neurosci* 24:3752–3761

Kalled SL (2006) Impact of the BAFF/BR3 axis on B cell survival, germinal center maintenance and antibody production. *Semin Immunol* 18:290–296

Kallo T, Nagata T, Yamamoto S, Okamura H, Nishizaki T (2004) Interleukin-18 stimulates synap-tically released glutamate and enhances post-synaptic AMPA receptor responses in the CA1 region of mouse hippocampal slices. *Brain Res* 25:190–193

Kerschensteiner M, Gallmeier E, Behrens L, Leal VV, Misgeld T, Klinkert WE, Kolbeck R, Hoppe E, Oropeza-Wekerle RL, Bartke I, Stadelmann C, Lassmann H, Wekerle H, Hohlfeld R (1999) Activated human T cells, B cells, and monocytes produce brain-derived neurotrophic factor in vitro and in inflammatory brain lesions: a neuroprotective role of inflammation. *J Exp Med* 189:865–870

Kerschensteiner M, Stadelmann C, Dechant G, Wekerle H, Hohlfeld R (2003) Neurotrophic cross-talk between the nervous and immune systems: implications for neurological diseases. *Ann Neurol* 53:292–304

Kieseier BC, Wiendl H, Hemmer B, Hartung HP (2007) Treatment and treatment trials in multiple sclerosis. *Curr Opin Neurol* 20:286–293

Killestein J, Polman CH (2005) Current trials in multiple sclerosis: established evidence and future hopes. *Curr Opin Neurol* 18:253–260

Kipnis J, Yoles E, Porat Z, Cohen A, Mor F, Sela M, Cohen IR, Schwartz M (2000) T cell immu-nity to copolymer 1 confers neuroprotection on the damaged optic nerve: possible therapy for optic neuropathies. *Proc Natl Acad Sci USA* 97:7446–7451

Kishimoto T (2005) Interleukin-6: from basic science to medicine – 40 years in immunology. *Ann Rev Immunol* 23:1–21

Kivisäkk P, Mahad DJ, Callahan MK, Trebst C, Tucky B, Wei T, Wu L, Baekkevold ES, Lassmann H, Staugaitis SM, Campbell JJ, Ransohoff RM (2003) Human cerebrospinal fluid central memory CD4+ T cells: evidence for trafficking through choroid plexus and meninges via P-selectin. *Proc Natl Acad Sci USA* 100:8389–8394

Knoblach SM, Fan L, Faden AI. (1999) Early neuronal expression of tumor necrosis factor-alpha after experimental brain injury contributes to neurological impairment. *J Neuroimmunol* 95:115–125

Kornek B, Storch MK, Weissert R, Wallstroem E, Stefferl A, Olsson T, Linington C, Schmidbauer M, Lassmann H (2000) Multiple sclerosis and chronic autoimmune encephalomyelitis: a com-parative quantitative study of axonal injury in active, inactive, and remyelinated lesions. *Am J Pathol* 157:267–276

Kucia M, Jankowski K, Reca R, Wysoczynski M, Bandura L, Allendorf DJ, Zhang J, Ratajczak J, Ratajczak MZ (2004) CXCR4-SDF-1 signalling, locomotion, chemotaxis and adhesion. *J Mol Histol* 35:233–245

Kuhlmann T, Remington L, Cognet I, Bourbonniere L, Zehntner S, Guihot F, Herman A, Guay-Giroux A, Antel JP, Owens T, Gauchat JF (2006) Continued administration of ciliary neurotrophic factor protects mice from inflammatory pathology in experimental autoimmune encephalomyelitis. *Am J Pathol* 169:584–598

Krams M, Lees KR, Hacke W, Grieve AP, Orgogozo JM, Ford GAASTIN Study Investigators(2003). Acute stroke therapy by inhibition of neutrophils (ASTIN): an adaptive dose-response study of UK-279 276 in acute ischemic stroke. *Stroke* 34:2543–2548

Krumbholz M, Theil D, Derfuss T, Rosenwald A, Schrader F, Monoranu CM, Kalled SL, Hess DM, Serafini B, Aloisi F, Wekerle H, Hohlfeld R, Meinl E (2005) BAFF is produced by astrocytes and up-regulated in multiple sclerosis lesions and primary central nervous system lymphoma. *J Exp Med* 201:195–200

Krumbholz M, Theil D, Cepok S, Hemmer B, Kivisäkk P, Ransohoff RM, Hofbauer M, Farina C, Derfuss T, Hartle C, Newcombe J, Hohlfeld R, Meinl E (2006) Chemokines in multiple sclerosis: CXCL12 and CXCL13 up-regulation is differentially linked to CNS immune cell recruitment. *Brain* 129:200–211

Krumbholz M, Theil D, Steinmeyer F, Cepok S, Hemmer B, Hofbauer M, Farina C, Derfuss T, Junker A, Arzberger T, Sinicina I, Hartle C, Newcombe J, Hohlfeld R, Meinl E (2007) CCL19 is constitutively expressed in the CNS, up-regulated in neuroinflammation, active and also inactive multiple sclerosis lesions. *J Neuroimmunol* 190:72–79

Kutzelnigg A, Lucchinetti CF, Stadelmann C, Brück W, Rauschka H, Bergmann M, Schmidbauer M, Parisi JE, Lassmann H (2005) Cortical demyelination and diffuse white matter injury in multiple sclerosis. *Brain* 128:2705–2712

Lauro C, Catalano M, Trettel F, Mainiero F, Ciotti MT, Eusebi F, Limatola C (2006) The chemokines CX3CL1 reduces migration and increases adhesion of neurons with mechanisms dependent on the beta1 integrin subunit. *J Immunol* 177:7599–7606

Lewin GR, Barde YA (1996) Physiology of the neurotrophins. *Annu Rev Neurosci* 19:289–317

Li G, Adesnik H, Li J, Long J, Nicoll RA, Rubenstein JL, Pleasure SJ (2008) Regional distribution of cortical interneurons and development of inhibitory tone are regulated by Cxcl12/Cxcr4 signaling. *J Neurosci* 28:1085–1098

Li M, Ransohoff RM (2008) Multiple roles of chemokine CXCL12 in the central nervous system: a migration from immunology to neurobiology. *Prog Neurobiol* 4:116–131

Liew FY (2002) T (H)1 and T(H)2 cells: a historical perspective. *Nat Rev Immunol* 2:55–60

Linker R, Lee DH, Siglienti I, Gold R (2007) Is there a role for neurotrophins in the pathology of multiple sclerosis. *J Neurol* 254:I/33–I/40

Linker R, Kruse N, Israel S, Wei T, Seubert S, Hombach A, Holtmann B, Luhder F, Ransohoff RM, Sendtner M, Gold R (2008) Leukemia inhibitory factor deficiency modulates the immune response and limits autoimmune demyelination: a new role for neurotrophic cytokines in neuroinflammation. *J Immunol* 180:2204–2213

Liu L, Li Y, Van Eldik LJ, Griffin WS, Barger SW (2005) S100B-induced microglial and neuronal IL-1 expression is mediated by cell type-specific transcription factors. *J Neurochem* 92:546–553

Losseff NA, Wang L, Lai HM, Yoo DS, Gawne-Cain ML, McDonald WI, Miller DH, Thompson AJ (1996) Progressive cerebral atrophy in multiple sclerosis. A serial MRI study. *Brain* 119:2009–2019

Lu B, Pang PT, Woo NH (2005) The yin and yang of neurotrophin action. *Nat Rev Neurosci* 6:603–614

Lu D, Goussev A, Chen J, Pannu P, Li Y, Mahmood A, Chopp M (2004) Atorvastatin reduces neurological deficits and increases synaptogenesis, angiogenesis and neuronal survival in rats subjected to traumatic brain injury. *J Neurotrauma* 21:21–32

Mantovani A, Sica A, Locati M (2005) Macrophage Polarization comes of age. *Immunity* 23:344–346

Manz RA, Hauser AE, Hiepe F, Radbruch A (2005) Maintenance of serum antibody levels. *Annu Rev Immunol* 23:367–386

Maroder M, Bellavia D, Meco D, Napolitano M, Stigliano A, Alesse E, Vacca A, Frati L, Gulinno A, Screpanti I (1996) Expression of trkB neurotrophin receptor during T cell development. Role of brain derived neurotrophic factor in immature thymocyte survival. *J Immunol* 157:2864–2872

McAllister AK, Katz LC, Lo DC (1999) Neurotrophins and synaptic plasticity. *Ann Rev Neurosci* 22:295–318

Meinl E, Krumbholz M, Hohlfeld R (2006) B lineage cells in the inflammatory central nervous system environment: migration, maintenance, local antibody production, and therapeutic modulation. *Ann Neurol* 59:880–892

Miller DH, Leary SM (2007) Primary-progressive multiple sclerosis. *Lancet Neurol* 6:903–912

Moalem G, Leibowitz-Amit R, Yoles E, Mor F, Cohen IR, Schwartz M. (1999) Autoimmune T cells protect neurons from secondary degeneration after central nervous system axotomy. *Nat Med* 5:49–55

Moalem G, Gdalyahu A, Shani Y, Otten U, Lazarovici P, Cohen IR, Schwartz M (2000) Production of neurotrophins by activated T cells: implications for neuroprotective autoimmunity. *J Autoimmun* 15:331–345

Mulcahy NJ, Ross J, Rothwell NJ, Loddick SA (2003) Delayed administration of interleukin-1 receptor antagonist protects against transient cerebral ischemia in the rat. *Br J Pharmacol* 140:471–476

Nagappan G, Woo NH, Lu B (2008) Ama "zinc" link between TrkB transactivation and synaptic plasticity. *Neuron* 57:477–479

Neumann H, Schmidt H, Wilharm E, Behrens L, Wekerle H (1997) Interferon gamma gene expression in sensory neurons: evidence for autocrine gene regulation. *J Exp Med* 186:2023–2031

Neumann H, Schweigreiter R, Yamashita T, Rosenkranz K, Wekerle H, Barde YA (2002) Tumor necrosis factor inhibits neurite outgrowth and branch formation of hippocampal neurons by a rho-dependent mechanism. *J Neurosci* 22:854–862

Neumann H, Misgeld T, Matsumuro K, Wekerle H (1998) Neurotrophins inhibit major histocompatibility class II inducibility of microglia: involvement of the p75 neurotrophin receptor. *Proc Natl Acad Sci USA* 95:5779–5784

Newman TA, Woolley ST, Hughes PM, Sibson NR, Anthony DC, Perry VH (2001) T-cell- and macrophage-mediated axon damage in the absence of a CNS- specific immune response: involvement of metalloproteinases. *Brain* 124:2203–2214

Ohshima Y, Kubo T, Koyama R, Ueno M, Nakagawa M, Yamashita T (2008) Regulation of axonal elongation and pathfinding from the entorhinal cortex to the dentate gyrus in the hippocampus by the cytokine stromal cell-derived factor 1alpha. *J Neurosci* 28:8344–8353

Omari KM, John GR, Sealfon SC, Raine CS (2005) CXC chemokine receptors on human oligodendrocytes: implications for multiple sclerosis. *Brain* 128:1003–1015

Polman CH, O'Connor PW, Havrdova E, Hutchinson M, Kappos L, Miller DH, Phillips JT, Lublin FD, Giovannoni G, Wajgt A, Toal M, Lynn F, Panzara MA, Sandrock AW; AFFIRM Investigators (2006) A randomized, placebo-controlled trial of natalizumab for relapsing multiple sclerosis. *N Engl J Med* 354:899–910

Prass K, Meisel C, Hoflich C, Braun J, Halle E, Wolf T, Ruscher K, Victorov IV, Priller J, Dirnagl U, Volk HD, Meisel A (2003) Stroke-induced immunodeficiency promotes spontanous bacterial infections and is mediated by sympathetic activation reversal by poststroke T helper cell type 1-like immunostimulation. *J Exp Med* 198:725–736

Ragozzino D, Di Angelantonio S, Trettel F, Bertollini C, Maggi L, Gross C, Charo IF, Limatola C, Eusebi F (2006) Chemokine fractalkine/CX3CL1 negatively modulates active glutamatergic synapses in rat hippocampal neurons. *J Neurosci* 26:10488–10498

Rapalino O, Lazarov-Spiegler O, Agranov E, Velan GJ, Yoles E, Fraidakis M, Solomon A, Gepstein R, Katz A, Belkin M, Hadani M, Schwartz M (1998) Implantation of stimulated homologous macrophages results in partial recovery of paraplegic rats. *Nat Med* 4:814–821

Rose CR, Blum R, Pichler B, Lepier A, Kafitz KW, Konnerth A (2003) Truncated TrkB-T1 mediates neurotrophin-evoked calcium signalling in glia cells. *Nature* 426:74–78

Serpe CJ, Kohm AP, Huppenbauer CB, Sanders VM, Jones KJ (1999) Exacerbation of facial motoneuron loss after facial nerve transection in severe combined immunodeficient (scid) mice. *J Neurosci* 19:RC7

Sicotte M, Tsatas O, Jeong SY, Cai CQ, He Z, David S (2003) Immunization withmyelin or recombinant Nogo-66/MAG in alum promotes axon regeneration and sprouting after corticospinal tract lesions in the spinal cord. *Mol Cell Neurosci* 23:251–263

Soriano SG, Amaravadi LS, Wang YF, Zhou H, Yu GX, Tonra JR, Fairchild-Huntress V, Fang Q, Dunmore JH, Huszar D, Pan Y. S. G. Soriano (2002) Mice deficient in fractalkine are less susceptible to cerebral ischemia-reperfusion injury. *J Neuroimmunol* 125:59–65

Stadelmann C, Kerschensteiner M, Misgeld T, Brück W, Hohlfeld R, Lassmann H (2002) BDNF and gp145trkB in multiple sclerosis brain lesions: neuroprotective interactions between immune and neuronal cells. *Brain* 125:75–85

Storch MK, Stefferl A, Brehm U, Weissert R, Wallström E, Kerschensteiner M, Olsson T, Linington C, Lassmann H (1998) Autoimmunity to myelin oligodendrocyte glycoprotein in rats mimics the spectrum of multiple sclerosis pathology. *Brain Pathol* 8:681–694

Sun JH, Yang B, Donnelly DF, Ma C, LaMotte RH (2006) MCP-1 enhances excitability of nociceptive neurons in chronically compressed dorsal root ganglia. *J Neurophysiol* 96:2189–2199

Sunnemark D, Eltayeb S, Nilsson M, Wallström E, Lassmann H, Olsson T, Berg AL, Ericsson-Dahlstrand A (2005) CX3CL1 (fractalkine) and CX3CR1 expression in myelin oligodendrocyte glycoprotein-induced experimental autoimmune encephalomyelitis: kinetics and cellular origin. *J Neuroinflammation* 2:17

Thoenen H, Sendtner M (2002) Neurotrophins: from enthusiastic expectations through sobering experiences to rational therapeutic approaches. *Nat Neurosci* 5S:1046–1050

Tomac A, Lindqvist E, Lin LF, Ogren SO, Young D, Hoffer BJ, Olson L (1995) Protection and repair of the nigrostriatal dopaminergic system by GDNF in vivo. *Nature* 373:335–339

Tomac AC, Agulnick AD, Haughey N, Chang CF, Zhang Y, Bäckman C, Morales M, Mattson MP, Wang Y, Westphal H, Hoffer BJ (2002) Effects of cerebral ischemia in mice deficient in Persephin. *Proc Natl Acad Sci USA* 99:9521–9526

Torcia M, Bracci-Laudiero L, Lucibello M, Nencioni L, Labardi D, Rubartelli A, Cozzolino F, Aloe L, Garaci E (1996) Nerve growth factor is an autocrine survival factor for memory B lymphocytes. *Cell* 85:345–356

Trapp BD, Peterson J, Ransohoff RM, Rudick R, Mörk S, Bö L (1998) Axonal transection in the lesions of multiple sclerosis. *N Engl J Med* 338:278–285

Tsai HH, Frost E, To V, Robinson S, Ffrench-Constant C, Geertman R, Ransohoff RM, Miller RH (2002) The chemokine receptor CXCR2 controls positioning of oligodendrocyte precursors in developing spinal cord by arresting their migration. *Cell* 110:373–383

Vargas-Leal V, Bruno R, Derfuss T, Krumbholz M, Hohlfeld R, Meinl E (2005) Expression and function of glial cell line-derived neurotrophic factor family ligands and their receptors on human immune cells. *J Immunol* 175:2301–2308

Vergote D, Butler GS, Ooms M, Cox JH, Silva C, Hollenberg MD, Jhamandas JH, Overall CM, Power C (2006) Proteolytic processing of SDF-1alpha reveals a change in receptor specificity mediating HIV-associated neurodegeneration. *Proc Natl Acad Sci USA* 103: 19182–19187

Villoslada P, Hauser SL, Bartke I, Unger J, Heald N, Rosenberg D, Cheung SW, Mobley WC, Fisher S, Genain CP (2000) Human nerve growth factor protects common marmosets against autoimmune encephalomyelitis by switching the balance of T helper cell type 1 and 2 cytokines within the central nervous system. *J Exp Med* 191:1799–1806

Wang LJ, Lu YY, Muramatsu S, Ikeguchi K, Fujimoto K, Okada T, Mizukami H, Matsushita T, Hanazono Y, Kume A, Nagatsu T, Ozawa K, Nakano I (2002) Neuroprotective effects of glial cell line-derived neurotrophic factor mediated by an adeno-associated virus vector in a transgenic animal model of amyotrophic lateral sclerosis. *J Neurosci* 22:6920–6928

Wang Q, Nan Tang X, Yenari MA (2007) The inflammatory response in stroke. *J Neuroimmunol* 184:53–68

Weber MS, Starck M, Wagenpfeil S, Meinl E, Hohlfeld R, Farina C (2004) Multiple sclerosis: glatiramer acetate inhibits monocyte reactivity in vitro and in vivo. *Brain* 127:1370–1378

Weber MS, Prod'homme T, Youssef S, Dunn SE, Rundle CD, Lee L, Patarroyo JC, Stüve O, Sobel RA, Steinman L, Zamvil SS (2007) Type II monocytes modulate T-cell mediated central nervous system autoimmune disease. *Nat Med* 13:935–943

Wong G, Goldshmit Y, Turnley AM (2004) Interferon-gamma but not TNF alpha promotes neuronal differentiation and neurite outgrowth of murine adult neuronal stem cells. *Exp Neurol* 187:171–177

Xia Y, Hu HZ, Liu S, Ren J, Zafirov DH, Wood JD (1999) IL-1beta and IL-6 excite neurons and suppress nicotinic and noradrenergic neurotransmission in guinea pig enteric nervous system. *J Clin Invest* 103:1309–1316

Yang GY, Zhao YJ, Davidson BL, Betz AL (1997) Overexpression of interleukin-1 receptor antagonist in the mouse brain reduces ischemic brain injury. *Brain Res* 751:181–188

Yoles E, Hauben E, Palgi O, Agranov E, Gothilf A, Cohen A, Kuchroo V, Cohen IR, Weiner H, Schwartz M (2001) Protective autoimmunity is a physiological response to CNS trauma. *J Neurosci* 21:3740–3748

Zhang K, McQuibban GA, Silva C, Butler GS, Johnston JB, Holden J, Clark-Lewis I, Overall CM, Power C (2003) HIV-induced metalloproteinase processing of the chemokine stromal cell derived factor-1 causes neurodegeneration. *Nat Neurosci* 6:1064–1071

Zhang RX, Liu B, Li A, Wang L, Ren K, Qiao JT, Berman BM, Lao L (2008) Interleukin 1beta facilitates bone cancer pain in rats by enhancing NMDA receptor NR-1 subunit phosphorylation. *Neuroscience* 154:1533–1538

Zhu Y, Yu T, Zhang XC, Nagasawa T, Wu JY, Rao Y. (2002) Role of the chemokine SDF-1 as the meningeal attractent for embryonic cerebellar neurons. *Nat Neurosci* 5:719–720

Ziemssen T, Kümpfel T, Klinkert WE, Neuhaus O, Hohlfeld R (2002) Glatiramer acetate-specific T-helper 1- and 2-type cell lines produce BDNF: implications for multiple sclerosis therapy. *Brain* 125:2381–2391

Ziv Y, Ron N, Butovsky O, Landa G, Sudai E, Greenberg N, Cohen H, Kipnis J, Schwartz M (2006) Immune cells contribute to the maintenance of neurogenesis and spatial learning abilities in adulthood. *Nat Neurosci* 9:268–275

Prospects for Antigen-Specific Tolerance Based Therapies for the Treatment of Multiple Sclerosis

Danielle M. Turley and Stephen D. Miller

Abstract A primary focus in autoimmunity is the breakdown of central and peripheral tolerance resulting in the survival and eventual activation of autoreactive T cells. As CD4$^+$ T cells are key contributors to the underlying pathogenic mechanisms responsible for onset and progression of most autoimmune diseases, they are a logical target for therapeutic strategies. One method for restoring self-tolerance is to exploit the endogenous regulatory mechanisms that govern CD4$^+$ T cell activation. In this review, we discuss tolerance strategies with the common goal of inducing antigen (Ag)-specific tolerance. Emphasis is given to the use of peptide-specific tolerance strategies, focusing on ethylene carbodiimide (ECDI)-peptide-coupled cells (Ag-SP) and nonmitogenic anti-CD3, which specifically target the T cell receptor (TCR) in the absence of costimulatory signals. These approaches induce a TCR signal of insufficient strength to cause CD4$^+$ T cell activation and instead lead to functional T cell anergy/deletion and activation of Ag-specific induced regulatory T cells (iTregs) while avoiding generalized long-term immunosuppression.

1 Introduction

This chapter will focus on tolerance mechanisms in multiple sclerosis (MS) and its mouse model, experimental autoimmune encephalomyelitis (EAE). An important goal of current research is to develop new therapies for autoimmune diseases by specifically inhibiting and/or tolerizing autoreactive CD4$^+$ T cells. Although this chapter focuses on tolerance strategies that directly target autoreactive T cells in MS and EAE, similar approaches are ongoing in other autoimmune diseases, as well as tissue transplantation.

D.M. Turley and S.D. Miller(✉)
Department of Microbiology-Immunology and the Interdepartmental Immunobiology Center
Northwestern University Medical School, Tarry 6-718, 303 E. Chicago Ave
Chicago, IL 60611 USA
e-mail: s-d-miller@northwestern.edu

Results Probl Cell Differ, doi:10.1007/400_2008_13

MS is an immune-mediated disease of the central nervous system (CNS) characterized by perivascular CD4$^+$ and CD8$^+$ T cell and mononuclear cell infiltration with subsequent primary demyelination of axonal tracks leading to progressive paralysis (Wekerle 1991). MS is generally understood to be an autoimmune disease characterized by T cell responses to myelin basic protein (MBP), proteolipid protein (PLP), and/or myelin-oligodendrocyte glycoprotein (MOG) (Bernard and de Rosbo 1991; de Rosbo et al. 1997; Ota et al. 1990); however a straightforward cause–effect relationship between myelin reactivity and disease pathology has not been demonstrated.

Characteristically there are four courses of clinical disease in MS: (1) relapsing–remitting, (2) secondary-progressive, (3) primary-progressive, and (4) progressive-relapsing. Correspondingly, there are relapsing–remitting and chronic mouse models of MS, i.e., EAE. Relapsing–remitting EAE (R–EAE) is characterized by transient ascending hind limb paralysis, perivascular mononuclear-cell infiltration, and fibrin deposition in the brain and spinal cord with adjacent areas of acute and chronic demyelination (Paterson and Swanborg 1988). Given that the etiology of MS is unknown, the inducing antigen (Ag) has yet to be identified, and the probability that CD4$^+$ T cells respond to multiple epitopes contained within several myelin proteins responsible for chronic disease progression, the use of Ag-specific tolerance-based immunotherapy targeting a single protein is challenging. Furthermore, a pathological role for epitope spreading is difficult to verify in human MS because the initiating Ag is not known. In contrast, animal models, such as EAE, have the advantage that the initiating Ag is known. For example, in the SJL model of disease in which mice are primed with PLP$_{139-151}$ in complete Freund's adjuvant (CFA), PLP$_{139-151}$-specific CD4$^+$ T cell reactivity in secondary lymphoid organs is maintained throughout the disease course. Beside the activation of CD4$^+$ T cells specific for the initiating antigen, PLP$_{178-191}$-specific CD4$^+$ T cell reactivity arises (intramolecular epitope spreading) during the first disease relapse, and CD4$^+$ T cells specific for a myelin basic protein epitope, MBP$_{84-104}$, arise (intermolecular epitope spreading) during the second disease relapse (McRae et al. 1995b; Vanderlugt et al. 2000). While Ag-specific tolerance can be induced in this experimental model as the self peptides are well characterized, this is not true for humans with MS.

Although the etiology of MS is unknown, both genetic (Ebers et al. 1995) and environmental factors appear to play a role in susceptibility and initiation of disease. Epidemiological studies provide strong circumstantial evidence for an environmental trigger, most likely viral, in the induction of MS (Kurtzke 1993; Olson et al. 2001b; Waksman 1995). CNS pathology may therefore result from bystander myelin damage mediated via T cells targeting a CNS-persisting virus; and/or from activation of autoreactive T cells secondary to an encounter with a pathogen directly by *molecular mimicry* (Fujinami and Oldstone 1985; Olson et al. 2001a; Wucherpfennig and Strominger 1995), or indirectly by *epitope spreading* resulting from the release of sequestered antigens secondary to virus-specific T cell-initiated myelin damage (McRae et al. 1995a; Miller and Karpus 1994; Miller et al. 1997). It is believed that combination of persistent CNS inflammation and resulting myelin/nerve damage produces clinical symptoms of MS. In addition to the EAE model of MS, another commonly used and virally induced model of MS is Theiler's murine encephalomyelitis virus (TMEV) induced demyelinating disease (TMEV–IDD) (Dal Canto et al. 1997). TMEV is a naturally occurring mouse pathogen, and disease is induced in recipient mice by intracerebral injection of the positive strand RNA virus that leads to primary progressive demyelinating disease (Miller et al. 1997).

As integral members of the adaptive immune system, CD4+ T cells are key mediators in multiple phases of the protective immune response by recognizing foreign Ags via their antigen-specific T cell receptor (TCR) complex during cognate interactions with antigen presenting cells (APCs) displaying peptide/major histocompatibility complex (MHC) class II complexes. Thus, an essential characteristic of intrathymic T cell development is the generation of TCR diversity enabling T cells to respond to an unlimited number of foreign antigens (Ags). However, one inevitable consequence of TCR diversity is the generation of self-reactive TCRs creating potential for autoimmune disease. To balance this, the immune system has developed regulatory checkpoints that govern lymphocyte development which includes biphasic processes of central tolerance which only permits generation of T cells with a functional TCR while deleting populations of T cells which express TCRs strongly reactive to self-peptides (Hogquist et al. 2005). Additionally, the immune system has created peripheral tolerance mechanisms to safe guard against autoreactive T cells that escape thymic deletion. Mechanisms of peripheral tolerance include T cell intrinsic mechanisms (ignorance, anergy, phenotypic skewing/immune deviation and deletion/apoptosis) as well as T cell extrinsic mechanisms (induction of tolerogenic dendritic cells (DCs) and/or regulatory T cells (Treg) (Walker and Abbas 2002). Thus, when functioning properly, the process of central and peripheral tolerance ensures selective generation and regulation of functional, non-self-reactive T cells.

Breakdown of immune tolerance resulting in the persistence and eventual activation of autoreactive T cells (Christen and von Herrath 2004) is a fundamental theme in autoimmunity. Since CD4+ T cells are key contributors to the underlying pathogenic mechanisms responsible for onset and progression of most autoimmune diseases, they are also a logical target for therapeutic intervention. However, as discussed above, these cells are critical to the induction of adaptive immunity, thus creating a complex functional dichotomy that underscores the necessity for active regulatory mechanisms, such as the two-signal hypothesis, and therapeutic interventions that both promote immunity against foreign Ags, while inhibiting self-directed responses.

One technique for restoring self-tolerance is to exploit the endogenous regulatory mechanisms that govern CD4+ T cell activation. Typically, endogenous ligation of the TCR by peptide/MHC class II alone produces a signal of insufficient strength to activate a CD4+ T cell and can instead induce functional anergy or deletion. As a consequence, additional APC-derived costimulatory signals (e.g., CD80/86 engagement of CD28) are required to lower the threshold required for successful T cell activation. This "two-signal" hypothesis predicts that TCR stimulation in the absence of costimulatory signals leads to CD4+ T cell anergy, tolerance, and/or depletion (Sharpe and Freeman 2002). Therefore, either TCR ligation in the absence of costimulatory signals or exogenous targeting of the costimulatory pathway would appear to be a logical target of therapeutic strategies to downregulate pathologic functions of autoreactive CD4$^+$ T cells. In light of this, various therapeutic approaches have been designed to block autoreactive CD4$^+$ T cell function during autoimmune disease, including administration of blocking antibodies directed against a variety of epitopes including CD3, CD4, CD28, CD40, CD80, CD86, CD154, ICOS, OX40, and 4–1BB, as well as CTLA4-Ig (Karandikar et al. 1998; Zhang et al. 2003). However, these treatment strategies, if administered over a long time period, often result in either nonspecific immune suppression or other undesirable side effects. In this chapter techniques with the common purpose of inducing Ag-specific tolerance by specifically targeting the TCR to avoid detrimental influences on nonspecific/bystander immune processes will be discussed.

2   Monoclonal Antibody Induced Tolerance

In an attempt to further test the two-signal hypothesis, several groups have investigated the therapeutic potential of anti-CD3 mAb treatment in the absence of costimulatory signals for treatment of various autoimmune diseases. However, treatment with an unaltered anti-CD3 mAb is potentially a double-edged sword. While treatment may modify activity of pathogenic autoreactive CD4$^+$ T cells, thereby ameliorating autoimmune disease progression, this therapy may also induce nonspecific side effects through activation of bystander T cells - for example, induction of general immunosuppression which increases the patient's susceptibility to opportunistic infection and common occurrence of high-dose syndrome in which treatment recipients suffer severe side-effects due to the nonspecific production of inflammatory cytokines including TNF-α. Furthermore, cross-linking of CD3 may in some cases initiate a signal of sufficient strength that eliminates the need for a costimulatory molecule-induced reduction in the signal threshold required for T cell activation. Due to the aforementioned complications associated with the use of mitogenic anti-CD3 mAb, structural alterations have been made to the Fc binding domain so that deleterious side effects may be avoided by lowering the level of nonspecific T cell signaling. Nonmitogenic anti-CD3 mAb treatments induce lower levels of TCR-mediated signaling (Herold et al. 2003), and it is believed that lower levels of TCR-mediated signaling favor immune deviation from a Th1/17 to a Th2 phenotype and activation of Tregs. In this scenario, the T cell-mediated immune response is changed from a Th1/17-like (disease-promoting) response, to a Th2-like (disease-regulating) response.

The therapeutic efficacy of bypassing costimulatory signals has been demonstrated using nonmitogenic anti-CD3 mAb therapy for treatment of both EAE and the nonobese diabetic (NOD) mouse model of type 1 diabetes (TID) (Chatenoud 2003; Chatenoud et al. 1994; Kohm et al. 2005). While anti-CD3 mAb treatment causes functional cross-linking and activation of the TCR in an Ag-nonspecific manner, a humanized form of anti-CD3 mAb, mutated to prevent binding to Fc receptors (OKT3 Ala–Ala), has had some success in phase I/II clinical trials for delaying onset of type I diabetes and treating psoriatic arthritis (Herold et al. 2002; Keymeulen et al. 2005; Pozzilli et al. 2000; Utset et al. 2002). This nonmitogenic anti-CD3 mAb induces a suboptimal level of TCR-mediated signalling (Herold et al. 2003) and the success of the therapy is thought to be due to multiple mechanisms including anergy induction, immune deviation and activation of Tregs (Belghith et al. 2003; Kohm et al. 2005). Administration of the humanized form of nonmitogenic anti-CD3 mAb in patients was however associated with side effects including a moderate cytokine release syndrome and symptoms of Epstein–Barr viral mononucleosis (Herold et al. 2005). Thus, broad-based TCR-directed therapies may have other undesired long-term effects on the immune system.

Other studies using monoclonal antibody therapies directed against molecules involved in lymphocyte/monocyte recruitment and activation include use of antibodies directed against $\alpha 4 \beta 1$-integrin (VLA4; Tysabri; natalizumab) (Miller et al. 2003; Yednock et al. 1992), chemokines such as CC-chemokine ligand 3 (CCL3) (Karpus et al. 1995), or proinflammatory cytokines such as lymphotoxin (Ruddle et al. 1990). These and other immunosuppressive strategies promote physical deletion or inactivation of entire subsets of T cells or cause nonspecific inhibition of Ag presentation, proinflammatory cytokine production, or T cell trafficking. All these strategies can compromise ability of the host to combat opportunistic pathogens and/or increase risk of neoplasia. This is illustrated by recent deaths from progressive multifocal leukoencephalopathy (PML), an infection of the CNS by JC virus, which destroys myelin-producing oligodendrocytes, of several participants in an MS clinical trial following treatment with Tysabri (Khalili et al. 2007). Thus, Ag-specific tolerance strategies have the best therapeutic potential, but their success requires a more precise understanding of the autoantigen(s) and epitope(s) involved in the ongoing pathogenesis of a particular autoimmune disease.

3   Antigen Specific Induced Tolerance Induction

Direct targeting of autoreactive T cells is the ideal treatment strategy for autoimmune disease, resulting in Ag-specific unresponsiveness without global immunosuppression. There are currently four different protocols employed for inducing peptide-specific immune tolerance: altered peptide ligand (APL)-induced tolerance, mucosal (oral–nasal)-induced tolerance, soluble-peptide-induced tolerance, and ECDI-coupled-cell-induced tolerance. Each of these methods of peptide-specific tolerance induction is briefly discussed in the following section, as well as their putative mechanisms of action.

3.1   Altered Peptide Ligand Induced Tolerance

APLs are peptide analogues that bind to the same TCR, but elicit different functional responses as compared to the native autoepitope (Anderton and Wraith

2002). APLs compete with the naïve peptide for TCR binding, altering the cascade of signalling events necessary for full T cell activation. APLs contain one or more amino-acid substitutions, typically binding with lower affinity to the TCR than native peptide and function as either antagonists or partial agonists. Antagonistic APLs induce T cell anergy, while partial-agonist APLs induce incomplete activation of T cells. This partial activation can precipitate cytokine production in the absence of proliferation, thereby inducing immune deviation from Th1- and Th17-cell dependent responses to Th2- and Th3-cell dependent responses, or bystander suppression through the induction of Tregs (Nicholson et al. 1997; Young et al. 2000). A number of groups have tested the therapeutic efficacy of APLs of various myelin epitopes in treating established CNS autoimmune disease. In vivo administration of these myelin APLs were reported to prevent or reverse clinical disease progression in EAE, and were effective regardless of the route of administration, including subcutaneous (s.c.), intraperitoneal (i.p.) in incomplete Freund's adjuvant, intranasal (i.n.), or intravenous (i.v.) routes (Nicholson and Kuchroo 1997; Samson and Smilek 1995; Wraith et al. 1989). Disease prevention with APLs of $PLP_{139-151}$ is associated with the induction of Th2-cell differentiation. In support of immune deviation as the mechanistic basis of APL treatment, therapeutic clones responsive to $PLP_{139-151}$-derived APLs have been shown to produce IL-4, IL-10, IL-13, and TGF-β, all of which are believed to suppress EAE, and antibodies directed against each of these cytokines attenuate the protective influence of APL-mediated treatment in EAE (Nicholson et al. 1995).

APLs with substitutions at amino-acid positions necessary for TCR engagement, which compete with the natural ligand and interfere with T cell activation, have also been tested in clinical trials. As discussed previously, APL therapy successfully inhibits EAE; however two separate MS Phase II clinical trials testing an APL of MBP_{83-99}, the immunodominant HLA-DR2-restricted MBP T cell epitope, were halted due to safety concerns (Bielekova et al. 2000; Kappos et al. 2000). The number of CNS lesions of patients undergoing therapy was assessed by MRI and incidence of clinical relapse and hypersensitivity reactions were monitored. Participants in the first study, which included only eight patients tested at a 50 mg dose, demonstrated a higher incidence of MS exacerbations, as determined by both MRI and clinical criteria, and the APL cross-stimulated self-antigen reactive Th1 cells (Bielekova et al. 2000). A second double-blind, placebo-controlled study included 142 patients receiving various APL doses (Kappos et al. 2000) and was halted because 9% of patients (mostly in the group receiving the highest dose) developed hypersensitivity reactions. A potential problem with this approach is that a particular APL may be antagonistic for certain T cell clones, but at the same time serve as an agonist or super-agonist for peptide-specific clones expressing different TCRs. These studies aptly demonstrate the difficulty in moving from animal model to human patient.

In order to induce peptide-specific tolerance, and for effective prevention and treatment of disease, it is hypothesized that either the peptide(s) responsible for disease induction or the dominant peptide(s) driving ongoing autoimmunity must be identified. However, the development of glatiramer acetate (GA, Teva Pharmaceuticals), a random mixture of glutamine, lysine, alanine, and tyrosine peptides of various lengths, is thought to act as an APL for treatment of patients with MS by simulating MBP reactive T cells (Bornstein et al. 1987; Duda et al. 2000; Neuhaus et al. 2000). Recent studies suggest that GA induces immune deviation from a Th1/Th17 cell-type response to a Th2 cell-type response and does not induce anergy or deletion of autoreactive T cells (Aharoni et al. 1999). In contrast to the aforementioned trials using the MBP APL, GA is the only approved semi-Ag-specific approved drug for the treatment of MS. Although GA appears to be well tolerated, 10% of patients experience a transient systemic postinjection reaction,, characterized by flushing, chest tightness, palpitations, dyspnea, and anxiety (Korczyn and Nisipeanu 1996). As with all approved MS therapies, treatment requires daily s.c. injections and is beneficial to only a minority of MS patients (Johnson et al. 1995).

3.2   Mucosal Tolerance

Tolerance induced by the mucosal (oral/nasal) route is biologically relevant, as individual foreign dietary Ags are normally tolerated by the host, except in the case of food allergy. T cells found within the gastrointestinal (GI) tract are exposed regularly to ingested foreign Ags, yet they remain largely unresponsive to these nonself Ags. GI surfaces are constantly exposed to exogenous foreign Ags and allow for protective tolerance against some (primarily food) Ags while at the same time serving as an immunological defense against other harmful (pathogenic) Ags. For this reason, induction of tolerance using the mucosal route for administration of soluble Ags is appealing as it is antigen-specific, is a relatively easy method of administration, and carries decreased risk of toxicity when compared with parenteral injection of soluble Ag.

Efficiency of oral tolerance is dependent on various factors including the animal model employed, type of Ag, and whether a high or low treatment dose is used (Faria and Weiner 1999; Mayer and Shao 2004; Mowat et al. 1982). High-dose oral tolerance results in the induction of anergy or deletion of peripheral Ag-specific T cells (Bitar and Whitacre 1988; Whitacre et al. 1991). At high doses, Ag can diffuse through the GI wall and into the systemic circulation, where it can induce T cell unresponsiveness via anergy and/or deletion. By contrast, low doses of oral Ag act by bystander suppression via activation of regulatory-cell-driven tolerance within the target organ (Chen et al. 1994; Khoury et al. 1992). Low-dose antigen is taken up by mucosa-associated APCs that activate Tregs to secrete suppressive cytokines, such as TGF-β, IL-4 and IL-10 (Miller et al. 1992a). Thus, induction of tolerance using high doses of Ag is believed to act by mechanisms that directly influence CD4$^+$ T cells, whereas use of low doses is believed to induce tolerance by indirect bystander suppression.

Oral Ag administration has been shown to suppress the initiation of disease in multiple animal models of autoimmune disease, including EAE, uveitis, and colitis, as well as asthma (Faria and Weiner 2006; Mowat et al. 2004; Weiner

2004). Multiple studies in EAE have shown that preadministration of soluble myelin peptides by the oral or nasal route protects against disease induction (Metzler and Wraith 1993); however attempts to treat EAE following onset of clinical symptoms with oral tolerance have been less successful (Bai et al. 1998, 1997; Benson et al. 1999; Karpus et al. 1996; Kennedy et al. 1997; Meyer et al. 1996) without the addition of other compounds such as soluble IL-10 delivered either orally or nasally (Slavin et al. 2001). In addition, treatment with orally administered bovine MBP in MS clinical trials proved unsuccessful as a therapy (Barnett et al. 1998; Faria and Weiner 2006; Weiner 2004; Weiner et al 1993). Thus, oral tolerance appears to be effective at preventing induction of EAE, but it is significantly less effective in treating preestablished EAE and MS. While the use of mucosal tolerance remains an attractive possibility for induction of tolerance, this therapy is currently limited in its ability to induce tolerance in ongoing disease, which restricts its potential for treating human autoimmune disease.

3.3 Soluble Peptide Tolerance

Injection of high doses of soluble peptides leads to a state of anergy by blocking T cell proliferation and/or IL-2 cytokine production upon restimulation with the cognate peptide (Burstein et al. 1992; Critchfield et al. 1994; Gaur et al. 1992). Upon encountering a high-dose of Ag, T cells undergo an initial burst of proliferation and on repeated encounter are rendered anergic or deleted due to activation induced cell death (AICD) (Critchfield et al. 1994; Racke et al. 1996). For this reason, it has been hypothesized that tolerance induced by soluble peptides may be useful for Ag-specific immunotherapy for human autoimmune diseases. For example, the induction of tolerance to MBP was examined in a Phase I clinical trial in patients with primary-progressive MS using a peptide that is immunodominant for MBP-specific T cells and B cells. The induction of tolerance was monitored by quantification of MBP-specific autoantibodies in cerebrospinal fluid (CSF). Following a single i.v. injection of MBP_{85-96} peptide, autoantibodies were undetectable for 3–4 months in the CSF of these patients and tolerance was more prolonged following a second injection (Warren et al. 1997).

A cautionary note for the use of i.v. injected soluble-myelin peptide monomers or oligomers as a therapy in mice with preestablished adjuvant-induced EAE, is that this treatment regimen was found to induce a fatal anaphylactic response in various mouse strains (Smith et al. 2005). Furthermore, i.v. administration of soluble MOG in a primate model of EAE was shown to exacerbate disease (Genain et al. 1996). Due to the highly variable outcome of treatment with soluble-peptide-induced tolerance, there is currently a significant level of uncertainty regarding its safety. Anaphylactic responses to i.v. soluble peptide administration appear to occur in an Ag-specific manner, such that the same Ag must be administered during both the initial sensitization phase and the rechallenge phase to induce the effect. In light of the fact that recurrent tolerogenic treatments may be required to ameliorate disease progression, soluble-peptide-induced anaphylaxis is a significant safety concern. Contributing to the complications of i.v. administered soluble-peptide-induced anaphylaxis is the observation that not all autoantigens induce anaphylaxis. Moreover, the capacity of specific Ags to induce anaphylactic response has been reported to directly correlate with the thymic expression of each Ag, such that if the self-peptide is expressed in the thymus and therefore is subject to central tolerance, anaphylaxis does not occur (Pedotti et al. 2001). This hypothesis is however not supported by findings from our laboratory that showed that equivalent levels of anaphylaxis were induced in mice with preestablished EAE regardless of whether or not the peptide was expressed in the thymus. For example, MBP_{Ac1-11} and MBP_{84-104} are expressed in the thymus, whereas $PLP_{139-151}$ and MOG_{35-55} are not; yet all four peptides equally induced anaphylaxis when the soluble peptide was administered to mice with actively-induced EAE (Smith et al. 2005). In addition, the capacity of a specific self-Ag to elicit an anaphylactic response failed to correlate with its ability to induce an antibody (Ab) response; $PLP_{178-191}$, which is a B cell epitope and induces IgG production, failed to promote anaphylactic shock during autoimmune disease treatment (Smith et al. 2005). Although i.v. administration of soluble peptides has been shown to ameliorate EAE in an Ag-specific manner, the technique has significant efficacy and safety concerns.

3.4 ECDI–Peptide-Coupled Cell Induced Tolerance

One of the more promising modes of tolerance induction for prevention and treatment of autoimmune diseases, as well as in preventing transplant rejection, is the i.v. treatment with Ag-coupled, ethylene carbodiimide (ECDI)-fixed splenocytes (referred to as Ag-coupled cells, Ag-SP). Treatment with Ag–SP is a powerful method to induce anergy in vitro and peripheral tolerance in vivo (Miller et al. 1995a, 1979), as i.v. injection of myelin-Ag-coupled cells induces rapid and long-lived Ag-specific tolerance in mice with EAE.Fixation of donor cells with ECDI in the presence of Ag results in the formation of peptide bonds between free amino and carboxyl groups, which binds the peptide to the cells. Specific peptides as well as intact proteins can be used to induce tolerance with this method.

Experimentally, this tolerogenic method not only prevents onset of EAE in mice, but is also an effective treatment for ameliorating progression of established disease in both the active and adoptive transfer models of EAE, by tolerizing host $CD4^+$ T cells that are specific for spread epitopes (Kennedy et al. 1990a, b; Su and Sriram 1991; Vandenbark et al. 1996; Vanderlugt et al. 2000). Induction of tolerance by Ag–SP treatment has also been shown to be an effective therapy in other disease models, including experimental autoimmune thyroiditis (Braley-Mullen et al. 1980), uveitis (Dua et al. 1992), and neuritis (Gregorian et al. 1993), and in the NOD model of diabetes (Fife et al. 2006). Ag–SP therapy appears to be nontoxic and well tolerated at all stages of disease progression. Unlike soluble-peptide therapy, in which the tolerizing Ag can induce an anaphylactic response resulting in the death of treated mice, Ag–SP therapy does not induce an anaphylactic response regardless of the Ag used (Pedotti et al. 2001; Smith et al. 2005).

I.v. administration of myelin antigen- or peptide-coupled splenocytes is a highly efficacious way to regulate EAE in mice and rats. Pretolerization of mice with the initiating myelin protein or epitope inhibits the induction of disease in various active R–EAE models. In addition, Ag–SP inhibit expression of R–EAE when administered shortly after the adoptive transfer of preactivated neuroantigen-peptide-specific T cells or during disease remission (Karpus et al. 1994; Miller and Karpus 1994; Miller et al. 1995a, 1992b; Tan et al. 1991; Vanderlugt et al. 2000). Induction of peripheral tolerance with Ag–SP has also been useful for defining immunodominant myelin proteins within the myelin sheath and immunodominant epitopes within myelin proteins. For example, MBP_{84-104}-specific tolerance significantly inhibits relapsing–remitting EAE initiated by MBP-primed lymph node-derived T cells and $PLP_{139-151}$-specific tolerance significantly inhibits active R–EAE induced by either mouse spinal cord homogenate (MSCH) or intact PLP. This not only indicates that the MBP_{84-104} and $PLP_{139-151}$ peptides are immunodominant in their respective proteins, but also that other epitopes also contribute to disease induced by the intact proteins (McRae et al. 1995a; Miller and Karpus 1994; Miller et al. 1995a, b, 1992b; Tan et al. 1992). Ag–SP tolerance has also helped define the specificity and pathological contribution of epitope spreading to endogenous myelin epitopes in clinical relapses. Tolerance studies showed a major pathological contribution of $PLP_{139-151}$-specific T cells to the relapses in MBP_{84-104}-induced R–EAE and of $PLP_{178-191}$-specific T cells to relapses in the $PLP_{139-151}$-induced EAE and that responses to the initiating epitope do not play a major role in the chronic disease phase (McRae et al. 1995a; Miller et al. 1995a; Tan et al. 1991; Vanderlugt et al. 2000; Vanderlugt and Miller 1996).

In a virally induced model of MS, TMEV, pretolerance with TMEV-coupled cells was found to induce "split-tolerance." This split-tolerance is due to tolerization of virus-specific Th1-cells, concomitant with activation of virus-specific Th2-cell responses (Karpus et al. 1994; Peterson et al. 1993). Split-tolerance was characterized by a decrease in virus-specific delayed-type hypersensitivity (DTH) and IgG2a Ab responses, but a normal to elevated IgG1 Ab response compared to sham tolerized control mice (Peterson et al. 1993). Likewise, the level of T cell-dependent IL-2 and IFNg produced were decreased upon rechallenge with viral epitopes, but no change was observed in Th2-cell-derived (IL-4) cytokine levels. ECDI-fixed peripheral-blood lymphocytes (PBLs) coupled with MBP peptides also selectively induced anergy in vitro in human Th1-cell but not Th2-cell clones (Vandenbark et al. 2000). While oral administration of soluble $PLP_{139-151}$ to SJL mice efficiently prevented acute and relapsing EAE induced by either $PLP_{139-151}$ or intact PLP (Karpus et al. 1996), tolerance induced with $PLP_{139-151}$-coupled cells significantly downregulated ongoing adoptive R–EAE when administered at either disease onset or the peak of acute disease. Oral tolerance is not effective at these time points (Kennedy et al. 1997).

The mechanism(s) underlying Ag–SP-induced tolerance has yet to be fully elucidated, however the route of administration, dosage, levels of costimulation (the two-signal hypothesis), Th cell polarization, and Treg cell induction are all likely factors that contribute to efficacy of treatment. A likely mechanism for Ag–SP tolerance is by suboptimal T cell activation through the engagement of the TCR in the absence of costimulation. In vitro studies revealed that ECDI-treated splenocytes pulsed with soluble peptide Ag are unable to deliver critical costimulatory signals for activation of Th1 clones leading to anergy (Jenkins and Schwartz 1987). Data from our laboratory demonstrates that the effectiveness of Ag–SP tolerance in PLP$_{139-151}$-induced EAE in the SJL mouse is dependent on having a low level of CD80 and CD86 expression on the fixed APCs (Eagar et al. 2002). This is further supported by the observation that blocking CTLA-4:CD80/CD86 interaction at the time of secondary antigen encounter reverses the tolerized state (Eagar et al. 2004). Programmed cell death ligand-1 (PDL-1)/PD1 engagement has also been demonstrated to be important for maintenance of insulin-coupled-cell-induced tolerance in the NOD diabetes model (Fife et al. 2006) and in islet graft transplant survival with allogeneic-coupled-cell-tolerance (Lou et al. 2008). Efficiency of this therapy is also critically dependent upon i.v. administration of antigen-coupled cells, whereas neither i.p. nor s.c. injection are effective at inducing tolerance - with the latter actually enhancing immune responses to the target Ag (Tan et al. 1992). Furthermore, ECDI fixation of cells is absolutely necessary for induction of tolerance; however, de novo Ag processing by the donor cells is not a contributing factor, as the inclusion of Ag-processing inhibitors in the coupling reaction does not reverse the tolerance phenotype (Pope et al. 1992).

The mechanism of Ag–SP tolerance is believed to be twofold through both direct and indirect interaction of the Ag–SP and recipient T cells (Fig. 1). We have hypothesized that direct induction of tolerance occurs via interaction of host autoreactive CD4$^+$ T cells with the donor Ag–SP cells and is strongly dependent on costimulation. Alternatively indirect tolerance can occur through representation of the bound Ag by host APCs such as splenic macrophages, B cells, and immature dendritic cells (DCs,) that phagocytize the donor Ag–SP and reprocess and represent the bound antigen to host T cells in a tolerogenic fashion in a costimulatory-deficient manner (Turley and Miller 2007). Ag–SP tolerance induction occurs even if the donor cells are coupled with intact proteins or if they are derived from donors that are deficient in MHC class I and/or MHC class II molecules; however twice the number of donor MHC deficient donor cells are required to induce levels of protection equivalent to that of syngeneic MHC-expressing donor cells. Additionally, ECDI fixation actively induces apoptosis of the donor cells (Turley and Miller 2007) likely assisting in the ability of the peptide-coupled allogeneic derived donor cells to tolerize recipient T cells. In addition to these findings, tolerance is inhibited in splenectomized recipients (Turley and Miller, unpublished observations) suggesting the requirement for a recipient splenic APC population for representation of Ag in a tolerance-inducing manner. Tolerizing Ag can also be coupled to donor red blood cells (RBCs) to induce tolerance capable of both preventing disease induction and treating ongoing EAE (Turley and Miller, unpublished observations). Use of RBCs as donor carrier cells greatly increases the clinical efficacy of Ag–SP as a therapy for MS, as RBCs are more readily available than purified APCs. Collectively this data suggests that ECDI

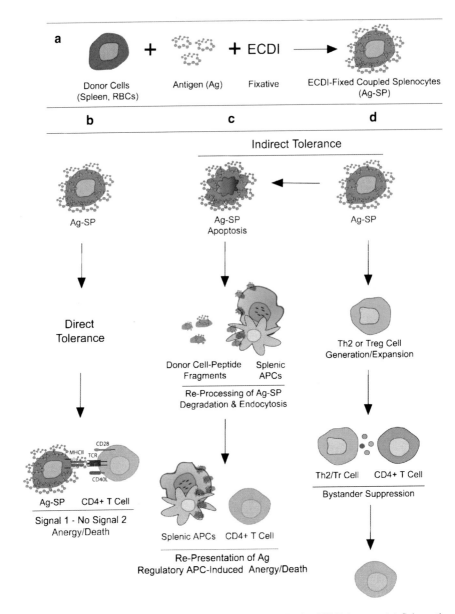

Fig. 1 Potential Mechanisms of ECDI-fixed Ag-coupled cell (Ag–SP) Tolerance. (**a**) Schematic representation of Ag–SP formation. Donor cells [bulk splenocytes or red blood cells (RBCs)] are fixed with ethylene carbodiimide (ECDI) in the presence of peptide for 1 h at 4°C to generate Ag–SP. (**b–d**) Possible mechanisms of Ag–SP-induced tolerance: Ag–SP can induce tolerance by direct (**b**) or indirect mechanisms (**c, d**). (**b**) Donor Ag–SP can directly interact with host antigen-specific T cells delivering signal 1 (MHC/peptide:TCR) without signal 2 (costimulation; B7–1,-2/CD28; CD40/CD40L), rendering the cells anergic. (**c**) Alternatively Ag–SP can induce indirect or cross-tolerance as the donor Ag–SP undergo apoptosis, leading to phagocytic uptake by host splenic antigen presenting cells (APCs), which then can present peptide fragments to host T cells inducing anergy. (**d**) Another possible mechanism for Ag–SP tolerance is the indirect generation/expansion of a Th2 or regulatory T cell (Treg) population, which through bystander suppression can also render antigen-specific CD4+ T cells tolerant

nonspecifically cross-links Ag to the cell surface while inducing apoptosis, which allows for the donor cells to be perceived by the host in a nonimmunogenic fashion in the spleen. This would aid in the ability of immature/tolerogenic host splenic APCs to reprocess and represent the coupled Ag in a tolerance-inducing manner.

Another interesting aspect of Ag–SP tolerance is the finding that multiple peptides can be coupled to a donor cell, allowing for simultaneous targeting of multiple T cell specificities. This may be critical for directed tolerance therapy of chronic autoimmune diseases, in which responses to multiple tissue Ags, activated by epitope spreading, are likely important in sustaining the destruction of self-tissue (Smith and Miller 2006). T cells removed from the CNS of mice with ongoing EAE that were tolerized at the peak of the acute clinical stage of disease, with Ag–SP also making increased amounts of the antiinflammatory cytokines IL-10 and TGFβ, compared with control mice, although the levels of these cytokines in the periphery remained roughly equivalent to control mice (Smith and Miller 2006). This pattern of increased regulatory cytokine production is suggestive of an increase in natural Treg function in ongoing EAE. Tregs appear to have a role in Ag–SP tolerance and are required for the long-term maintenance of tolerance, but are not necessary for tolerance induction (Feeny and Miller, unpublished observations). However the opposite was found in the allogeneic-coupled-cell tolerance in islet transplant survival - Tregs were found to be required for tolerance induction (Lou et al. 2008). Additionally, necessity of Treg cells in adoptively transferred tolerance also suggests that a regulatory population of cells is induced, activated and/or expanded by the administration of Ag–SP (Smith and Miller, unpublished observations) and necessary for the transfer of tolerance. Based on current success of Ag–SP cell tolerance in treating EAE, this therapy is in the final stages of toxicity testing for an initial Phase I/IIa clinical trial designed to test safety and efficacy of this therapy in treating new-onset RRMS patients. It is proposed to intravenously reinfuse autologous peripheral blood leukocytes (PBLs), collected from patients using leukocytaphaeresis, ECDI-coupled with a cocktail of 5–7 previously identified immunodominant myelin peptides including MBP_{13-32}, $MBP_{111-129}$, MBP_{83-99}, $MBP_{146-170}$, MOG_{1-20}, MOG_{35-55} and $PLP_{139-154}$ which cause T cell expansion in peripheral blood T cells of MS patients (Bielekova et al. 2004). The goal is to induce long-term tolerance in effector autoreactive T cells as well as prevent future relapses by tolerizing naïve T cells specific for potential endogenously released myelin epitopes, without compromising immune responses to foreign pathogens. Our data demonstrating use of RBCs as efficient donor ECDI-fixed carrier cells to induce tolerance lends support to the clinical efficacy of Ag–SP tolerance induction for treatment of human autoimmune disease.

4   Conclusions

When moving research studies from the lab to the clinic, it is necessary to keep in mind that a contributing factor to the variable results gained from Ag-specific tolerance approaches which are initially used for treatment of disease models in inbred mice, and those observed to date in humans in the clinical setting, is likely the effect of the diverse nature of human MHC polymorphisms. Continued research to better understand underlying molecular mechanisms of tolerance and enhance specificity and efficacy of each of these treatment strategies, perhaps using combinatorial approaches, is thus necessary to deal with the complexity of the human immune system.

Development/identification of therapeutics that either inhibit signalling intermediates necessary for T cell activation or activate T cell anergy-associated signalling intermediates, when used in combination with therapies such as Ag–SP, presents a possible combinatorial strategy that may increase therapeutic efficacy. Therefore, continued research to enhance specificity and efficacy of treatment in Ag–SP as well as other aforementioned antigen-specific strategies is necessary in both animal models as well as in the clinic with advanced patient screening using modern genomic and pharmacogenomic techniques. Use of these tolerogenic approaches in combination with non Ag-specific therapies also has the potential to one day provide "tailored therapy" to deal with the complexity of the human immune system.

References

Aharoni R, Teitelbaum D, Arnon R, Sela M (1999) Copolymer 1 acts against the immunodominant epitope 82–100 of myelin basic protein by T cell receptor antagonism in addition to major histocompatibility complex blocking. Proc Natl Acad Sci U S A 96:634–639

Anderton SM, Wraith DC (2002) Selection and fine-tuning of the autoimmune T-cell repertoire. Nat Rev Immunol 2:487–498

Bai XF, Li HL, Shi FD, Liu JQ, Xiao BG, van der Meide PH, Link H (1998) Complexities of applying nasal tolerance induction as a therapy for ongoing relapsing experimental autoimmune encephalomyelitis (EAE) in DA rats. Clin Exp Immunol 111:205–210

Bai XF, Shi FD, Xiao BG, Li HL, van der Meide PH, Link H (1997) Nasal administration of myelin basic protein prevents relapsing experimental autoimmune encephalomyelitis in DA rats by activating regulatory cells expressing IL-4 and TGF-beta mRNA. J Neuroimmunol 80:65–75

Barnett ML, Kremer JM, St Clair EW, Clegg DO, Furst D, Weisman M, Fletcher MJ, Chasan-Taber S, Finger E, Morales A et al. (1998) Treatment of rheumatoid arthritis with oral type II collagen. Results of a multicenter, double-blind, placebo-controlled trial. Arthritis Rheum 41:290–297

Belghith M, Bluestone JA, Barriot S, Megret J, Bach JF, Chatenoud L (2003) TGF-beta-dependent mechanisms mediate restoration of self-tolerance induced by antibodies to CD3 in overt autoimmune diabetes. Nat Med 9:1202–1208

Benson JM, Stuckman SS, Cox KL, Wardrop RM, Gienapp IE, Cross AH, Trotter JL, Whitacre CC (1999) Oral administration of myelin basic protein is superior to myelin in suppressing established relapsing experimental autoimmune encephalomyelitis. J Immunol 162:6247–6254

Bernard CC, de Rosbo NK (1991) Immunopathological recognition of autoantigens in multiple sclerosis. Acta Neurol (Napoli) 13:171–178

Bielekova B, Goodwin B, Richert N, Cortese I, Kondo T, Afshar G, Gran B, Eaton J, Antel J, Frank JA et al. (2000) Encephalitogenic potential of the myelin basic protein peptide (amino acids 83–99) in multiple sclerosis: results of a phase II clinical trial with an altered peptide ligand. Nat Med 6:1167–1175

Bielekova B, Sung MH, Kadom N, Simon R, McFarland H, Martin R (2004) Expansion and functional relevance of high-avidity myelin-specific CD4+ T cells in multiple sclerosis. J Immunol 172:3893–3904

Bitar DM, Whitacre CC (1988) Suppression of experimental autoimmune encephalomyelitis by the oral administration of myelin basic protein. Cell Immunol 112:364–370

Bornstein MB, Miller A, Slagle S, Weitzman M, Crystal H, Drexler E, Keilson M, Merriam A, Wassertheilsmoller S, Spada V et al. (1987) A Pilot Trial of Cop-1 in Exacerbating Remitting Multiple-Sclerosis. N Engl J Med 317:408–414

Braley-Mullen H, Tompson JG, Sharp GC, Kyriakos M (1980) Suppression of experimental autoimmune thyroiditis in guinea pigs by pretreatment with thyroglobulin-coupled spleen cells. Cell Immunol 51:408–413

Burstein HJ, Shea CM, Abbas AK (1992) Aqueous antigens induce in vivo tolerance selectively in IL-2- and IFN-gamma-producing (Th1) cells. J Immunol 148:3687–3691

Chatenoud L (2003) CD3-specific antibody-induced active tolerance: from bench to bedside. Nat Rev Immunol 3:123–132

Chatenoud L, Thervet E, Primo J, Bach JF (1994) Anti-CD3 antibody induces long-term remission of overt autoimmunity in nonobese diabetic mice. Proc Natl Acad Sci USA. 91:123–127

Chen Y, Kuchroo VK, Inobe J, Hafler DA, Weiner HL (1994) Regulatory T cell clones induced by oral tolerance: suppression of autoimmune encephalomyelitis. Science 265:1237–1240

Christen U, von Herrath MG (2004) Initiation of autoimmunity. Curr Opin Immunol 16:759–767

Critchfield JM, Racke MK, Zuniga-Pflucker JC, Cannella B, Raine CS, Goverman J, Lenardo MJ (1994) T cell deletion in high antigen dose therapy of autoimmune encephalomyelitis. Science 263:1139–1143

Dal Canto MC, Kim BS, Miller SD, Melvold RW (1997) Theiler's murine encephalomyelitis virus (TMEV)-induced demyelination: A model for human multiple sclerosis. Neuroprotocols 10:453–461

de Rosbo NK, Hoffman M, Mendel I, Yust I, Kaye J, Bakimer R, Flechter S, Abramsky O, Milo R, Karni A, Ben-Nun A (1997). Predominance of the autoimmune response to myelin oligodendrocyte glycoprotein (MOG) in multiple sclerosis: reactivity to the extracellular domain of MOG is directed against three main regions. Eur J Immunol 27:3059–3069

Dua HS, Gregerson DS, Donoso LA (1992) Inhibition of experimental autoimmune uveitis by retinal photoreceptor a ntigens coupled to spleen cells. Cell Immunol 139:292–305

Duda PW, Schmied MC, Cook SL, Krieger JI, Hafler DA (2000) Glatiramer acetate (Copaxone (R)) induces degenerate, Th2-polarized immune responses in patients with multiple sclerosis. J Clin Invest 105:967–976

Eagar TN, Karandikar NJ, Bluestone J, Miller SD (2002) The role of CTLA-4 in induction and maintenance of peripheral T cell tolerance. Eur J Immnol 32:972–981

Eagar TN, Turley DM, Padilla J, Karandikar NJ, Tan LJ, Bluestone JA, Miller SD (2004) CTLA-4 regulates expansion and differentiation of Th1 cells following induction of peripheral T cell tolerance. J Immunol 172:7442–7450

Ebers GC, Sadovnick AD, Risch NJ (1995) A genetic basis for familial aggregation in multiple sclerosis. Nature 377:150–151

Faria AM, Weiner HL (1999) Oral tolerance: mechanisms and therapeutic applications. Adv Immunol 73:153–264

Faria AM, Weiner HL (2006) Oral tolerance: therapeutic implications for autoimmune diseases. Clin Dev Immunol 13:143–157

Fife BT, Guleria I, Gubbels Bupp M, Eagar TN, Tang Q, Bour-Jordan H, Yagita H, Azuma M, Sayegh MH, Bluestone JA (2006) Insulin-induced remission in new-onset NOD mice is maintained by the PD-1-PD-L1 pathway. J Exp Med 203:2737–2747

Fujinami RS, Oldstone MB (1985) Amino acid homology between the encephalitogenic site of myelin basic protein and virus: mechanism for autoimmunity. Science 230:1043–1045

Gaur A, Wiers B, Liu A, Rothbard J, Fathman CG (1992) Amelioration of autoimmune encephalomyelitis by myelin basic protein synthetic peptide-induced anergy. Science 258:1491–1494

Genain CP, Abel K, Belmar N, Villinger F, Rosenberg DP, Linington C, Raine CS, Hauser SL (1996) Late complications of immune deviation therapy in a nonhuman primate. Science 274:2054–2057

Gregorian SK, Clark L, Heber-Katz E, Amento EP, Rostami A (1993) Induction of peripheral tolerance with peptide-specific anergy in experimental autoimmune neuritis. Cell Immunol 150:298–310

Herold KC, Burton JB, Francois F, Poumian-Ruiz E, Glandt M, Bluestone JA (2003) Activation of human T cells by FcR nonbinding anti-CD3 mAb, hOKT3gamma1(Ala-Ala). J Clin Invest 111:409–418

Herold KC, Gitelman SE, Masharani U, Hagopian W, Bisikirska B, Donaldson D, Rother K, Diamond B, Harlan DM, Bluestone JA (2005) A single course of anti-CD3 monoclonal antibody hOKT3gamma1(Ala-Ala) results in improvement in C-peptide responses and clinical parameters for at least 2 years after onset of type 1 diabetes. Diabetes 54:1763–1769

Herold KC, Hagopian W, Auger JA, Poumian-Ruiz E, Taylor L, Donaldson D, Gitelman SE, Harlan DM, Xu D, Zivin RA, Bluestone JA (2002) Anti-CD3 monoclonal antibody in new-onset type 1 diabetes mellitus. N Engl J Med 346:1692–1698

Hogquist KA, Baldwin TA, Jameson SC (2005) Central tolerance: learning self-control in the thymus. Nat Rev Immunol 5:772–782

Jenkins MK, Schwartz RH (1987) Antigen presentation by chemically modified splenocytes induces antigen-specific T cell unresponsiveness in vitro and in vivo. J Exp Med 165:302–319

Johnson KP, Brooks BR, Cohen JA, Ford CC, Goldstein J, Lisak RP, Myers LW, Panitch HS, Rose JW, Schiffer RB (1995) Copolymer 1 reduces relapse rate and improves disability in relapsing-remitting multiple sclerosis: results of a phase III multicenter, double-blind placebo-controlled trial. The Copolymer 1 Multiple Sclerosis Study Group. Neurology 45:1268–1276

Kappos L, Comi G, Panitch H, Oger J, Antel J, Conlon P, Steinman L (2000) Induction of a non-encephalitogenic type 2 T helper-cell autoimmune response in multiple sclerosis after administration of an altered peptide ligand in a placebo-controlled, randomized phase II trial. The Altered Peptide Ligand in Relapsing MS Study Group. Nat Med 6:1176–1182

Karandikar NJ, Vanderlugt CL, Bluestone JA, Miller SD (1998) Targeting the B7/CD28:CTLA-4 costimulatory system in CNS autoimmune disease. J Neuroimmunol 89:10–18

Karpus WJ, Kennedy KJ, Smith WS, Miller SD (1996) Inhibition of relapsing experimental autoimmune encephalomyelitis in SJL mice by feeding the immunodominant PLP139–151 molecule. J Neurosci Res 45:410–423

Karpus WJ, Lukacs NW, McRae BL, Streiter RM, Kunkel SL, Miller SD (1995) An important role for the chemokine macrophage inflammatory protein-1 alpha in the pathogenesis of the T cell-mediated autoimmune disease, experimental autoimmune encephalomyelitis. J Immunol 155:5003–5010

Karpus WJ, Peterson JD, Miller SD (1994) Anergy in vivo: Down-regulation of antigen-specific CD4+ Th1 but not Th2 cytokine responses. Int Immunol 6:721–730

Kennedy KJ, Smith WS, Miller SD, Karpus WJ (1997) Induction of antigen-specific tolerance for the treatment of ongoing, relapsing autoimmune encephalomyelitis – A comparison between oral and peripheral tolerance. J Immunol 159:1036–1044

Kennedy MK, Tan LJ, Dal Canto MC, Miller SD (1990a) Regulation of the effector stages of experimental autoimmune encephalomyelitis via neuroantigen-specific tolerance induction. J Immunol 145:117–126

Kennedy MK, Tan LJ, Dal Canto MC, Tuohy VK, Lu ZJ, Trotter JL, Miller SD (1990b) Inhibition of murine relapsing experimental autoimmune encephalomyelitis by immune tolerance to proteolipid protein and its encephalitogenic peptides. J Immunol 144:909–915

Keymeulen B, Vandemeulebroucke E, Ziegler AG, Mathieu C, Kaufman L, Hale G, Gorus F, Goldman M, Walter M, Candon S et al. (2005) Insulin needs after CD3-antibody therapy in new-onset type 1 diabetes. N Engl J Med 352:2598–2608

Khalili K, White MK, Lublin F, Ferrante P, Berger JR (2007) Reactivation of JC virus and development of PML in patients with multiple sclerosis. Neurology 68:985–990

Khoury SJ, Hancock WW, Weiner HL (1992) Oral tolerance to myelin basic protein and natural recovery from experimental autoimmune encephalomyelitis are associated with downregulation of inflammatory cytokines and differential upregulation of transforming growth factor beta, interleukin 4, and prostaglandin E expression in the brain. J Exp Med 176:1355–1364

Kohm AP, Williams JS, Bickford AL, McMahon JS, Chatenoud L, Bach JF, Bluestone JA, Miller SD (2005) Treatment with nonmitogenic anti-CD3 monoclonal antibody induces CD4+ T cell unresponsiveness and functional reversal of established experimental autoimmune encephalomyelitis. J Immunol 174:4525–4534

Korczyn AD, Nisipeanu P (1996) Safety profile of copolymer 1: Analysis of cumulative experience in the United States and Israel. J Neruol 243:S23–S26

Kurtzke JF (1993) Epidemiologic evidence for multiple sclerosis as an infection. Clin Microbiol Rev 6:382–427

Luo X, Pothoven KL, McCarthy D, DeGutes M, Martin A, Getts DR, Xia G, He J, Zhang X, Kaufman DB, Miller SD (2008) ECDI-fixed allogeneic splenocytes induce donor-specific tolerance for long-term survival of islet transplants via two distinct mechanisms. Proc Natl Acad Sci U S A 105:14527–14532

Mayer L, Shao L (2004) Therapeutic potential of oral tolerance. Nat Rev Immunol 4:407–419

McRae BL, Vanderlugt CL, Dal Canto MC, Miller SD (1995a) Functional evidence for epitope spreading in the relapsing pathology of experimental autoimmune encephalomyelitis. J Exp Med 182:75–85

McRae BL, Vanderlugt CL, Dal Canto MC, Miller SD (1995b) Functional evidence for epitope spreading in the relapsing pathology of experimental autoimmune encephalomyelitis. J Exp Med 182:75–85

Metzler B, Wraith DC (1993) Inhibition of experimental autoimmune encephalomyelitis by inhalation but not oral administration of the encephalitogenic peptide: influence of MHC binding affinity. Int Immunol 5:1159–1165

Meyer AL, Benson JM, Gienapp IE, Cox KL, Whitacre CC (1996) Suppression of murine chronic relapsing experimental autoimmune encephalomyelitis by the oral administration of myelin basic protein. J Immunol 157:4230–4238

Miller A, Lider O, Roberts AB, Sporn MB, Weiner HL (1992a) Suppressor T cells generated by oral tolerization to myelin basic protein suppress both in vitro and in vivo immune responses by the release of transforming growth factor after antigen-specific triggering. Proc Nat Acad Sci 89:421–425

Miller DH, Khan OA, Sheremata WA, Blumhardt LD, Rice GPA, Libonati MA, Willmer-Hulme AJ, Dalton CM, Miszkiel KA, O'Connor PW (2003) A controlled trial of natalizumab for relapsing multiple sclerosis. N Engl J Med 348:15–23

Miller SD, Karpus WJ (1994) The immunopathogenesis and regulation of T-cell mediated demyelinating diseases. Immunol Today 15:356–361

Miller SD, McRae BL, Vanderlugt CL, Nikcevich KM, Pope JG, Pope L, Karpus WJ (1995a) Evolution of the T cell repertoire during the course of experimental autoimmune encephalomyelitis. Immunol Rev 144:225–244

Miller SD, Tan LJ, Pope L, McRae BL, Karpus WJ (1992b) Antigen-specific tolerance as a therapy for experimental autoimmune encephalomyelitis. Int Rev Immunol 9:203–222

Miller SD, Vanderlugt CL, Begolka WS, Pao W, Yauch RL, Neville KL, Katz-Levy Y, Carrizosa A, Kim BS (1997) Persistent infection with Theiler's virus leads to CNS autoimmunity via epitope spreading. Nat Med 3:1133–1136

Miller SD, Vanderlugt CL, Lenschow DJ, Pope JG, Karandikar NJ, Dal Canto MC, Bluestone JA (1995b) Blockade of CD28/B7-1 interaction prevents epitope spreading and clinical relapses of murine EAE. Immunity 3:739–745

Miller SD, Wetzig RP, Claman HN (1979) The induction of cell-mediated immunity and tolerance with protein antigens coupled to syngeneic lymphoid cells. J Exp Med 149:758–773

Mowat AM, Parker LA, Beacock-Sharp H, Millington OR, Chirdo F (2004) Oral tolerance: overview and historical perspectives. Ann N Y Acad Sci 1029:1–8

Mowat AM, Strobel S, Drummond HE, Ferguson A (1982) Immunological responses to fed protein antigens in mice. I. Reversal of oral tolerance to ovalbumin by cyclophosphamide. Immunology 45:105–113

Neuhaus O, Farina C, Yassouridis A, Wiendl H, Bergh FT, Dose T, Wekerle H, Hohlfeld R (2000) Multiple sclerosis: Comparison of copolymer-1-reactive T cell lines from treated and untreated subjects reveals cytokine shift from T helper 1 to T helper 2 cells. Proc Nat Acad Sci U S A 97:7452–7457

Nicholson LB, Greer JM, Sobel RA, Lees MB, Kuchroo VK (1995) An altered peptide ligand mediates immune deviation and prevents autoimmune encephalomyelitis. Immunity 3:397–405

Nicholson LB, Kuchroo VK (1997) T cell recognition of self and altered self antigens. Crit Rev Immunol 17:449–462

Nicholson LB, Murtaza A, Hafler BP, Sette A, Kuchroo VK (1997) A T cell receptor antagonist peptide induces T cells that mediate bystander suppression and prevent autoimmune encephalomyelitis induced with multiple myelin antigens. Proc Nat Acad Sci U S A 94:9279–9284

Olson JK, Croxford JL, Calenoff M, Dal Canto MC, Miller SD (2001a) A virus-induced molecular mimicry model of multiple sclerosis. J Clin Invest 108:311–318

Olson JK, Croxford JL, Miller SD (2001b) Virus-induced autoimmunity: Potential role of viruses in initiation, perpetuation, and progression of T cell-mediated autoimmune diseases Viral Immunol 14 227–250

Ota K, Matsui M, Milford EL, Mackin GA, Weiner HL, Hafler DA (1990) T-cell recognition of an immunodominant myelin basic protein epitope in multiple sclerosis. Nature 346:183–187

Paterson PY, Swanborg RH (1988) Demyelinating diseases of the central and peripheral nervous systems. In: Sampter M, Talmage DW, Frank MM, Austen KF, Claman HN (eds) Immunological diseases. Brown, Boston, Little, pp 1877–1916

Pedotti R, Mitchell D, Wedemeyer J, Karpuj M, Chabas D, Hattab EM, Tsai M, Galli, SJ, Steinman L (2001) An unexpected version of horror autotoxicus: anaphylactic shock to a self-peptide. Nat Immunol 2:216–222

Peterson JD, Karpus WJ, Clatch RJ, Miller SD (1993) Split tolerance of Th1 and Th2 cells in tolerance to Theiler's murine encephalomyelitis virus. Eur J Immunol 23:46–55

Pope L, Paterson PY, Miller SD (1992) Antigen-specific inhibition of the adoptive transfer of experimental autoimmune encephalomyelitis in Lewis rats. J Neuroimmunol 37:177–190

Pozzilli P, Pitocco D, Visalli N, Cavallo MG, Buzzetti R, Crino A, Spera S, Suraci C, Multari G, Cervoni M, et al. (2000) No effect of oral insulin on residual beta-cell function in recent-onset type I diabetes (the IMDIAB VII). IMDIAB Group. Diabetologia 43:1000–1004

Racke MK, Critchfield JM, Quigley L, Cannella B, Raine CS, McFarland Hf, Lenardo MJ (1996) Intravenous antigen administration as a therapy for autoimmune demyelinating disease. Ann Neurol 39:46–56

Ruddle NH, Bergman CM, McGrath KM, Lingenheld EG, Grunnet ML, Padula SJ, Clark RB (1990) An antibody to lymphotoxin and tumor necrosis factor prevents transfer of experimental allergic encephalomyelitis. J Exp Med 172:1193–1200

Samson MF, Smilek DE (1995) Reversal of acute experimental autoimmune encephalomyelitis and prevention of relapses by treatment with a myelin basic protein peptide analogue modified to form long-lived peptide-MHC complexes. J Immunol 155:2737–2746

Sharpe AH, Freeman GJ (2002) The B7-CD28 superfamily. Nat Rev Immunol 2:116–126

Slavin AJ, Maron R, Weiner HL (2001) Mucosal administration of IL-10 enhances oral tolerance in autoimmune encephalomyelitis and diabetes. Int Immunol 13:825–833

Smith CE, Eagar TN, Strominger JL, Miller SD (2005) Differential induction of IgE-mediated anaphylaxis after soluble vs. cell-bound tolerogenic peptide therapy of autoimmune encephalomyelitis. Proc Natl Acad Sci U S A 102:9595–9600

Smith CE, Miller SD (2006) Multi-peptide coupled-cell tolerance ameliorates ongoing relapsing EAE associated with multiple pathogenic autoreactivities. J Autoimmunity 27:218–231

Su XM, Sriram S (1991) Treatment of chronic relapsing experimental allergic encephalomyelitis with the intravenous administration of splenocytes coupled to encephalitogenic peptide 91–103 of myelin basic protein. J Neuroimmunol 34:181–190

Tan LJ, Kennedy MK, Dal Canto MC, Miller SD (1991) Successful treatment of paralytic relapses in adoptive experimental autoimmune encephalomyelitis via neuroantigen- specific tolerance. J Immunol 147:1797–1802

Tan LJ, Kennedy MK, Miller SD (1992) Regulation of the effector stages of experimental autoimmune encephalomyelitis via neuroantigen-specific tolerance induction. II. Fine specificity of effector T cell inhibition. J Immunol 148:2748–2755

Turley DM, Miller SD (2007) Peripheral tolerance Induction using ethylenecarbodiimide-fixed APCs uses both direct and indirect mechanisms of antigen presentation for prevention of experimental autoimmune encephalomyelitis. J Immunol 178:2212–2220

Utset TO, Auger JA, Peace D, Zivin RA, Xu D, Jolliffe L, Alegre ML, Bluestone JA, Clark MR (2002) Modified anti-CD3 therapy in psoriatic arthritis: a phase I/II clinical trial. J Rheumatol 29:1907–1913

Vandenbark AA, Barnes D, Finn T, Bourdette DN, Whitham R, Robey I, Kaleeba J, Bebo BF Jr, Miller SD, Offner H, Chou YK (2000) Differential susceptibility of human T(h)1 versus T (h) 2 cells to induction of anergy and apoptosis by ECDI/antigen-coupled antigen- presenting cells. Int Immunol 12:57–66

Vandenbark AA, Vainiene M, Ariail K, Miller SD, Offner H (1996) Prevention and treatment of relapsing autoimmune encephalomyelitis with myelin peptide-coupled splenocytes. J Neurosci Res 45:430–438

Vanderlugt CL, Eagar TN, Neville KL, Nikcevich KM, Bluestone JA, Miller SD (2000) Pathologic role and temporal appearance of newly emerging autoepitopes in relapsing experimental autoimmune encephalomyelitis. J Immunol 164:670–678

Vanderlugt CL, Miller SD (1996) Epitope spreading. Curr Opin Immunol 8:831–836

Waksman BH (1995) Multiple sclerosis: More genes versus environment. Nature 377:105–106

Walker LS, Abbas AK (2002) The enemy within: keeping self-reactive T cells at bay in the periphery. Nat Rev Immunol 2:11–19

Warren KG, Catz I, Wucherpfennig KW (1997) Tolerance induction to myelin basic protein by intravenous synthetic peptides containing epitope P85 VVHFFKNIVTP96 in chronic progressive multiple sclerosis. J Neurol Sci 152:31–38

Weiner HL (2004) Current issues in the treatment of human diseases by mucosal tolerance. Ann N Y Acad Sci 1029:211–224

Weiner HL, Mackin GA, Matsui M, Orav EJ, Khoury SJ, Dawson DM, Hafler DA (1993) Double-blind pilot trial of oral tolerization with myelin antigens in multiple sclerosis. Science 259:1321–1324

Wekerle H (1991) Immunopathogenesis of multiple sclerosis. Acta Neurol (Napoli) 13:197–204

Whitacre CC, Gienapp IE, Orosz CG, Bitar DM (1991) Oral tolerance in experimental autoimmune encephalomyelitis: III. Evidence for clonal anergy. J Immunol 147:2155–2163

Wraith DC, Smilek DE, Mitchell DJ, Steinman L, McDevitt HO (1989) Antigen recognition in autoimmune encephalomyelitis and the potential for peptide-mediated immunotherapy. Cell 59:247–255

Wucherpfennig KW, Strominger JL (1995) Molecular mimicry in T cell-mediated autoimmunity: viral peptides activate human T cell clones specific for myelin basic protein. Cell 80:695–705

Yednock TA, Cannon C, Fritz LC, Sanchez-Madrid F, Steinman L, Karin N (1992) Prevention of experimental autoimmune encephalomyelitis by antibodies against α4β1 integrin. Nature 356:63–66

Young DA, Lowe LD, Booth SS, Whitters MJ, Nicholson L, Kuchroo VK, Collins M (2000) IL-4, IL-10, IL-13, and TGF-beta from an altered peptide ligand-specific Th2 cell clone down-regulate adoptive transfer of experimental autoimmune encephalomyelitis. J Immunol 164:3563–3572

Zhang X, Hupperts R, De Baets M (2003) Monoclonal antibody therapy in experimental allergic encephalomyelitis and multiple sclerosis. Immunol Res 28:61–78

Immuno-Therapeutic Potential of Haematopoietic and Mesenchymal Stem Cell Transplantation in MS

Paolo A. Muraro and Antonio Uccelli

Abstract In the last few years there has been extraordinary progress in the field of stem cell research. Two types of stem cells populate the bone marrow: haematopoietic stem/progenitor cells (HSC) and mesenchymal stem cells (MSC). The capacity of HSC to repopulate the blood has been known and exploited thera-peutically for at least four decades. Today, haematopoietic stem cell transplantation (HSCT) holds a firm place in the therapy of some haematological malignancies, and a potential role of HSCT for treatment of severe autoimmune diseases has been explored in small-scale clinical studies. Multiple sclerosis (MS) is the noncancerous immune mediated disease for which the greatest number of transplants has been performed to date. The results of clinical studies are double-faced: on the one hand, HSCT has demonstrated powerful effects on acute inflammation, arresting the development of focal CNS lesions and clinical relapses; on the other hand, the treatment did not arrest chronic worsening of disability in most patients with secondary progressive MS, suggesting limited or no beneficial effects on the chronic processes causing progressive disability. MSC are a more recent addition to the range of experimental therapies being developed to treat MS. While interest in MSC usage was originally raised by their potential capacity to differentiate into different cell lineages, recent work showing their interesting immunological properties has led to a revised concept, envisioning their utilization for immuno-modulatory purposes. In this review we will summarize the current clinical and experimental evidence on HSC and MSC and outline some key questions warranting further investigation in this exciting research area.

P.A. Muraro(✉)
Division of Neuroscience and Mental Health, Faculty of Medicine,
Imperial College Building 560/Burlington Danes, Room E415,
Imperial College London, 160 Du Cane Road, London
e-mail: p.muraro@imperial.ac.uk

A. Uccelli
Department of Neurosciences, Ophthalmology and Genetics, University of Genoa, Italy
Centre of Excellence for Biomedical Research, University of Genoa, Italy

Results Probl Cell Differ, doi:10.1007/400_2008_14
© Springer-Verlag Berlin Heidelberg 2009

1   Bone-Marrow-Derived Stem Cells

Stem cells are defined by their capacity to self-renew and differentiate into mature cell types. While embryonic stem cells exhibit the capacity to differentiate into any possible cell type and regenerate most tissues and organs, adult stem cells have more limited capabilities but their procurement and isolation are more straightforward in ethical and practical terms. In adults, the bone marrow is the home of two types of stem cells that have relevant immuno-therapeutic potential: haematopoietic stem/progenitor cells (HSC) and mesenchymal stem/stromal cells (MSC).

1.1   The Haematopoietic Niche

Within the bone cavity the developing hematopoietic cells are surrounded by stromal cells contributing to the formation of the HSCs niche where hematopoiesis takes place (Schofield 1978; Fig. 1). The stromal cells inside the "niche" create a

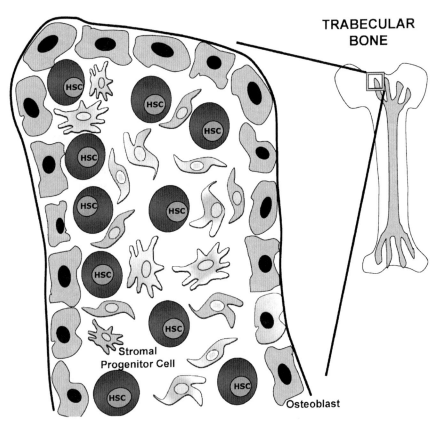

Fig. 1 Stromal cells in the hematopoietic stem cells niche within the bone marrow of trabecular bone. Multipotential stromal progenitor cells at different stages of maturation, from which MSCs originate upon in vitro culture, contribute to the formation of the HSCs niche

sheltering microenvironment that supports HSCs maintenance and self-renewal by shielding them from differentiation and apoptotic stimuli that would challenge stem cell reserves. On the other hand, the niche controls HSCs differentiation and release of mature progenies in the vascular system.

1.2   Haematopoietic Stem Cells and Their Properties

Large numbers of HSCs are generated during life to populate the blood and immune system, and to maintain homeostasis. HSCs can differentiate into the main haemato- and lymphopoietic precursors which in turn evolve into mature cell lineages. The differentiation of HSCs into T lymphocytic lineage is schematically depicted in Fig. 2. At the haematopoietic precursor stage, HSCs are identifiable by the expression of cell surface markers such as the glycoproteins CD34 and CD133 and the absence of expression of mature lineage markers. HSCs can be further characterized by their clonogenic colony-forming capacity in vitro. Definitive demonstration of HSC "stemness" is the capacity to reconstitute all haemato- and lymphopoietic lineages following myeloablative treatment of a recipient. The capability of HSCs to repopulate the blood and immune organs has a multitude of clinical applications. HSCs can indeed rescue a subject from marrow failure that can be

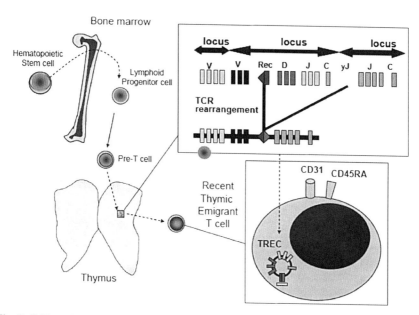

Fig. 2 Differentiation of HSC into T cells. Hematopoietic stem cells commit to the lymphoid progenitor lineage. Precursors of T cells home to the thymus where they undergo maturation and receptor rearrangement. TCR rearrangement generates excision circles (TREC) that are useful for the identification of recent thymic emigrant (RTE) cells. RTEs coexpress CD31 and CD45RA on the cell membrane

caused by marrow disorders, chemotherapy or irradiation. In haematology HSCs are routinely utilized to reconstitute blood and immune system following high-dose myelo- and/or immuno-ablative chemotherapy. HSCs have traditionally been obtained by direct aspiration of the bone marrow from the hip bones; today they are most frequently mobilized from the haematopoietic niche using immunosuppressive agents, growth factors such as G-CSF, or both. In response to the mobilization treatment, HSCs move into the blood stream and can be collected from a large vein by leukoapheresis. Umbilical cord blood is also a source of HSCs and a growing number of public or commercial cord blood banks have been established to provide an alternative source for haematopoietic transplants.

Purification of HSCs, if required by the clinical protocol, can be accomplished by CD34 selection using automated clinical-grade cell concentrators. When HSCs are harvested and reinfused into the same patient the procedure is termed autologous haematopoietic stem cell transplantation. Transplantation of HSC from a healthy, HLA-matching donor (frequently a relative, typically a sibling) is termed allogeneic transplantation. Autologous and allogeneic HSC transplantation are indicated for treatment of different malignant haematological disorders including myelodysplastic syndrome/myeloma, leukemia, and lymphoma of specific types and stages. In addition, therapeutic potential for nonneoplastic conditions including severe autoimmune diseases has been established or explored. We will discuss the role of HSCT for treatment of MS later in this chapter.

1.3 Identification of Mesenchymal Stem Cells and Their Role in Bone Marrow

Multipotential stromal precursor cells were first identified by Friedenstein and colleagues who reported on the isolation from the bone marrow (BM) of fibroblastoid, clonogenic cells capable of transferring the microenvironment of hematopoietic tissues upon transplantation to an ectopic site (Schofield 1978). From Friedenstein's pioneer observation originated more than two decades of studies showing that BM derived stromal cells are the common ancestors of mesenchymal tissues such as cartilage, fat, bone and other connective tissues (Bianco and Gehron Robey 2000). More recently such "orthodox view" of marrow stromal cell plasticity was challenged by several studies reporting their capability to differentiate also into cells from unrelated germ lineages including neural cells (Kopen et al. 1999; Pittenger et al. 1999).

BM stromal cells are not a homogeneous population as they contain an admixture of progenitors at different stages of mesodermal lineage commitment, fibroblasts and only a small number of true stem cells expressing MCAM/CD146 residing in the subadventitial space of the BM sinusoids and, upon in vivo transplantation, capable of regenerating the full hematopoietic microenvironment (Pedemonte et al. 2007; Sacchetti et al. 2007). This heterogeneity is reflected by a complex transcriptome encoding a wide array of proteins involved in a large number of diverse biological processes (Phinney et al. 2006; Pedemonte et al. 2007).

On the basis of their transdifferentiation potential, adult stromal precursor cells acquired over the years the "status" of stem cells, namely mesenchymal stem cells (MSCs), defined on the basis of their capability to grow as adherent cells on plastic displaying a fibroblast-like morphology and expressing stromal markers while lacking hematopoietic markers, and to differentiate into fat, bone, and cartilage (Horwitz et al. 2005; Uccelli et al. 2008). Nevertheless, despite several evidences showing that MSCs can transdifferentiate into multiple cell types in vitro and in vivo, their real contribution to tissue repair through significant engraftment and differentiation into biologically and functionally relevant tissue-specific cell types is still elusive (Phinney and Prockop 2007).

Although stromal cells apparently fulfilling the criteria for MSCs definition were isolated virtually from every connective tissue including fat and amniotic fluid (Meirelles et al. 2006), MSCs have been mostly characterized upon isolation from the BM. Thus, in this chapter we focus exclusively on BM – derived MSCs.

2 Haematopoietic Stem Cell Transplantation as a Therapy for Autoimmune Disease

Interest in investigating HSCT for treatment of autoimmune disease were elicited by (1) clinical observations in patients with autoimmune disease and cancer who following HSCT administered the malignancy experienced remission of the autoimmune disorder; and (2) preclinical studies of marrow transplantation in animal models of autoimmune disease such as collagen-induced arthritis and experimental autoimmune encephalomyelitis. Clinical studies of autologous HSCT for autoimmune disease were initiated in the mid-1990s and a number of Phase I and Phase II trials have been reported.

2.1 Autologous HSCT for MS: Clinical Results

Amongst nonhaematological autoimmune diseases, multiple sclerosis is the condition for which the largest number of autologous transplants has been carried out to date (>350). A detailed analysis of the clinical studies is beyond the scope of this chapter, and the subject has been reviewed in recent publications (Muraro et al. 2008; Mancardi and Saccardi 2008). Here a succinct summary of the major aspects emerging from the clinical experience will be provided. No randomized controlled studies of HSCT for MS have been completed to date and the treatment remains an experimental procedure. Early reports of clinical trials have included small number patients and follow-up duration was limited. These studies were unsuitable to establish efficacy. However, in the majority of treated subjects profound suppressive effects were documented on indices of inflammatory disease activity such as acute clinical exacerbations and new lesions. Abrogation of inflammatory disease activity after

HSCT was best demonstrated in studies that included systematic neuroimaging assessments (Mancardi et al. 2001; Saiz et al. 2001; Saccardi et al. 2005). In addition, a stabilization of the clinical course of disease posttherapy was observed in majority of patients with the exception of one study (Samijn et al. 2006). Therefore, these experiences provided proof of the principle that HSCT might have a role in the management of severe MS and raised the interesting prospect of achieving long-term disease remission by means of a one-time radical immuno-therapeutic intervention. Unfortunately, continued worsening of disability posttherapy was often observed in patients who had chronically high degree of disability before HSCT (EDSS 6.0 and above), suggesting that chronic deterioration in those patients was more likely related to secondary axonal and oligodendroglial degeneration rather than to failure to suppress an active inflammatory process, similar to observations in Campath-1H treated patients (Coles et al. 2005). However, a study showing ongoing CNS demyelination and axonal injury in association with activated microglia in spite of suppressed inflammation postHSCT in a small number of autoptic cases has raised the question that CNS resident immune cells' (i.e., microglial) activation may not be controlled by HSCT and may continue to promote CNS damage (Metz et al. 2006). This concern needs to be addressed by studying a larger number of cases, ideally by measuring microglial activation in vivo (Banati et al. 2000) posttherapy in the vast majority of patients who do not develop a fatal adverse event and survive long-term.

2.2 Unraveling the Effects of Autologous HSCT on the Immune System

Current understanding of autoimmune disorders attributes a key pathogenic role to T and B cells inappropriately recognizing self antigens and initiating a cell-mediated or humoral reaction, or both, which leads to inflammatory tissue damage (Shlomchik et al. 2001; Sospedra and Martin 2005). HSCT is probably the most radical strategy that can be applied to eliminate disease-inducing autoimmune cells. The fundamental notion of the mode of action of autologous HSCT is that immune ablation (or maximal suppression) has the capability to radically eliminate the pathogenic autoreactive cells together with all other immune cells. This "take no prisoners" approach does not require prior knowledge of the pathogenic cell specificity.

A further assumption is that a healthy immune system can be regenerated from HSCs during immune reconstitution. Obviously autologous HSCT confers no correction of any genetic defect since the cells and genetic material of the HSCs are from the same individual. However, even though genetic predisposing factors cannot be modified by autologous HSCT, there is strong evidence that MS is not a purely genetic defect and that an environmental component, possibly a microbial agent (virus or bacteria) is implicated. Therefore, it is reasonable to postulate that disease recapitulation will require a contact with the same or similar environmental agents to which the genetically predisposed individuals have been exposed prior to initial disease development. Several lines of evidence converge at suggesting that the primary infection with viruses associated with MS such as EBV infection are followed by several years of latency before the development of clinical disease (DeLorenze et al. 2006). We surmise that the latency in the development of immunopathology may extend the duration of disease remission after autologous HSCT even in patients treated with and who remain in contact with ubiquitous environmental agents that have the potential to reinitiate the immune disturbances required for disease induction, such as EBV.

Immunological work has been aimed at understanding the mechanism of action of autologous HSCT in order to provide a biological basis to explain the observed clinical and neuroradiological remissions. The demonstration of a contribution of immune renewal to posttransplant self-tolerance, however, remained elusive for almost a decade after the initiation of clinical studies. In the absence of proof of immune renewal immuno-suppression (defined as unspecific reduction in number and function of immune cells) remained the dogma in explaining the action of autologous HSCT against focal brain inflammation and relapses. Early, short-term follow-up studies of immune reconstitution after HSCT had shown a profound lymphopoenia in the first year after transplantation. The cytopoenia was observed to affect in different measure the lymphocyte subsets with the kinetics of reconstitution being related to different timing of recovery for each cell type. In the lymphocytic compartment, B cells, natural killer (NK) cells, and CD8+ T cells showed a rapid reconstitution to pretransplantation levels. In contrast, the recovery of CD4+ T cells has consistently been shown to be delayed, and often incomplete (Burt et al. 1998; Carreras et al. 2003; Farge et al. 2005; Fassas et al. 2002; Kozak et al. 2000; Nash et al. 2003; Sun et al. 2004; Guillaume et al. 1998; Passweg et al. 2004; Verburg et al. 2001). It is only by extending the longitudinal follow-up of the patients that recovery of CD4+ T cell numbers was shown to require 2 years postHSCT in young adults with MS (Muraro et al. 2005) and rheumatoid arthritis (Snowden et al. 2004), and after 12 months in children with juvenile idiopathic arthritis (JIA) (De Kleer et al. 2004). Since recovery of lymphocytes was not correlated to disease relapse, these studies demonstrated that quantitative immune depletion was an inadequate mechanistic explanation.

2.3 Immune Regeneration After HSCT

Addressing the question whether HSCT is actually followed by regeneration/renewal of the immune system required (1) a systematic analysis of the qualitative reconstitution of the immune repertoire after autologous HSCT; and (2) a reliable estimate of generation of new T cells in the thymus. The premise that transplanted HPC could generate ex novo the T cell repertoire through a revival of thymopoiesis (thymic rebound) was previously in doubt according to the notion that the thymus involutes into a vestigial, nonfunctional organ in adults. This notion has now been revisited since estimates of thymic output, such as the T cell receptor excision circle (TREC) assay to measure recent thymic emigrant (RTE) T cell numbers and volumetric assessment of thymic size, allowed to show that maturation of new T cells in the thymus occurs even in older adults, even if at a decreased rate (Douek and Koup 2000; Douek et al. 2000), and that thymus size can significantly increase during lymphocytic recovery from cyto- (and lympho-) toxic chemotherapy (Hakim et al. 2005).

New naïve CD4+ and CD8+ T cells are almost exclusively generated from lymphoid precursors via intrathymic maturation of double-positive (CD4+ CD8+) thymocytes (Berzins et al. 2002). Naïve CD4+ T cells can be identified based on the expression of cell surface markers such as CD45RA and CD45RO isoforms together with other markers of cell activation. Flow cytometric staining for naïve and memory phenotypic T cell markers provided an apparently straightforward way to start looking at qualitative changes in the T cell repertoire occurring after HSCT. Paradoxically, the size of the CD4+ naïve subset was consistently found to be diminished for at least 1 year after transplant (Saccardi et al. 2005; Carreras et al. 2003; Farge et al. 2005; Kozak et al. 2000; Verburg et al. 2001; Muraro et al. 2005). Reciprocally, several studies found an expansion of memory T cells at early posttranplantation time points (Saccardi et al. 2005; Verburg et al. 2001; Koehne et al. 1997; Rosen et al. 2000). The interpretation of these findings was difficult until the methods for analysis and the duration of the follow-up of the patients were extended and the notion of homeostatic proliferation was appreciated in this context.

It has now been shown that the naïve CD4+ T cell population can recover or even exceed pretransplant size 2 years after a myeloablative conditioning (Muraro et al. 2005; Hakim et al. 2005). The expansion of T cells with memory phenotype in the first year posttransplantation that we also observed is a consequence of homeostatic proliferation that occurs in a host with absolute (initially) and relative (further on) lymphopoenia (Muraro and Douek 2006). After allowing sufficient time for lymphocytic repopulation, naïve and memory cell frequency can be interpreted more reliably. At steady state, the observation of an increase of the naïve population suggested a thymic involvement in immune reconstitution. Naïve T cells, however, do not exclusively derive from the thymus, but could also expand homeostatically in a lymphopenic host, either having survived the immunosuppressive conditioning regimen or having been reintroduced with the autologous graft, similar to what had been shown after allogeneic transplantation (Storek et al. 2003). The true thymic origin of the reconstituted naïve cells is supported by the finding that an increase in naïve T cells was found after reconstitution with CD34+ -selected, but not with CD34+-unselected peripheral blood stem cells (Malphettes et al. 2003). Two studies utilizing specific assays have provided definitive evidence for reactivation of the thymic output during the long-term regeneration of the T-cell repertoire after autologous HSCT was obtained in MS patients undergoing a myeloablative irradiation regimen (Muraro et al. 2005) as well as in cancer patients receiving intensive chemotherapy (Hakim et al. 2005). In the MS study, a recent thymic origin of the naïve T cells expanded after immune reconstitution was demonstrated by the coexpression of CD31 and CD45RA in combination with high levels of TRECs. Evaluation of the antigen-receptor diversity by TCRBV CDR3-spectratyping and sequencing demonstrated substantial renewal of the T cell repertoire (Muraro et al. 2005), a finding that can only be reasonably explained by de novo intrathymic TCR rearrangement. Although some preexisting clones persisted after transplant, their degree of expansion was usually reduced, and other newly appeared clones became dominant. Similar demonstration of TCR diversification was provided by the spectratyping data reported by Hakim et al. in patients with cancer during immune recovery postchemotherapy. Furthermore, the National Cancer Institute study showed the patients' thymic enlargement by high-resolution computed tomography (Hakim et al. 2005). Neither the study by Hakim et al. nor ours allowed to examine potential correlations between thymic rebound and response to treatment. Interestingly, a more pronounced initial recovery of TRECs was observed at 1 year after HSCT in a subgroup of patients with systemic sclerosis who benefited from sustained clinical responses, compared to a subgroup that had poor clinical response (Farge et al. 2005).

A number of factors influence the reactivation of the thymus and the extent of thymic naïve cell recovery after HSCT; most crucially patient age, previous cytotoxic/immunosuppressive treatments and the transplantation protocol itself. Both the intensity of the conditioning regimen (myeloablative or reduced intensity/nonmyeloablative) and the manipulation of the haematopoietic graft (graft purification and/or in vivo T cell purging, and no graft or in vivo T cell depletion) are expected to influence the reconstitution mechanisms. Depletion of T cells in the graft seemed to be a factor that delayed T cell reconstitution, but promoted thymic regeneration in adults who had undergone total body irradiation (Malphettes et al. 2003). Comparative analysis of immune reconstitution following different treatment schemes are needed to better understand the rules governing the quality and kinetics of immune reconstitution.

A recent study provided important insight into antibody and T cell responses to alloantigens following HSCT. Immuno-ablative conditioning and autologous HSCT eliminated immunological memory for a neo-antigen to which patients were immunized after the graft harvest, and diminished the immunological memory for a recall antigen boosted before HSC mobilization (Brinkman et al. 2007). These observations are in line with our T cell repertoire data suggesting a regeneration of adaptive immunity (Muraro et al. 2005). What are the therapeutic implications of immune renewal resulting in ablation or attenuation of immunological memory and long-term naïve T cell augmentation? Naïve cells are recognized as an important component of immune homeostasis (Berzins et al. 2002), and their increase results in a diversification of the immune repertoire and a dilution of the self-reactive memory immune response. This "passive" tolerogenic role relies on the widely held notion that central tolerance mechanisms eliminate already in the thymus the vast majority of cells bearing receptors recognizing self antigens with high affinity. Autoantigen-reactive cells are found at similar frequency in healthy individuals and patients with autoimmune disease. The naïve T cell compartment is actually a major source of myelin basic protein (MBP)-specific T cells in both healthy and MS affected humans (Muraro et al. 2000). However, high-affinity autoreactive cells that have escaped negative selection will be rare in the thymic output but can become highly expanded in peripheral immune system following activation and clonal expansion. MBP-responsive T cells in MS patients were shown to have a greater functional avididity/affinity for cognate antigen than cells of healthy donors, even though the quantity of autoreactive cells was not found to be significantly different (Bielekova et al. 2004). In patients with MS who were treated with autologous HSCT, Sun et al (Sun et al. 2004) detected an initial decrease of MBP-specific cells but reactivity returned to baseline levels after 12 months. Further investigation into the epitope recognition patterns using MBP peptides showed that a more heterogeneous pattern of reactivity to MBP was elicited at 12 months. Preexisting dominant responses to specific peptides in individual patients were also reconstituted, but at reduced frequency and new epitope specificities had emerged. These changes resulted in a broader repertoire and a different hierarchy of immunodominant peptides of MBP than pretransplantation.

We are not aware of published studies addressing whether the reemerging MBP- (or any other self antigen-) specific cells have different functional avidity for the stimulatory ligands. In addition, not only the specificity of the cells, but also their functional subtype should be investigated since it would be important in determining their pathogenic vs tolerogenic role. Data characterizing the profile of cytokine secretion and other lymphocyte effector functions after autologous HSCT are scant, particularly for antigen-specific cells. This uncertainty currently precludes a definitive interpretation of the reemergence of potentially autoreactive T cells after autologous transplant.

Increasing evidence implicates failure of immuno-regulatory mechanisms in the development of autoimmune disease (Shevach 2000). Increase of CD4+ CD25+ regulatory T cells in an animal model of BMT (Herrmann et al. 2005) and in JIA patients after syngeneic HSCT (de Kleer et al. 2005) has suggested that recovery of immune regulation is involved in induction and/or maintenance tolerance post-transplant. Similar observations are emerging from an ongoing study in MS patients who received a reduced intensity conditioning regimen (Abrahamsson and Muraro, World Congress on Treatment and Research in Multiple Sclerosis, Montreal 2008 and unpublished data).

Thus far, changes in B cell population and autoantibodies have received less attention in MS than in systemic autoimmune disorders with better documented B cell involvement such as SLE. Intrathecal synthesis of IgG and detection of oligo-clonal bands (OCBs) in the cerebrospinal fluid (CSF) are nonspecific but typical (~90%) findings in MS. In most studies oligoclonal banding has been shown to persist in MS patients CSF at 1 year after HSCT (Carreras et al. 2003; Healey et al. 2004; Openshaw et al. 2000; Saiz et al. 2001). Given the long half-life of autoantibodies and plasma cells, reassessment after a longer follow-up is required to understand the evolution of CSF OBs after HSCT. Since CSF OBs, however, are not a surrogate marker of disease activity the interpretation of their disappearance or persistence even after long-term follow-up post-transplant requires caution. Further studies are warranted on B cell reconstitution and the possible role of modifications of B cell repertoire and function in the mechanism of action of HSCT for autoimmune diseases.

3. MSCs: Potential Role in Immunotherapy of MS

3.1 MSCs Effect on Immune cells

There is general consensus that MSCs are capable of mediating immunosuppression both on cells of the innate and adaptive immunity (Keating 2006; Nauta and Fibbe 2007; Rasmusson et al. 2005; Uccelli et al. 2006; Uccelli et al. 2007; Uccelli et al. 2008). Although it was initially proposed that the immunoregulatory functions of MSCs in vivo may be related to the immunoprivileged status of these cells, such assumption has been challenged by recent studies (Eliopoulos et al. 2005; Nauta et al. 2006). MSCs impair in vitro maturation of monocyte-derived myeloid dendritic cells (DC) through the downregulation of MHC class II, CD11c, CD83 and costimulatory molecules resulting in an impaired antigen presentation as well as IL-12 production (Aggarwal and Pittenger 2005; Beyth et al. 2005; Jiang et al. 2005; Maccario et al. 2005; Ramasamy et al. 2007).

Recently, MSCs have been demonstrated to closely interact with natural killer cells in vitro. MSCs significantly inhibit proliferation of resting NK cells, NK-mediated cytotoxicity and cytokine production (Poggi et al. 2005; Sotiropoulou et al. 2006; Spaggiari et al. 2008; Spaggiari et al. 2006). Conversely IL-2-activated NK cells efficiently lyse autologous and allogeneic MSCs. While the interplay between MSCs and NK cells should be considered when designing allogeneic MSC-based cell therapy protocols, the impact of NK-mediated cytotoxicity of MSCs in vivo is not known.

MSCs also inhibit the proliferation of T lymphocytes stimulated with polyclonal mitogens, allogeneic cells or specific antigens through the induction of "cell division arrest" and the consequent accumulation of T cells in the G0 phase of the cell cycle (Aggarwal and Pittenger 2005; Bartholomew et al. 2002; Di Nicola et al. 2002; Gieseke et al. 2007; Glennie et al. 2005; Krampera et al. 2003; Meisel et al. 2004; Rasmusson et al. 2005; Tse et al. 2003; Zappia et al. 2005). MSCs have been also reported to promote the generation of CD4$^+$ T regulatory cells (Treg) (Prevosto et al. 2007; Selmani et al. 2008; Maccario et al. 2005) and to switch CD4$^+$ T cell responses from a Th1 to a Th2 polarized phenotype (Aggarwal and Pittenger 2005; Maccario et al. 2005).

Recent studies showed also that MSCs inhibited B cell proliferation, differentiation and constitutive expression of chemokine receptors (Augello et al. 2005; Corcione et al. 2006; Glennie et al. 2005). However, in vivo and in vitro studies have shown that MSC-mediated inhibition of B cell function is mainly T cell dependent. Thus, in an animal model of multiple sclerosis (MS), i.e., the SJL/J mouse injected with proteolipid protein peptide (PLP), the production of antigen specific antibodies in vivo was abolished by MSCs infusion with significant therapeutic efficacy and downregulation of PLP specific T cell responses (Gerdoni et al. 2007).

Although a large number of studies have addressed the immunosuppressive activities of MSCs, the underlying mechanisms are only partially known. Usually, both cell contact and soluble factors released as a consequence of a cross-talk between MSCs and target cells are responsible for immunosuppression, including prostaglandin E2 (PGE-2), interleukin-10 (IL-10), transforming growth factor-β1

(TGF-β1), hepatocyte growth factor (HGF), indoleamine 2,3-dyoxigenase (IDO), nitric oxide (NO), heme oxygenase-1 and HLA-G (Aggarwal and Pittenger 2005; Chabannes et al. 2007; Di Nicola et al. 2002; Meisel et al. 2004; Sato et al. 2006).

Regardless of the mediators of the immunomodulatory activity of MSCs it is likely that different molecules act through diverse and possibly overlapping mechanisms leading to multiple in vivo effect including (1) freezing of DC in an immature state precluding efficient antigen presentation to T cells, (2) inhibition of T cell division and possibly, (3) generation of CD4+ Treg cells. As consequence, impaired CD4+ T cell activation results into defective T helper activity on B cell proliferation and differentiation in to antibody secreting cells. These effects may be reinforced by the direct inhibitory activities of MSCs on B cells (Fig. 3).

3.2 MSCs In vivo Effects in Preclinical Model of CNS Diseases

Current evidence suggests that, upon in vivo infusion, MSCs home with different efficiency to different organs including the CNS (Meisel et al. 2004). Importantly, MSCs seem to migrate preferentially to the site of injury through the release of metalloproteinases and upon chemotactic signals suggesting that they can sense danger cues in the local microenvironment where they promote functional recovery.

The initial in vivo studies with MSCs focused mainly on their capability to facilitate the engraftment of transplanted HSCs (Almeida-Porada et al. 2000) and promote structural (Pereira et al. 1995) and functional repair of damaged tissues

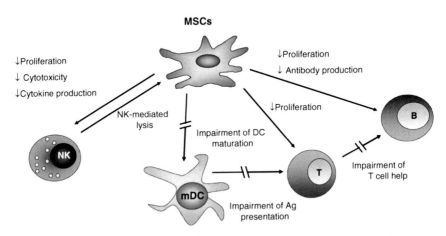

Fig. 3 A simplified scheme of the major interactions between MSCs and immune cells. The immunosuppressive effect of MSCs leads to (1) inhibition of NK cell proliferation, cytotoxicity and cytokine production, (2) accumulation of immature DC (ImDCs), (3) impairment of T cell responses occurring through freezing of DC in an immature state precluding efficient antigen presentation to T cells and T cell division arrest anergy, (4) impairment of T cell help to B cell, (5) inhibition of B cell proliferation and differentiation to ASC

(Orlic et al. 2001). The ability of MSCs to modulate in vivo immune responses was first investigated by Bartholomew and colleagues, who demonstrated that, upon injection into nonhuman primates, allogeneic MSCs prolonged skin graft survival (Bartholomew et al. 2002). We have extensively evaluated the effect of MSCs on EAE, the experimental model of MS. We demonstrated that MSCs infused i.v. following the onset of disease ameliorated myelin oligodendrocyte glycoprotein (MOG) induced EAE, homed into lymphoid organs and reduced infiltration of the CNS by T cells, B cells, and macrophages. Importantly, T cells from the lymph nodes of MSCs-treated mice did not proliferate upon in vitro rechallenge with MOG suggesting the induction of tolerance (Zappia et al. 2005). More recently we extended these results showing for the first time that MSCs can inhibit pathogenic B cells in vivo. Upon i.v. infusion of MSCs in mice with PLP-induced EAE we observed a striking inhibition of the production of antigen-specific antibodies. We also showed that MSCs suppress the encephalitogenic potential of PLP-specific T cells in passive-transfer experiments. Finally, we demonstrated that green fluorescent protein-labeled MSCs home into the inflamed CNS but do not transdifferentiate into neural cells. These findings are in agreement with most in vivo studies challenging in vitro results demonstrating MSCs transdifferentiation potential (Phinney and Prockop, Stem Cells 2007). However, we observed a decreased level of inflammation and demyelination associated with decreased axonal loss (Gerdoni et al. 2007). These results suggest that MSCs exert their therapeutic activity on inflamed CNS through the peripheral control of encephalitogenic T and B cells and a protective effect in situ (Uccelli et al. 2007). A protective effect of infused MSCs on injured neurons is also supported by the results obtained in other neurological diseases besides EAE (Chopp and Li 2002; Hofstetter et al. 2002). In these studies, regardless of the type of injury, the therapeutic effect is unrelated to MSCs engraftment in the damaged organ and transdifferentiation into neural cells but appears to depend on the release of antiapoptotic, antiinflammatory and trophic molecules as in the case of stroke in rats (Li et al. 2002), and, possibly, on the recruitment of local progenitors and their subsequent induction to differentiate into neural cells (Munoz et al. 2005). On this matter, MSCs have been shown to favor oligodendrogenic fate decision by neural precursor cells (Rivera et al. 2006).

3.3 Clinical Applications of MSCs

At first, the stem cell-like ability of MSCs prompted their exploitation as treatment for tissue repair, for example, in severe osteogenesis imperfecta (Horwitz et al. 1999) metachromatic leukodystrophy and Hurler syndrome (Koc et al. 2002/8) and in stroke (Bang et al. 2005). However, the in vivo immunosuppressive effect of infused MSCs has been tested so far mainly in acute severe GVHD where these cells have been demonstrated to reverse acute disease (Le Blanc et al. 2004; Le Blanc K et al. 2008) and in haploidentical HSCs transplantation where they can modulate host alloreactivity while promoting the engraftment of donor hematopoiesis (Ball et al. 2007). In addition,

MSCs are currently being tested for the treatment of Crohn disease based on the observation that transplanted BM cells can repopulate the human gastrointestinal tract and that BM-derived epithelial cells are remarkably increased during the epithelial regeneration after GVHD and ulcer formation (Okamoto et al. 2002).

On the basis of their immunomodulatory properties, ability of releasing growth factors and antiapoptotic molecules, capacity of migrating at the site of inflammation inside the CNS upon intravenous administration and ameliorating EAE, it would be reasonable to exploit MSCs for the treatment of severe, rapidly evolving cases of MS where a high inflammatory load is present and inexorably moves toward tissue destruction. Despite these encouraging perspectives the limited numbers of treated patients mostly in the frame of noncontrolled studies (Slavin et al. 2008) do not allow to draw firm conclusions as to the safety of MSCs in chronic diseases such as MS. A potential risk of MSCs treatment may ironically arise from their capability to modulate immune responses if this results in tumor engraftment and metastasis (Djouad et al. 2003; Karnoub et al. 2007). A further risk associated with MSCs expansion in vitro is the induction of cytogenetic abnormalities (Rubio et al. 2005) and subsequent differentiation into tumor cells upon in vivo administration as demonstrated in rodents (Tolar et al. 2006). Even so, there is general agreement that BM-derived MSCs can be safely expanded in vitro with very limited risk of malignant transformation (Bernardo et al. 2007) and, until now, there are no reports on generation of tumors by culture-expanded cells, thus rendering these cells still amenable for transplantation purposes.

Finally, some reports have recently challenged the tenet that allogeneic MSC are immunoprivileged (Eliopoulos et al. 2005; Nauta et al. 2006), suggesting that in some clinical conditions such as autoimmune diseases an autologous MSCs source could be preferable. On this matter, recent studies have shown that MSCs from autoimmune patients display normal ability to support hematopoiesis (Papadaki et al. 2005), immunomodulatory activity (Bocelli-Tyndall et al. 2006) and surface and molecular phenotype (Kastrinaki et al. 2007; Mazzanti et al. 2008) thus supporting the possibility of utilizing autologous MSCs for transplantation purposes.

4 Conclusions

Our review has attempted to cover relevant immunotherapeutic features of the bone marrow-derived cell populations, HSCs and MSCs. Recent studies have demonstrated that HSC transplantation not only is the most powerful immunosuppressive therapy that can be used to induce disease remission in cases of severely exacerbating/ rapidly progressive MS, but can also induce long-term qualitative changes in the composition of the immune system that favor immune tolerance. These modifications include a regeneration of the immune system sustained by enhanced naïve T cell output from the thymus. The observed increase in the number of T cells with regulatory T phenotypes suggests that the immune system reformatting may also rebuild impaired immuno-regulatory circuits. Further research is required to fully elucidate the mechanisms of HSCT in MS. A better understanding of the mode of action will help optimize the safety and efficacy of HSCT regimens and may reveal important aspects of the immune disturbances in MS.

Overall, current data suggest that bone marrow derived MSCs are endowed with several features that support their utilization for the treatment of MS. While the biological relevance of MSCs transdifferentiation into neural cells is questionable, it is now clear that their therapeutic effect relies on other characteristics, such as their anti-inflammatory, immunosuppressive, antiapoptotic and trophic properties. The compelling evidence of clinical efficacy in EAE mostly during the acute phase of disease and the relatively limited evidence of engraftment in the CNS suggest that MSCs are unlikely to be effective through the regeneration of damaged tissues. In contrast, MSCs immunosuppressive activity provides a tool for inducing peripheral tolerance and substantially modifying the local microenvironment through bystander mechanisms.

It is tempting to speculate that the properties of HSCs and MSCs could be exploited in combination or sequentially for treatment of severe and highly active MS: HSCs to support haematopoietic recovery and induce immune regeneration after intense lymphodepletion; and MSCs to maintain immune tolerance and promote CNS regeneration via their immunomodulatory and trophic actions.

Acknowledgements This work was partially supported by the National Institute for Health Research and by the Hammersmith Hospitals Research Trustees Committee (to P.A.M.).

References

Aggarwal S, Pittenger MF (2005) Human mesenchymal stem cells modulate allogeneic immune cell responses. Blood 105:1815–1822

Almeida-Porada G, Porada CD, Tran N, Zanjani ED (2000) Cotransplantation of human stromal cell progenitors into preimmune fetal sheep results in early appearance of human donor cells in circulation and boosts cell levels in bone marrow at later time points after transplantation. Blood 95:3620–3627.

Augello A, Tasso R, Negrini SM, Amateis A, Indiveri F, Cancedda R, Pennesi G (2005) Bone marrow mesenchymal progenitor cells inhibit lymphocyte proliferation by activation of the programmed death 1 pathway. Eur J Immunol 35:1482–1490.

Ball LM, Bernardo ME, Roelofs H, Lankester A, Cometa A, Egeler RM, Locatelli F, Fibbe WE (2007) Cotransplantation of ex vivo expanded mesenchymal stem cells accelerates lymphocyte recovery and may reduce the risk of graft failure in haploidentical hematopoietic stem-cell transplantation. Blood 110(7):2764–2767

Banati RB, Newcombe J, Gunn RN, Cagnin A, Turkheimer F, Heppner F, Price G, Wegner F, Giovannoni G, Miller DH et al (2000) The peripheral benzodiazepine binding site in the brain in multiple sclerosis: quantitative in vivo imaging of microglia as a measure of disease activity. Brain 123(Pt 11):2321–2337

Bang OY, Lee JS, Lee PH, Lee G (2005) Autologous mesenchymal stem cell transplantation in stroke patients. Ann Neurol 57(6):874–882

Bartholomew A, Sturgeon C, Siatskas M, Ferrer K, McIntosh K, Patil S, Hardy W, Devine S, Ucker D, Deans R (2002) Mesenchymal stem cells suppress lymphocyte proliferation in vitro and prolong skin graft survival in vivo. Exp Hematol 30(1):42–48

Bernardo ME, Zaffaroni N, Novara F, Cometa AM, Avanzini MA, Moretta A, Montagna D, Maccario R, Villa R, Daidone MG, Zuffardi O, Locatelli F (2007) Human bone marrow derived mesenchymal stem cells do not undergo transformation after long-term in vitro culture and do not exhibit telomere maintenance mechanisms. Cancer Res 67:9142–9149.

Berzins SP, Uldrich AP, Sutherland JS, Gill J, Miller JF, Godfrey DI, Boyd RL (2002) Thymic regeneration: teaching an old immune system new tricks. Trends Mol Med 8(10):469–476

Beyth S, Borovsky Z, Mevorach D, Liebergall M, Gazit Z, Aslan H, Galun E, Rachmilewitz J (2005) Human mesenchymal stem cells alter antigen-presenting cell maturation and induce T-cell unresponsiveness. Blood 105:2214–2219.

Bianco P, Gehron Robey P (2000) Marrow stromal stem cells. J Clin Invest 105(12):1663–1668

Bielekova B, Sung MH, Kadom N, Simon R, McFarland H, Martin R (2004) Expansion and functional relevance of high-avidity myelin-specific CD4+ T cells in multiple sclerosis. J Immunol 172(6):3893–3904

Bocelli-Tyndall C, Bracci L, Spagnoli G, Braccini A, Bouchenaki M, Ceredig R, Pistoia V, Martin I, Tyndall A (2006) Bone marrow mesenchymal stromal cells (BM-MSCs) from healthy donors and auto-immune disease patients reduce the proliferation of autologous- and allogeneic-stimulated lymphocytes in vitro. Rheumatology 46(3):403–408

Brinkman DM, Jol-van der Zijde CM, Ten Dam MM, Te Boekhorst PA, Ten Cate R, Wulffraat NM, Hintzen RQ, Vossen JM, van Tol MJ (2007) Resetting the adaptive immune system after autologous stem cell transplantation: lessons from responses to vaccines. J Clin Immunol 27:647–658

Burt RK, Traynor AE, Pope R, Schroeder J, Cohen B, Karlin KH, Lobeck L, Goolsby C, Rowlings P, Davis FA et al (1998) Treatment of autoimmune disease by intense immuno-suppressive conditioning and autologous hematopoietic stem cell transplantation. Blood 92(10):3505–3514

Carreras E, Saiz A, Marin P, Martinez C, Rovira M, Villamor N, Aymerich M, Lozano M, Fernandez-Aviles F, Urbano-Izpizua A et al (2003) CD34 + selected autologous peripheral blood stem cell transplantation for multiple sclerosis: report of toxicity and treatment results at one year of follow-up in 15 patients. Haematologica 88(3):306–314

Chabannes D, Hill M, Merieau E, Rossignol J, Brion R, Soulillou JP, Anegon I, Cuturi MC (2007) A role for heme oxygenase - 1 in the immunosuppressive effect of adult rat and human mes-enchymal stem cells. Blood 110:3691–3694.

Chopp M, Li Y (2002) Treatment of neural injury with marrow stromal cells. Lancet Neurol 1(2):92–100

Coles AJ, Cox A, Le Page E, Jones J, Trip SA, Deans J, Seaman S, Miller DH, Hale G, Waldmann H et al (2005) The window of therapeutic opportunity in multiple sclerosis Evidence from mono-clonal antibody therapy. J Neurol 27:27

Corcione A, Benvenuto F, Ferretti E, Giunti D, Cappiello V, Cazzanti F, Risso M, Gualandi F, Mancardi GL, Pistoia V, Uccelli A (2006) Human mesenchymal stem cells modulate B-cell functions. Blood 107:367–372.

de Kleer I, Vastert B, Klein M, Teklenburg G, Arkesteijn G, Puga Yung G, Albani S, Kuis W, Wulffraat N, Prakken B (2005) Autologous stem cell transplantation for autoimmunity induces immunologic self-tolerance by reprogramming autoreactive T-cells and restoring the CD4+ CD25+ immune regulatory network. Blood 1:1

De Kleer IM, Brinkman DM, Ferster A, Abinun M, Quartier P, Van Der Net J, Ten Cate R, Wedderburn LR, Horneff G, Oppermann J et al (2004) Autologous stem cell transplantation for refractory juvenile idiopathic arthritis: analysis of clinical effects, mortality, and transplant related morbidity. Ann Rheum Dis 63(10):1318–1326

DeLorenze GN, Munger KL, Lennette ET, Orentreich N, Vogelman JH, Ascherio A (2006) Epstein-Barr virus and multiple sclerosis: evidence of association from a prospective study with long-term follow-up. Arch Neurol 63(6):839–844

Di Nicola M, Carlo-Stella C, Magni M, Milanesi M, Longoni PD, Matteucci P, Grisanti S, Gianni, AM (2002) Human bone marrow stromal cells suppress T-lymphocyte proliferation induced by cellular or nonspecific mitogenic stimuli. Blood 99:3838–3843.

Djouad F, Plence P, Bony C, Tropel P, Apparailly F, Sany J, Noel D, Jorgensen C (2003) Immunosuppressive effect of mesenchymal stem cells favors tumor growth in allogeneic animals. Blood 102(10):3837–3844

Douek DC, Koup RA (2000) Evidence for thymic function in the elderly. Vaccine 18(16):1638–1641

Douek DC, Vescio RA, Betts MR, Brenchley JM, Hill BJ, Zhang L, Berenson JR, Collins RH, Koup RA (2000) Assessment of thymic output in adults after haematopoietic stem-cell transplantation and prediction of T-cell reconstitution. Lancet 355(9218):1875–1881

Eliopoulos N, Stagg J, Lejeune L, Pommey S, Galipeau J (2005) Allogeneic marrow stromal cells are immune rejected by MHC class I- and class II-mismatched recipient mice. Blood 106: 4057–4065.

Farge D, Henegar C, Carmagnat M, Daneshpouy M, Marjanovic Z, Rabian C, Ilie D, Douay C, Mounier N, Clave E et al (2005) Analysis of immune reconstitution after autologous bone marrow transplantation in systemic sclerosis. Arthritis Rheum 52(5):1555–1563

Fassas A, Passweg JR, Anagnostopoulos A, Kazis A, Kozak T, Havrdova E, Carreras E, Graus F, Kashyap A, Openshaw H et al (2002) Hematopoietic stem cell transplantation for multiple sclerosis. A retrospective multicenter study. J Neurol 249(8):1088–1097

Friedenstein AJ, Chailakhyan RK, Latsinik NV, Panasyuk AF, Keiliss-Borok IV (1974) Stromal cells responsible for transferring the microenvironment of the hemopoietic tissues. Cloning in vitro and retransplantation in vivo. Transplantation 17:331–340.

Gerdoni E, Gallo B, Casazza S, Musio S, Bonanni I, Pedemonte E, Mantegazza R, Frassoni F, Mancardi G, Pedotti R et al (2007) Mesenchymal stem cells effectively modulate pathogenic immune response in experimental autoimmune encephalomyelitis. Ann Neurol 61(3):219–227

Gieseke F, Schutt B, Viebahn S, Koscielniak E, Friedrich W, Handgretinger R, Muller I (2007) Human multipotent mesenchymal stromal cells inhibit proliferation of PBMCs independently of IFNgammaR1 signaling and IDO expression. Blood 110:2197–2200.

Glennie S, Soeiro I, Dyson PJ, Lam EW, Dazzi F (2005) Bone marrow mesenchymal stem cells induce division arrest anergy of activated T cells. Blood 105:2821–2827.

Guillaume T, Rubinstein DB, Symann M (1998) Immune reconstitution and immunotherapy after autologous hematopoietic stem cell transplantation. Blood 92(5):1471–1490

Hakim FT, Memon SA, Cepeda R, Jones EC, Chow CK, Kasten-Sportes C, Odom J, Vance BA, Christensen BL, Mackall CL et al (2005) Age-dependent incidence, time course, and consequences of thymic renewal in adults. J Clin Invest 115(4):930–939

Healey KM, Pavletic SZ, Al-Omaishi J, Leuschen MP, Pirruccello SJ, Filipi ML, Enke C, Ursick MM, Hahn F, Bowen JD, Nash RA (2004) Discordant functional and inflammatory parameters in multiple sclerosis patients after autologous haematopoietic stem cell transplantation. Mult Scler 10:284–289.

Herrmann MM, Gaertner S, Stadelmann C, van den Brandt J, Boscke R, Budach W, Reichardt HM, Weissert R (2005) Tolerance induction by bone marrow transplantation in a multiple sclerosis model. Blood 106(5):1875–1883

Hofstetter CP, Schwarz EJ, Hess D, Widenfalk J, El Manira A, Prockop DJ, Olson L (2002) Marrow stromal cells form guiding strands in the injured spinal cord and promote recovery. Proc Natl Acad Sci U S A 99:2199–2204.

Horwitz E, Le Blanc K, Dominici M, Mueller I, Slaper-Cortenbach I, Marini F, Deans R, Krause D, Keating A (2005) Clarification of the nomenclature for MSC: The International Society for Cellular Therapy position statement. Cytotherapy 7(5):393–395

Horwitz EM, Prockop DJ, Fitzpatrick LA, Koo WW, Gordon PL, Neel M, Sussman M, Orchard P, Marx JC, Pyeritz RE, Brenner MK (1999) Transplantability and therapeutic effects of bone marrow-derived mesenchymal cells in children with osteogenesis imperfecta. Nat Med 5:309–313.

Jiang XX, Zhang Y, Liu B, Zhang SX, Wu Y, Yu XD, Mao N (2005) Human mesenchymal stem cells inhibit differentiation and function of monocyte-derived dendritic cells. Blood 105:4120–4126.

Karnoub AE, Dash AB, Vo AP, Sullivan A, Brooks MW, Bell GW, Richardson AL, Polyak K, Tubo R, Weinberg RA (2007) Mesenchymal stem cells within tumour stroma promote breast cancer metastasis. Nature 449(7162):557–563

Kastrinaki MC, Sidiropoulos P, Roche S, Ringe J, Lehmann S, Kritikos H, Vlahava VM, Delorme B, Eliopoulos G, Jorgensen C et al (2007) Functional, molecular and proteomic characterization of bone marrow mesenchymal stem cells in rheumatoid arthritis. Ann Rheum Dis 67:741–749

Keating A (2006) Mesenchymal stromal cells. Curr Opin Hematol 13:419–425.

Koc ON, Day J, Nieder M, Gerson SL, Lazarus HM, Krivit W (2002/8) Allogeneic mesenchymal stem cell infusion for treatment of metachromatic leukodystrophy (MLD) and Hurler syndrome (MPS-IH). Bone Marrow Transplant 30:215–222.

Koehne G, Zeller W, Stockschlaeder M, Zander AR (1997) Phenotype of lymphocyte subsets after autologous peripheral blood stem cell transplantation. Bone Marrow Transplant 19(2):149–156

Kopen GC, Prockop DJ, Phinney DG (1999) Marrow stromal cells migrate throughout forebrain and cerebellum, and they differentiate into astrocytes after injection into neonatal mouse brains. Proc.Natl.Acad.Sci U.S.A 96:10711–10716.

Kozak T, Havrdova E, Pit'ha J, Gregora E, Pytlik R, Maaloufova J, Mareckova H, Kobylka P, Vodvarkova S (2000) High-dose immunosuppressive therapy with PBPC support in the treatment of poor risk multiple sclerosis. Bone Marrow Transplant 25(5):525–531

Krampera M, Glennie S, Dyson J, Scott D, Laylor R, Simpson E, Dazzi F (2003) Bone marrow mesenchymal stem cells inhibit the response of naive and memory antigen-specific T cells to their cognate peptide. Blood 101:3722–3729.

Le Blanc K, Rasmusson I, Sundberg B, Gotherstrom C, Hassan M, Uzunel M, Ringden O (2004) Treatment of severe acute graft-versus-host disease with third party haploidentical mesenchymal stem cells. Lancet 363:1439–1441.

Le Blanc K, Frassoni F, Ball L, Locatelli F, Roelofs H, Lewis I, Lanino E, Sundberg B, Bernardo ME, Remberger M, Dini G, Egeler RM, Bacigalupo A, Fibbe W, Ringden O (2008) Mesenchymal stem cells for treatment of steroid-resistant, severe, acute graft-versus-host disease: a phase II study. Lancet 371:1579–1586.

Li Y, Chen J, Chen XG, Wang L, Gautam SC, Xu YX, Katakowski M, Zhang LJ, Lu M, Janakiraman N et al (2002) Human marrow stromal cell therapy for stroke in rat: Neurotrophins and functional recovery. Neurology 59(4):514–523

Maccario R, Podesta M, Moretta A, Cometa A, Comoli P, Montagna D, Daudt L, Ibatici A, Piaggio G, Pozzi S et al (2005) Interaction of human mesenchymal stem cells with cells involved in alloantigen-specific immune response favors the differentiation of CD4+ T-cell subsets expressing a regulatory/suppressive phenotype. Haematologica 90(4):516–525

Malphettes M, Carcelain G, Saint-Mezard P, Leblond V, Altes HK, Marolleau JP, Debre P, Brouet JC, Fermand JP, Autran B (2003) Evidence for naive T-cell repopulation despite thymus irradiation after autologous transplantation in adults with multiple myeloma: role of ex vivo CD34+ selection and age. Blood 101(5):1891–1897

Mancardi G, Saccardi R (2008) Autologous haematopoietic stem-cell transplantation in multiple sclerosis. Lancet Neurol 7(7):626–636

Mancardi GL, Saccardi R, Filippi M, Gualandi F, Murialdo A, Inglese M, Marrosu MG, Meucci G, Massacesi L, Lugaresi A et al (2001) Autologous hematopoietic stem cell transplantation suppresses Gd-enhanced MRI activity in MS. Neurology 57(1):62–68

Mazzanti B, Aldinucci A, Biagioli T, Barilaro A, Urbani A, Dal Pozzo S, Amato M, Siracusa G, Crescioli C, Manuelli C et al (2008) Differences in mesenchymal stem cell cytokine profiles between MS patients and healthy donors: Implication for assessment of disease activity and treatment. J Neuroimmunol 199(1–2):142–150

Meirelles LdS, Chagastelles PC, Nardi NB (2006) Mesenchymal stem cells reside in virtually all post-natal organs and tissues. J Cell Sci 119(11):2204–2213

Meisel R, Zibert A, Laryea M, Gobel U, Daubener W, Dilloo D (2004) Human bone marrow stromal cells inhibit allogeneic T-cell responses by indoleamine 2,3-dioxygenase mediated tryptophan degradation. Blood 103(12):4619–4621

Metz I, Lucchinetti CF, Openshaw H, Garcia-Merino A, Lassmann H, Freedman M, Azzarelli B, Kolar OJ, Atkins HL, Bruck W (2006) Multiple sclerosis pathology after autologous stem cell transplantation: ongoing demyelination and neurodegeneration despite suppressed inflammation. Mult Scler 12:S9

Munoz JR, Stoutenger BR, Robinson AP, Spees JL, Prockop DJ (2005) Human stem/progenitor cells from bone marrow promote neurogenesis of endogenous neural stem cells in the hippocampus of mice. Proc Natl Acad Sci U S A 102(50):18171–18176

Muraro PA, Douek DC (2006) Renewing the T cell repertoire to arrest autoimmune aggression. Trends Immunol 27:61–67.

Muraro PA, Douek DC, Packer A, Chung K, Guenaga FJ, Cassiani-Ingoni R, Campbell C, Memon S, Nagle JW, Hakim FT et al (2005) Thymic output generates a new and diverse TCR repertoire after autologous stem cell transplantation in multiple sclerosis patients. J Exp Med 201(5):805–816

Muraro PA, Pette M, Bielekova B, McFarland HF, Martin R (2000) Human Autoreactive CD4+ T Cells from Naive CD45RA+ and Memory CD45RO+ Subsets Differ with Respect to Epitope Specificity and Functional Antigen Avidity. J Immunol 164:5474–5481

Muraro PA, Van Laar JM, Illei G, Pavletic S (2008) Hematopoietic stem cell transplantation for autoimmune disorders. In: Barrett AJ, Treleaven JG (eds) Hematopoietic stem cell transplantation in clinical practice. Churchill Livingstone, Oxford, pp 197–210

Nash RA, Bowen JD, McSweeney PA, Pavletic SZ, Maravilla KR, Park MS, Storek J, Sullivan KM, Al-Omaishi J, Corboy JR et al (2003) High-dose immunosuppressive therapy and autologous peripheral blood stem cell transplantation for severe multiple sclerosis. Blood 102(7):2364–2372

Nauta AJ, Fibbe WE (2007) Immunomodulatory properties of mesenchymal stromal cells. Blood 110:3499–3506.

Nauta AJ, Westerhuis G, Kruisselbrink AB, Lurvink EG, Willemze R, Fibbe WE (2006) Donor-derived mesenchymal stem cells are immunogenic in an allogeneic host and stimulate donor graft rejection in a non-myeloablative setting. Blood 108:2114–2120.

Okamoto R, Yajima T, Yamazaki M, Kanai T, Mukai M, Okamoto S, Ikeda Y, Hibi T, Inazawa J, Watanabe M (2002/9) Damaged epithelia regenerated by bone marrow-derived cells in the human gastrointestinal tract. Nat Med 8(9):1011–1017

Openshaw H, Lund BT, Kashyap A, Atkinson R, Sniecinski I, Weiner LP, Forman S (2000) Peripheral blood stem cell transplantation in multiple sclerosis with busulfan and cyclophosphamide conditioning: report of toxicity and immunological monitoring. Biol Blood Marrow Transplant 6:563–575.

Orlic D, Kajstura J, Chimenti S, Jakoniuk I, Anderson SM, Li B, Pickel J, McKay R, Nadal-Ginard B, Bodine DM, Leri A, Anversa P (2001) Bone marrow cells regenerate infarcted myocardium. Nature 410:701–705.

Papadaki HA, Tsagournisakis M, Mastorodemos V, Pontikoglou C, Damianaki A, Pyrovolaki K, Stamatopoulos K, Fassas A, Plaitakis A, Eliopoulos GD (2005) Normal bone marrow hematopoietic stem cell reserves and normal stromal cell function support the use of autologous stem cell transplantation in patients with multiple sclerosis. Bone Marrow Transplant 36:1053–1063.

Passweg JR, Rabusin M, Musso M, Beguin Y, Cesaro S, Ehninger G, Espigado I, Iriondo A, Jost L, Koza V et al (2004) Haematopoetic stem cell transplantation for refractory autoimmune cytopenia. Br J Haematol 125(6):749–755

Pedemonte E, Benvenuto F, Casazza S, Mancardi G, Oksenberg JR, Uccelli A, Baranzini SE (2007) The molecular signature of therapeutic mesenchymal stem cells exposes the architecture of the hematopoietic stem cell niche synapse. BMC Genomics 8:65.

Pereira RF, Halford KW, O'Hara MD, Leeper DB, Sokolov BP, Pollard MD, Bagasra O, Prockop DJ (1995) Cultured adherent cells from marrow can serve as long-lasting precursor cells for bone, cartilage, and lung in irradiated mice. Proc Natl Acad Sci U S A 92(11):4857–4861

Phinney DG, Hill K, Michelson C, Dutreil M, Hughes C, Humphries S, Wilkinson R, Baddoo M, Bayly E (2006) Biological activities encoded by the murine mesenchymal stem cell transcriptome provide a basis for their developmental potential and broad therapeutic efficacy. Stem Cells 24(1):186–198

Phinney DG, Prockop DJ (2007) Concise Review: Mesenchymal Stem/Multi-Potent Stromal Cells (MSCs): The State of Transdifferentiation and Modes of Tissue Repair – Current Views. Stem Cells 25(11):2896–2902

Pittenger MF, Mackay AM, Beck SC, Jaiswal RK, Douglas R, Mosca JD, Moorman MA, Simonetti DW, Craig S, Marshak DR (1999) Multilineage potential of adult human mesenchymal stem cells. Science 284:143–147.

Poggi A, Prevosto C, Massaro AM, Negrini S, Urbani S, Pierri I, Saccardi R, Gobbi M, Zocchi MR (2005) Interaction between human NK cells and bone marrow stromal cells induces NK cell triggering: role of NKp30 and NKG2D receptors. J.Immunol 175:6352–6360.

Prevosto C, Zancolli M, Canevali P, Zocchi MR, Poggi A (2007) Generation of CD4+ or CD8+ regulatory T cells upon mesenchymal stem cell-lymphocyte interaction. Haematologica 92: 881–888.

Ramasamy R, Fazekasova H, Lam EW, Soeiro I, Lombardi G, Dazzi F (2007) Mesenchymal stem cells inhibit dendritic cell differentiation and function by preventing entry into the cell cycle. Transplantation 83:71–76.

Rasmusson I, Ringden O, Sundberg B, Le Blanc K (2005) Mesenchymal stem cells inhibit lymphocyte proliferation by mitogens and alloantigens by different mechanisms. Exp.Cell Res. 305:33–41.

Rivera FJ, Couillard-Despres S, Pedre X, Ploetz S, Caioni M, Lois C, Bogdahn U, Aigner L (2006) Mesenchymal stem cells instruct oligodendrogenic fate decision on adult neural stem cells. Stem Cells 24:2209–2219.

Rosen O, Thiel A, Massenkeil G, Hiepe F, Haupl T, Radtke H, Burmester GR, Gromnica-Ihle E, Radbruch A, Arnold R (2000) Autologous stem-cell transplantation in refractory autoimmune diseases after in vivo immunoablation and ex vivo depletion of mononuclear cells. Arthritis Res 2:327–336

Rubio D, Garcia-Castro J, Martin MC, de la FR, Cigudosa JC, Lloyd AC, Bernad A (2005) Spontaneous human adult stem cell transformation. Cancer Res. 65:3035–3039.

Saccardi R, Mancardi GL, Solari A, Bosi A, Bruzzi P, Di Bartolomeo P, Donelli A, Filippi M, Guerrasio A, Gualandi F et al (2005) Autologous HSCT for severe progressive multiple sclerosis in a multicenter trial: impact on disease activity and quality of life. Blood 105(6):2601–2607

Sacchetti B, Funari A, Michienzi S, Di Cesare S, Piersanti S, Saggio I, Tagliafico E, Ferrari S, Robey PG, Riminucci M, Bianco P (2007) Self-renewing osteoprogenitors in bone marrow sinusoids can organize a hematopoietic microenvironment. Cell 131:324–336.

Saiz A, Carreras E, Berenguer J, Yague J, Martinez C, Marin P, Rovira M, Pujol T, Arbizu T, Graus F (2001) MRI and CSF oligoclonal bands after autologous hematopoietic stem cell transplantation in MS. Neurology 56(8):1084–1089

Samijn JP, te Boekhorst PA, Mondria T, van Doorn PA, Flach HZ, van der Meche FG, Cornelissen J, Hop WC, Lowenberg B, Hintzen RQ (2006) Intense T cell depletion followed by autologous bone marrow transplantation for severe multiple sclerosis. J Neurol Neurosurg Psychiatry 77(1):46–50

Sato K, Ozaki K, Oh I, Meguro A, Hatanaka K, Nagai T, Muroi K, Ozawa K (2006) Nitric oxide plays a critical role in suppression of T cell proliferation by mesenchymal stem cells. Blood 109:228–234.

Schofield R (1978) The relationship between the spleen colony-forming cell and the haemopoietic stem cell. Blood Cells 4(1–2):7–25

Selmani Z, Naji A, Zidi I, Favier B, Gaiffe E, Obert L, Borg C, Saas P, Tiberghien P, Rouas-Freiss N, Carosella ED, Deschaseaux F (2008) Human leukocyte antigen-G5 secretion by human mesenchymal stem cells is required to suppress T lymphocyte and natural killer function and to induce CD4+CD25highFOXP3+ regulatory T cells. Stem Cells 26:212–222.

Shevach EM (2000) Regulatory T cells in autoimmmunity*. Annu Rev Immunol 18:423–449

Shlomchik MJ, Craft JE, Mamula MJ (2001) From T to B and back again: positive feedback in systemic autoimmune disease. Nat Rev Immunol 1(2):147–153

Slavin S, Kurkalli BG, Karussis D (2008) The potential use of adult stem cells for the treatment of multiple sclerosis and other neurodegenerative disorders. Clin Neurol Neurosurg 110: 943–946.

Snowden JA, Passweg J, Moore JJ, Milliken S, Cannell P, Van Laar J, Verburg R, Szer J, Taylor K, Joske D et al (2004) Autologous hemopoietic stem cell transplantation in severe rheumatoid arthritis: a report from the EBMT and ABMTR. J Rheumatol 31(3):482–488

Sospedra M, Martin R (2005) Immunology of multiple sclerosis. Annu Rev Immunol 23:683–747

Sotiropoulou PA, Perez SA, Gritzapis AD, Baxevanis CN, Papamichail M (2006) Interactions Between Human Mesenchymal Stem Cells and Natural Killer Cells. Stem Cells 24:74–85.

Spaggiari GM, Capobianco A, Abdelrazik H, Becchetti F, Mingari MC, Moretta L (2008) Mesenchymal stem cells inhibit natural killer-cell proliferation, cytotoxicity, and cytokine production: role of indoleamine 2,3-dioxygenase and prostaglandin E2. Blood 111: 1327–1333.

Spaggiari GM, Capobianco A, Becchetti S, Mingari MC, Moretta L (2006) Mesenchymal stem cell-natural killer cell interactions: evidence that activated NK cells are capable of killing MSCs, whereas MSCs can inhibit IL-2-induced NK-cell proliferation. Blood 107: 1484–1490.

Storek J, Dawson MA, Maloney DG (2003) Correlation between the numbers of naive T cells infused with blood stem cell allografts and the counts of naive T cells after transplantation. Biol Blood Marrow Transplant 9(12):781–784

Sun W, Popat U, Hutton G, Zang YC, Krance R, Carrum G, Land GA, Heslop H, Brenner M, Zhang JZ (2004) Characteristics of T-cell receptor repertoire and myelin-reactive T cells reconstituted from autologous haematopoietic stem-cell grafts in multiple sclerosis. Brain 127(Pt 5):996–1008

Tolar J, Nauta AJ, Osborn MJ, Panoskaltsis Mortari A, McElmurry RT, Bell S, Xia L, Zhou N, Riddle M, Schroeder TM et al (2006) Sarcoma Derived from Cultured Mesenchymal Stem Cells. Stem Cells 25:371–379

Uccelli A, Moretta L, Pistoia V (2006) Immunoregulatory function of mesenchymal stem cells. Eur.J.Immunol 36:2566–2573.

Uccelli A, Moretta L, Pistoia V (2008) Mesenchymal stem cells: role in health and disease. Nat Rev Immunol 8:726–736

Uccelli A, Pistoia V, Moretta L (2007) Mesenchymal stem cells: a new strategy for immunosuppression? Trends Immunol 28:219–226

Verburg RJ, Kruize AA, van den Hoogen FH, Fibbe WE, Petersen EJ, Preijers F, Sont JK, Barge RM, Bijlsma JW, van de Putte LB et al (2001) High-dose chemotherapy and autologous hematopoietic stem cell transplantation in patients with rheumatoid arthritis: results of an open study to assess feasibility, safety, and efficacy. Arthritis Rheum 44(4):754–760

Zappia E, Casazza S, Pedemonte E, Benvenuto F, Bonanni I, Gerdoni E, Giunti D, Ceravolo A, Cazzanti F, Frassoni F et al (2005) Mesenchymal stem cells ameliorate experimental autoimmune encephalomyelitis inducing T cell anergy. Blood 106(5):1755–1761

Multiple Sclerosis Therapies: Molecular Mechanisms and Future

Paulo Fontoura and Hideki Garren

Abstract The current treatments for multiple sclerosis (MS) are, by many measures, not satisfactory. The original interferon-β therapies were not necessarily based on an extensive knowledge of the pathophysiological mechanisms of the disease. As more and more insight has been acquired about the autoimmune mechanisms of MS and, in particular, the molecular targets involved, several treatment approaches have emerged. In this chapter, we highlight both promising preclinical approaches and therapies in late stage clinical trials that have been developed as a result of the improved understanding of the molecular pathophysiology of MS. These clinical stage therapies include oral agents, monoclonal antibodies, and antigen-specific therapies. Particular emphasis is given to the molecular targets when known and any safety concerns that have arisen because, despite the need for improved efficacy, MS remains a disease in which the safety of any agent remains of paramount importance.

1 Introduction

1.1 Multiple Sclerosis as an Immune-Mediated Disease

Multiple Sclerosis (MS) is the most prevalent demyelinating disease and the principal cause of neurological disability in young adults. Although cases have been studied and documented for more than 150 years, we are still unsure what mechanisms underlie the appearance of the hallmark inflammatory demyelinating lesions in the central nervous system (CNS), and their evolution into disseminated sclerotic plaques. Nowadays, the diagnostic process has become straightforward, and the appearance of typical clinical manifestations in a young (especially female) adult

P. Fontoura
Roche Pharmaceuticals, CNS Translational Medicine Group, Basel, Switzerland
and
Department of Immunology, Faculty of Medical Sciences, New University of Lisbon, Lisbon, Portugal

H. Garren (✉)
Bayhill Therapeutics, Inc., Palo Alto, CA, USA
e-mail: hgarren@bayhilltx.com

Results Probl Cell Differ, DOI 10.1007/400_2010_36
© Springer-Verlag Berlin Heidelberg 2010

leads rapidly to the documentation of CNS white matter lesions by magnetic resonance imaging (MRI), which also serves to monitor in vivo the evolution of the disease, as reflected by currently accepted diagnostic criteria (McDonald criteria) (McDonald et al. 2001). Unfortunately, this does not mean we have come to understand the underlying causation of this disease.

Any pathophysiological model of MS needs to be able to explain the generation of acute demyelinating lesions, their evolution into chronic sclerotic plaques, as well as an unpredictable clinical course, initially characterized by recurrent relapses and later by steady progression. The generally accepted pathophysiological model for MS is centered on an immune-mediated attack against CNS myelin antigens, in which several components of the immune system generate an inflammatory response that damages myelin and axons, leading to the formation of an acute plaque (Frohman et al. 2006; Hafler et al. 2005; Steinman 1996). It is less clear how such an acute lesion evolves into a chronic sclerotic plaque. In recent years, other aspects of this disease have become evident, such as the presence of widespread axonal loss and continuing CNS atrophy despite a reduction in classic inflammatory markers, leading some to propose an underlying – possibly independent – degenerative component to this disease. Therefore, it is possible to view MS as composed of an initial inflammatory stage, characterized by acute relapses, followed by a later degenerative stage in which relapse unrelated progression becomes apparent (Steinman 2001).

1.2   Molecular Mechanisms of MS Immunopathogenesis

Even early on in the history of MS, the presence of immune system cells in the plaques was evident, and comparisons with postinfectious and vaccinial encephalomyelitis provided ample evidence for the possibility of immune-mediated myelin destruction. Today, major MS etiological theories include autoimmune, infectious, and degenerative causes, and a mixture of these, with an emphasis on immune processes, provides our best working hypothesis (Hemmer et al. 2002). In this view, MS can be seen as a disease in which genetically susceptible individuals, encountering an environmental stimulus such as an infection, generate an autoimmune attack against CNS myelin, based on molecular mimicry between infectious and myelin antigens. If normal regulatory mechanisms are overcome, this response can become autonomous, possibly leading to epitope spreading to other CNS antigens, and the generation of a chronic recurrent condition such as MS (Bar-Or 2008; Sospedra and Martin 2005). Data in favor of this model come from genetics, pathology (macro, micro, and molecular), response to immune-based therapies, and mainly from analogies with the animal model experimental autoimmune encephalomyelitis (EAE).

As for most autoimmune diseases, genetic studies have revealed an association of MS with particular HLA haplotypes, especially the DR15 haplotype (DRB1*1501, DRB5*0101, DQA1*0102, DQB1*0602) for the European and North American

Caucasian population. Recently, this link was once again confirmed in the largest genome-wide association study done to date, which also revealed two other immune-related targets, the receptors for the interleukins IL-2 and IL-7 (Hafler et al. 2007; Lundmark et al. 2007). Pathological studies gave us the first insights into MS pathogenesis, and in recent years have once again transformed our view of this disease. Namely, the concept of MS as a pathologically homogeneous entity has been challenged by work from the International MS Lesion Project collaboration, which has shown that MS lesions can be grouped into four main patterns, of which pattern I and II are classical immune-mediated lesions, but patterns III and IV appear to be caused by oligodendroglial dysfunction (Lassmann et al. 2001; Lucchinetti et al. 2000). The existence of these latter patterns highlights both the limitations of our current MS model and the need for further human immunopathological studies from MS patients, but does not negate the important role of the immune system in generating demyelinating lesions. Response to immune-modulatory and suppressing therapies is also a strong indication that despite what the ultimate cause of the disease may be, several of its clinical manifestations can be attributed to immune system activation. Cases in point include the interferons and glatiramer acetate, as well as the recently introduced monoclonal antibody, natalizumab. Although mechanisms of action for the interferons and glatiramer continue to be discovered, the natalizumab experience proves that blocking lymphocyte migration into the CNS is sufficient to significantly reduce relapse rate, new lesion formation, and disability progression (Steinman 2005).

The animal model EAE has now been in existence for 75 years, and its versatility and robustness continue to make it useful both as a tool for dissecting immunopathological mechanisms and as a proving ground for new MS therapies (Steinman and Zamvil 2005). In many ways, EAE has shaped the evolution of the MS pathogenesis model, setting paradigms that are still useful today, including the focus on a Th1-cell autoimmune response to myelin antigens (Zamvil and Steinman 1990). In this model, acute lesions are caused by a perivascular inflammatory reaction coordinated by Th1 cells, which can directly and indirectly cause tissue destruction through the secretion of cytokines and recruitment of other immune cells, such as macrophages and microglia (Bar-Or 2008). These T cells are generated in peripheral lymphoid organs after being presented with an encephalitogenic antigen. The actual antigen is unknown in MS patients, but can presumably be derived from an infectious microorganism by molecular mimicry. This antigen must be presented by a mature professional antigen-presenting cell (APC), which has been stimulated to express high levels of both MHC class II and costimulatory molecules, implying activation by molecular danger signals sensed by receptors, such as the Toll-like receptor (TLR) family (Kielian 2006). Conditions present during T-cell priming, such as local cytokines (IL-12 or IL-4), costimulatory molecules, and intrinsic antigen characteristics, lead to differentiation along antagonistic Th1 or Th2 phenotypes. The Th1-Th2 paradigm dominated our perceptions up until recently, and therapies for MS and autoimmune diseases, in general, were developed and selected based on their capacity to suppress Th1 and promote Th2 responses (Hafler et al. 2005).

After differentiation, Th1 cells migrate to the CNS by crossing the blood–brain barrier, in a tightly regulated process that involves chemokine signaling, cell-adhesion molecules, and matrix metalloproteinases (Engelhardt and Ransohoff 2005; Szczucinski and Losy 2007). As said earlier, despite the complexity of this signaling, the $\alpha 4\beta 1$ integrin (targeted by natalizumab) has been shown both in EAE and in MS as a key component in transendothelial migration (Steinman 2005). Reactivation inside the CNS depends on local antigen presentation by resident APCs, such as microglia, that may then become the main effectors of tissue damage by generating reactive oxygen and nitrogen species, as well as proinflammatory cytokines (Becher et al. 2006). The list of potential myelin antigens has been steadily growing through the years, and now includes several major and minor proteins (MBP, PLP, MOG, MAG, OSP, MOBP) as well as lipids (Kanter et al. 2006; Sospedra and Martin 2005). The main antigenic determinants for these proteins have also been discovered, and in some cases, the structural requirements for MHC presentation and T-cell recognition have been studied in detail, such as for the MBP83-99 major epitope (Krogsgaard et al. 2000). Based on the trimolecular complex DR15-MBP-TCR, therapies such as altered peptide ligands generated from the MBP sequence were tested in clinical trials, proving both the existence of real biologic effects on MS patients and the inadvertent generation of high numbers of anti-MBP Th1 cell clones that can lead to disease exacerbation (Bielekova et al. 2000). This is a direct proof that in MS such cells play important roles in the genesis of new lesions.

This model of MS pathogenesis has gradually changed as the importance of other components of the immune response has emerged (Lassmann 2008; McFarland and Martin 2007). New T-cell differentiation phenotypes have been characterized, including proinflammatory Th17 cells and regulatory T cells (Treg). Th17 cell generation is dependent on TGFβ, IL-6, and IL-23, and is suppressed by IL-27 (McGeachy and Cua 2008). These cells may have important roles in disease progression and the generation of parenchymal lesions; recently, human Th17 cells were shown to be present in MS plaques, to have increased migratory capacity to the CNS, and to express granzyme B (Kebir et al. 2007). There are several regulatory immune cell populations, of which the natural and induced Treg CD4+ CD25+ FoxP3+ phenotype is the most studied (Jiang and Chess 2004); there are some indications that this population may be reduced, and its function deficient, in MS patients (Viglietta et al. 2004).

Besides Th cells, other immune system cells play relevant roles in MS patho-genesis. Cytotoxic CD8+ T cells have been found to be much more abundant in MS plaques than Th cells, and to show signs of local clonal expansion and reactivity to myelin antigens. Moreover, these cells have intrinsic destructive capability and are not dependent on MHC class II presentation, making them better candidates for generating demyelinating lesions (Neumann et al. 2002). B lymphocytes and antibodies are clearly present in MS plaques, and the presence of B-cell clonal expansion as well as oligoclonal immunoglobulins in the cerebrospinal fluid (CSF) is characteristic of this disease. Several targets for these immunoglobulins have been identified, including proteins from promising infectious candidates such as Epstein–Barr virus (EBV) (Cepok et al. 2005). Besides being antibody producers, B cells may also be involved in MS pathogenesis in other roles; recently, ectopic B-cell lymphoid tissues have been found in the meninges of progressive MS patients, and its presence was correlated with earlier onset and more aggressive disease course (Magliozzi et al. 2007). Also, these ectopic follicles were found to be associated with local expression of EBV proteins, presumably due to a chronic infectious process (Serafini et al. 2007). In this case, these follicles might function as a source of inflammatory stimuli, such as cytokines and antibodies, leading to immune activation of other resident cells such as microglia, and the generation of MS lesions. In fact, cortical lesions were found in the vicinity of such ectopic follicles, giving strength to this hypothesis.

Cortical pathology has also been found to be much more prevalent than previously thought, especially in progressive MS cases (Kutzelnigg et al. 2005). However, cortical lesions are associated mainly with signs of diffuse microglial activation instead of classical perivascular inflammation and are therefore undetectable by conventional MRI (Pirko et al. 2007). These findings, together with the newly recognized role for B cells, indicate that even in the progressive and neurodegenerative stage of MS, the underlying disease process is immune driven, therefore bridging the gap with the initial inflammatory and autoimmune stage. Microglial activation has also been found in acute pattern III MS lesions characterized by myelin disaggregation, oligodendrocyte apoptosis, and the expression of tissue protective proteins such as HIF-1α (Marik et al. 2007). Findings comparable to these led some authors to propose the existence of a prephagocytic lesion, in which a primary oligodendrocyte injury would lead to microglial activation, with the adaptive T- and B-cell response appearing as an epiphenomenon (Barnett et al. 2006; Barnett and Prineas 2004). It may be that the innate immune system plays a much more fundamental role than previously thought in MS lesion pathogenesis and that therapies aimed at this side of the immune response will prove effective, including in progressives stages of the disease. In fact, it has recently been shown in EAE that the therapeutic benefits of interferon-β are dependent mainly on myeloid lineage cells such as macrophages and microglia (Prinz et al. 2008).

Taken together, all these findings have led to a much more complex model of MS immunopathology in which several components of the adaptive and innate immune system contribute to the genesis of MS lesions, perhaps in different proportions depending on the disease stage.

1.3 Emerging Molecular Targets for MS Therapeutics

Developments in the model for MS pathogenesis have naturally led to the discovery of relevant molecules that may be useful as therapeutic targets. There are several new therapeutic agents in advanced clinical development, to be discussed in detail below, which can be grouped into three categories: oral agents, monoclonal antibodies, and antigen-specific therapies. Some of these drugs are classic nonspecific immunosuppressors, whereas others take advantage of newly discovered biological mechanisms, such as the role of sphingosine 1-phosphate receptors (fingolimod) (Brinkmann 2007) or the natural antioxidant Nrf2 pathway (BG-12) (Schreibelt et al. 2007).

Apart from these, there are several potentially relevant therapeutic targets that should be explored (Fontoura et al. 2006). Most of these have been discovered in EAE and therefore should be validated in MS studies before attempts are made to develop agents against them. Nonspecific regulators of the inflammatory response are naturally relevant targets for MS therapy. Many members of the nuclear receptor family including the glucocorticoid, sex hormone, peroxisome proliferator-activated receptors (PPARs – PPARα, PPARβ/δ, PPARγ), and liver-X receptors have been found to be the major regulators of inflammation (Glass and Ogawa 2006), and ligands for these receptors have been used to treat EAE with success. Early trials of these agents (estriol, fibrates, thiazolidinediones) are currently underway in MS. Regulatory enzymes such as indoleamine deoxygenase (IDO) (Platten et al. 2005) and heme oxygenase-1 (Chora et al. 2007) have also proven to be the key players in EAE pathogenesis, and induction of these molecules might prove beneficial in MS.

Besides modulating the immune response, effector mechanisms for myelin destruction and axonal degeneration including glutamate excitotoxicity and sodium channel function make promising targets for neuroprotective therapies. Glutamate AMPA and NMDA receptor antagonists have been shown effective in preventing myelin damage in animal models (Karadottir et al. 2005; Smith et al. 2000), and existing nonselective antagonists such as riluzole or memantine have been tried with success in EAE, but have yet to prove themselves in MS. Sodium channel regulation of intracellular calcium has also emerged as a central component in axonal degeneration (Waxman 2006). Phenytoin and carbamazepine, well-known sodium channel antiepileptic drugs, were effective in the treatment of EAE, but a severe rebound effect appeared after the drugs were stopped, raising concerns about their safety for human trials (Black et al. 2007).

It is inevitable that beyond the goals of stopping inflammation and tissue destruction, neural regeneration and remyelination are necessary to repair existing lesions. Axonal regeneration in the CNS is impeded by several factors, of which the myelin-associated inhibitors, MAG, OMgp, and Nogo-A, have attracted the most attention (Buchli and Schwab 2005; Filbin 2003). In animal models of CNS injury and early nonhuman primate trials, antibodies against Nogo-A have been shown to induce axonal growth and functional recovery (Freund et al. 2006). Anti-Nogo antibodies are currently in early clinical development for acute spinal cord injury (Novartis). Remyelination has been shown to occur in a significant percentage of MS lesions, and the presence of oligodendrocyte progenitors has been well documented (Chang et al. 2002; Patrikios et al. 2006). Differentiation of these cells into mature oligodendrocytes and myelin production is dependent on transcription factors such as Olig-1, activation of Jagged-Notch-Hes pathway signaling, and overcoming the inhibitory effects of the recently described LINGO-1 component of the Nogo-A receptor complex (Arnett et al. 2004; John et al. 2002; Mi et al. 2005). Blocking

LINGO-1 has been shown to promote remyelination in the EAE model (Mi et al. 2007), and anti-LINGO therapies are in late stage preclinical development (Biogen-Idec).

Finally, a different strategy to promote tissue regeneration is develolped based on stem-cell technology. Neural precursor and glial precursor cells have both been shown to be able to migrate to the CNS parenchyma and differentiate into useful phenotypes; neural precursor cells (neurospheres) have also been proven to migrate to inflamed areas of the CNS, where they differentiate into oligodendroglial precursors, promote remyelination and clinical recovery, and to abrogate the autoimmune T-cell response by inducing apoptosis (Martino and Pluchino 2006; Pluchino et al. 2005) and IDO (Matysiak et al. 2008).

In the future, regenerative strategies might be applied in individualized combinations with one or more therapies aimed at immune regulation and tissue protection, to provide a comprehensive treatment program for MS. For now, several molecularly targeted, immune-mediated therapies are in late stage clinical development.

2 Promising Molecular Target-Based MS Therapies

2.1 Oral Therapies

Oral therapies have the obvious advantage of a preferred route of administration. This advantage, however, does not infinitely outweigh the need to maintain a favorable safety profile. Five of these oral therapies in late stage clinical development are described in detail. In addition to describing the efficacy of these therapies, particular attention is given to the molecular targets of these therapies as well as to their side effects.

2.1.1 FTY720

FTY720 (Fingolimod, Novartis) is an oral sphingosine-1-phosphate (S1P) receptor modulator that appears to alter lymphocyte trafficking (Kappos et al. 2006a). FTY720 shares similarities in chemical structure to the sphingolipid, sphingosine. Like sphingosine, FTY720 is phosphorylated by sphingosine kinases, and the phosphorylated form inhibits the G-protein-coupled receptors for S1P, a molecule that is active in a number of cellular functions including morphogenesis, proliferation, and cytoskeletal rearrangement (Brinkmann 2007). In this way, FTY720 has been shown to prevent the egress of both T cells and B cells from secondary lymphoid tissues into the circulation, thus inhibiting the trafficking of lymphocytes to the CNS. This mechanism is, of course, similar to the approved monoclonal antibody natalizumab (Tysabri, Biogen-Idec) in that lymphocyte trafficking is inhibited, but FTY720 acts at the source of the lymphocytes, whereas natalizumab acts at the blood–brain barrier. By virtue of possessing this similar mechanism of action to natalizumab, there exists the potential of untoward effects including PML (progressive multifocal leukoencephalopathy), although no cases have so far been reported with FTY720. Because FTY720 is a small molecule with a short half-life, a potential advantage of FTY-720 over natalizumab is that its adverse effects may be rapidly reversible.

In a placebo-controlled phase 2 trial, FTY720 was shown to have significant efficacy in reducing lesions on brain MRI as well as reduction in clinical relapse rates (Kappos et al. 2006a). Two hundred and eighty one relapsing–remitting MS (RRMS) patients were randomized to receive daily oral FTY720 at either 1.25 or 5.0 mg, or placebo daily for 6 months. A total of 255 patients completed 6 months of dosing, and within these patients, the median total number of gadolinium-enhancing lesions on brain MRI was lower with 1.25 and 5.0 mg of FTY720 (1 lesion, $p < 0.001$ and 3 lesions, $p = 0.006$, respectively) than with placebo (5 lesions). The annualized relapse rates were also lower with 1.25 and 5.0 mg of FTY720 (0.35, $p = 0.009$ and 0.36, $p = 0.01$, respectively) than with placebo (0.77). After the 6-month time point, patients continued dosing into a 6-month extension phase where patients and investigators remained blinded to the dose assignments. Patients who received placebo were rerandomized to one of the FTY720 doses. A total of 227 patients completed the extension phase, and the number of gadolinium-enhancing lesions and relapse rates remained low in the groups that continued on FTY720. At month 12, more than 80% of patients who were treated with FTY720 were free of gadolinium-enhancing lesions. Furthermore, brain lesions and relapse activity decreased in patients who were rerandomized from placebo to FTY720.

The most frequent adverse events that were noted included nasopharyngitis, dyspnea, headache, diarrhea, and nausea. Elevations in liver enzymes were noted more frequently with FTY720 (10–12%) compared to placebo (1%). One case of posterior reversible encephalopathy syndrome was reported in the 5.0 mg group after 10 weeks of treatment. This patient's clinical symptoms and MRI findings began to improve 72 h after discontinuation of FTY720, but a residual right homonymous hemianopsia remained unchanged at 15 months of follow-up. Within 6 h after the first dose of FTY720, a reduction in heart rate of a mean of 13.8 and 16.6 beats per minute was noted with the 1.25 and 5.0 mg groups, respectively, with spontaneous resolution. Forced expiratory volume at 1 s was also reduced at month 6 compared to baseline to an average of 8.8% ($p = 0.003$) with 5.0 mg of FTY720, and 2.8% ($p = 0.68$) with 1.25 mg of FTY720, compared to 1.9% in the placebo group. Finally, in the 3-year extension study, the results of which were reported at the 60th annual meeting of the American Academy of Neurology (AAN), seven patients developed skin malignancies on FTY720.

Following the encouraging phase 2 results, FTY720 was tested in a 1,292-patient phase 3 trial of 12 months duration, comparing either a 0.5 or a 1.25 mg dose of FTY720 to intramuscular interferon beta-1a (Cohen et al. 2010). A total of 1,153 patients completed the trial, and the annualized relapse rate was significantly lower in both dosing groups compared to interferon beta-1a (0.16 in the lower dose, 0.20 in the higher dose, as compared to 0.33 in the interferon beta-1a group, $p < 0.001$). No effect on disability scores was seen. The most frequent adverse events reported with FTY720 include herpesvirus infections, bradycardia, hypertension, macular edema, and elevated liver enzyme levels. Two cases of fatal infections occurred during dosing in the 1.25 mg group – disseminated primary varicella zoster and herpes simplex encephalitis. Two other fatalities occurred 6–10 months after discontinuation of the 1.25 mg dose – progressive neurological deterioration and metastatic breast cancer. Multiple cancers were reported with FTY720 – ten cases of skin cancer and four cases of breast cancer.

In parallel, a 24-month placebo-controlled phase 3 study of FTY720 was also conducted (Kappos et al. 2010). A total of 1,272 patients were enrolled and 1,033 completed the study comparing 0.5 or 1.25 mg of FTY720 to placebo. The annualized relapse rates were significantly lower with both doses of FTY720 (0.18 in the lower dose, 0.16 in the higher dose, as compared to 0.40 with placebo, $p < 0.001$). There was also a significant reduction in disability progression for both doses compared to placebo ($p = 0.02$). Adverse events were similar to the other studies and included bradycardia, macular edema, elevated liver enzyme levels, and hypertension.

Based on the positive results of these two phase 3 studies, it is expected that Novartis will seek approval of FTY720 in 2010.

2.2   Teriflunomide

Teriflunomide (Sanofi-Aventis), an active metabolite of leflonimide (Arava), is an oral dihydro-orotate dehydrogenase inhibitor that inhibits T-cell proliferation (Cherwinski et al. 1995). By blocking dihydro-orotate dehydrogenase, an enzyme in the pyrimidine synthesis pathway, pyrimidine synthesis and thus cell division is inhibited in rapidly dividing cells such as T cells. Furthermore, in animal studies, teriflunomide was shown to suppress tyrosine kinases involved in signal transduction pathways, thus altering calcium mobilization and T-cell activation (Korn et al. 2001). Therefore, there may be multiple mechanisms involved in the drug's action to suppress autoimmune T cells; although since these pathways are not unique to autoimmune T cells, the drug would be expected to act on lymphocytes generally.

Teriflunomide was tested in a placebo-controlled phase 2 trial and was shown to reduce brain MRI lesions significantly in relapsing MS patients (O'Connor et al. 2006). In this trial, 157 RRMS patients and 22 secondary progressive MS (SPMS) patients with relapses were randomized to placebo, teriflunomide at 7 mg/day, or teriflunomide at 14 mg/day for 36 weeks. The primary endpoint was based on the number of combined unique active lesions on brain MRI, which were defined as new and persistent gadolinium-enhancing lesions and new or enlarging T2 lesions. There were significantly fewer median numbers of combined unique active lesions with teriflunomide 7 mg/day (0.2 lesions, $p < 0.03$) and teriflunomide 14 mg/day (0.3 lesions, $p < 0.01$) compared to placebo (0.5 lesions). Furthermore, patients on teriflunomide had nonsignificant trends toward lower annualized relapse rates compared to placebo (0.58 on 7 mg/day, 0.55 on 14 mg/day, and 0.81 on placebo). The proportion of patients with disability increases as measured by the EDSS

(Expanded Disability Status Scale) was lower in the higher 14 mg/day dosed group of teriflunomide compared to placebo (7.4% vs. 21.3%, $p < 0.04$).

The treatment appeared safe and well tolerated with similar incidence of adverse events with teriflunomide and placebo. Nasopharyngitis, alopecia, nausea, alanine aminotransferase increase, paraesthesia, back pain, limb pain, diarrhea, and arthralgia were more commonly reported by patients in the teriflunomide treatment groups than in the placebo group, though none of these symptoms reportedly reached a statistically significantly difference. Concerns, however, have been raised about the teratogenic capacity of teriflunomide's parent compound leflunomide. In animal studies, leflunomide at 30 mg/kg was shown to induce craniofacial malformations and other skeletal and cardiac deformities in the fetus (Fukushima et al. 2007). As a result, leflunomide has been ascribed a pregnancy category X, and strict contraceptive measures have been imposed with clinical trials involving leflunomide.

Currently, teriflunomide is being tested in a phase 3 trial in relapsing MS. This is a 2-year placebo-controlled study testing two doses of teriflonomide once daily with accumulation of disability and frequency of relapses as endpoints.

2.2.1 BG-12

BG-12 (Biogen-Idec) is an oral fumarate derivative, a second-generation formulation of the original compound Fumaderm that is approved for use in Germany for psoriasis (Wakkee and Thio 2007). Fumaderm is an enteric-coated tablet that contains a defined mixture of dimethylfumarate and salts of monoethylfumarate. It was later determined that dimethylfumarate was the principal component that provided antipsoriatic activity, and led to the development of BG-12 that contains dimethylfumarate as enteric-coated tablets. These fumaric acid ester compounds have been shown to have several mechanisms of action of relevance to the treatment of MS. These compounds are able to induce T-helper 2 (Th2) cytokines that are anti-inflammatory in nature, to induce apoptosis in activated T cells, and to reduce the expression of adhesion molecules (de Jong et al. 1996; Ockenfels et al. 1998; Vandermeeren et al. 1997). Furthermore, in the EAE animal model, a fumaric acid ester has been shown to reduce the severity of disease by increasing the expression of the anti-inflammatory cytokine IL-10 and decreasing the expression of proinflammatory cytokines TNFα and IL-6 (Schilling et al. 2006). All of these factors could contribute to the downregulation of the inflammatory autoimmune process as well as to the inhibition of lymphocyte infiltration of the CNS.

The results of two clinical trials with BG-12 in MS patients have been reported. In the initial open label study in ten RRMS patients, the trial was divided into four phases: 6-week baseline, 18-week treatment phase (Fumaderm 720 mg/day orally), 4-week washout, and another 48-week treatment phase (Fumaderm 360 mg/day orally) (Schimrigk et al. 2006). In this study, Fumaderm produced a significant reduction from baseline in the number (from a mean of 11.28 at baseline to 1.5 at week 18, $p < 0.05$) and volume (from a mean of 244.5 mm^3 at baseline to 26.1 mm^3 at week 18, $p < 0.018$) of gadolinium-enhancing lesions after the 18-week treatment phase, which persisted during the 48-week treatment phase. Clinical measures of disease remained stable or slightly improved. The most common adverse events were gastrointestinal symptoms and flushing. Almost all patients experienced mild-to-moderate gastrointestinal adverse events at the beginning of the study, but these events decreased continually during the study. One patient was discontinued because of the gastrointestinal symptoms. Transient, up to twofold elevation of liver enzyme levels occurred in four patients, but none were significant enough to withdraw from the study.

The results of a subsequent randomized, double-blind, placebo-controlled phase 2 study were reported at the Sixteenth Meeting of the European Neurological Society in May of 2006 at Lausanne, Switzerland (Kappos et al. 2006b). A total of 257 RRMS patients were enrolled in this study and were assigned to one of four oral BG-12 treatment groups for 24 weeks: 120 mg once daily, 120 mg three times daily (360 mg/day), 240 mg three times daily (720 mg/day), or placebo. The treatment period was followed by a 24-week dose-blinded extension study in which all patients continued to receive their assigned treatment except that the placebo patients were switched to 720 mg/day BG-12. During the 24-week double-blind stage, patients who received the 720 mg/day dose experienced a significant 69% reduction in the mean number of new gadolinium-enhancing lesions in the brain compared to placebo ($p < 0.001$). This group also had a 32% reduction in relapse rates compared to placebo, but this was not statistically significant. The results of the 120 mg/day and 360 mg/day groups were not statistically significant. The most common adverse events were flushing, gastrointestinal disorders, headache, and nasopharyngitis. Transient elevations of liver enzyme levels of an incidence that ranged between 2 and 8% in the three active treatment groups were observed, compared with 5% in the placebo group.

In early 2007, two large phase 3 trials of BG-12 were initiated in RRMS patients. These trials will enroll more than 2,000 patients in a 2-year, placebo-controlled, dose-comparison trial. The doses being tested are 240 mg twice a day (480 mg/day) or 240 mg three times a day (720 mg/day). One of these trials will also include glatiramer acetate in a comparator arm. Endpoints on both of these trials will include measurements of clinical relapses, disability progression, and brain MRI measures. These trials are expected to be completed in mid-2010.

2.2.2 Laquinimod

Laquinimod (Teva Pharmaceuticals and Active Biotech) is an oral immunomodulator that is structurally similar to roquinimex (Linomide). Roquinimex was shown to reduce the degree of new lesions on brain MRI in phase 2 and phase 3 MS studies, but its development was discontinued because of untoward adverse effects including serositis and myocardial infarction (Andersen et al. 1996; Karussis et al. 1996; Noseworthy et al. 2000; Tan et al. 2000; Wolinsky et al. 2000). Because of these issues, an effort was undertaken to chemically modify roquinimex to reduce its toxicity and improve its efficacy. More than 50 derivatives were tested in the mouse

EAE model for efficacy, and select compounds were tested for toxicity in the beagle for induction of a proinflammatory reaction (Jonsson et al. 2004). From this screen, laquinimod was chosen as a compound with improved potency and lower toxicity compared to roquinimex. As far as the mechanism of action of laquinimod in auto-immune disease is concerned, it has been shown that in animal models to inhibit T-cell and macrophage entry into the CNS as well as to shift the cytokine profile of T cells from Th1 to Th2/Th3 (Yang et al. 2004).

Because of its similarity with roquinimex, there has been a concern about the potential for cardiac side effects in clinical trials. In a double-blind, placebo-controlled phase 2 trial, the higher dose of laquinimod tested (0.3 mg) was shown to be effective in reducing active lesions on MRI (Polman et al. 2005). In this study, 209 relapsing MS patients were randomized to receive either 0.1 or 0.3 mg laquinimod or placebo tablets for 24 weeks. The cumulative number of active lesions on brain MRI (defined as the sum of new gadolinium-enhancing lesions and new or enlarging T2 lesions) was reduced by 44% on 0.3 mg of laquinimod compared to placebo ($p = 0.0498$). There was no significant effect of the 0.1 mg group, and no effects on clinical measures were noted with either dose. Laquinimod was well tolerated and caused relatively few side effects, with similar proportions of patients with adverse events in all three treatment groups. Of the four treatment-emergent serious adverse events, one was in the laquinimod 0.1 mg group (brain contusion), and two were in the laquinimod 0.3 mg group (iritis, burning sensation). Two patients withdrew from the study because of adverse events, one patient in the placebo group (myocardial ischemia) and one patient in the laquinimod 0.3 mg group (muscle tightness and depression). There were no clinically relevant changes in laboratory measurements.

In a subsequent phase 2b trial, an even higher dose of laquinimod, 0.6 mg, was tested, and it showed significant improvement in active lesions compared with placebo, with a similar safety profile as in previous trials. The results of this trial were presented at the 59th Annual Meeting of the AAN in April 2007. The phase 2b trial was conducted in 306 RRMS patients who were randomized to placebo, 0.3 or 0.6 mg/day of laquinimod treatment for 36 weeks. Patients who received the 0.6 mg daily dose experienced a 38% reduction in the cumulative number of enhancing lesions on brain MRI compared to placebo ($p = 0.0048$). There was a favorable trend toward reduction in clinical relapses compared to placebo. The 0.3 mg dose showed no statistically significant difference compared with placebo. Treatment with both doses of laquinimod was well tolerated with only transient and dose-dependent increases in liver enzymes.

The 0.6 mg dose of laquinimod is currently being tested in a phase 3 study of RRMS patients, which is expected to enroll approximately 1,000 patients. Patients are randomized to 0.6 mg of laquinimod daily or placebo for 24 months. The primary outcome measure is the number of confirmed relapses, with secondary assessment of disability and MRI lesions. The study began in late 2007, and results are expected in early 2011. The drug is considered by many to have modest efficacy, in part because there were no significant effects on relapses or disability, but concern remains about potential cardiac toxicity.

2.2.3 Cladribine

If approved, cladribine (Merck-Serono) would become the first oral agent available for MS as the results of a phase 3 trial are due in early 2009. Cladribine is a purine analog that was originally indicated for the treatment of hairy-cell leukemia. Cladribine is phosphorylated by deoxycytidine kinase to a triphosphate deoxynucleotide, but because it is resistant to deamination, it achieves high intracellular concentrations (Beutler 1992). It is thought that the accumulation of this deoxynucleotide may interfere with some aspects of DNA repair, and thus damage to DNA occurs with eventual ensuing cell death. Lymphocytes, in particular, have high concentrations of deoxycytidine kinase and, therefore, are relatively selectively sensitive to cell death by cladribine (Seto et al. 1985). Because of this lymphocyte selectivity, it was theorized that cladribine might be effective in T- and B-cell mediated autoimmune disorders. The first indication that this theory may be true came from results in which about 50% of patients with treatment-resistant autoimmune hemolytic anemia responded to cladribine (Beutler 1992). Subsequently, several trials in MS were begun, initially testing the parenteral formulation of cladribine.

In 1994, an intravenous formulation of cladribine was tested in a phase 2 trial of 51 chronic-progressive MS patients, which included both primary and secondary progressive MS patients (Sipe et al. 1994). This was a 2-year placebo-controlled, double-blind, crossover study in which patients were initially to receive cladribine 0.1 mg/kg/day or placebo for seven consecutive days every month for 6 months. However, because of profound thrombocytopenia, the regimen was modified to seven consecutive daily doses every 4 months. The crossover occurred at 1 year with the placebo group switching to cladribine at one-half of the original dose. The results at both 1 and 2 years show that cladribine-treated patients had statistically significant improvement in disability scores compared to placebo, as well as reductions in enhancing lesions on brain MRI compared to placebo.

Based on the results of the phase 2 study, a phase 3 study was conducted in chronic progressive MS patients (Rice et al. 2000). This was a 1 year, double-blind, placebo-controlled study of 159 patients. Patients were randomized to placebo or subcutaneous cladribine 0.07 mg/kg/day for five consecutive days every 4 weeks for either two or six cycles (total dose, 0.7 mg/kg or 2.1 mg/kg, respectively), followed by placebo, for a total of eight cycles. The results at 1 year show no difference in disability scores among any of the groups, which was the primary endpoint of the study. There was, however, a 90% reduction in the mean number of enhancing lesions and a greater than 87% reduction in the mean volume of enhancing lesions in both cladribine treatment groups. Although the clinical effect of cladribine in this study was disappointing, it has been pointed out that 30% of the patients in the study were diagnosed with primary progressive MS, which is felt to be more refractory to treatment than other forms of MS.

A total of three studies of cladribine in RRMS have been published. In the first study, ten patients with RRMS were dosed with cladribine either a 5 mg subcutaneous formulation or 10 mg oral formulation for five consecutive days, and repeated every 5 weeks for a total of six courses (Grieb et al. 1994). Slight improvements in disability scores were observed with an improvement in EDSS scores from 4.6 at baseline to 2.8 at 30 weeks. Mild leukopenia with about a 50% reduction of lymphocyte counts was noted at 30 weeks. In the second study, 31 RRMS patients were treated subcutaneously with the same dose and schedule of cladribine as the first study, and 40 patients were treated with placebo. Data at 30 weeks were reported and a reduction in the EDSS scores of 0.63 points in the cladribine arm were demonstrated, but only a 0.13 point reduction in the placebo arm (Grieb et al. 1994).

The third study was an 18-month, placebo-controlled, double-blind study conducted on 52 RRMS patients (Romine et al. 1999). Patients were randomized to either 8 months of placebo ($n = 25$) or cladribine 0.07 mg/kg/day ($n = 27$) by subcutaneous administration for five consecutive doses each month for 6 months followed by 2 months of placebo. The results demonstrated a statistically significant improvement in the joint frequency and severity of relapses with cladribine treatment compared to placebo during months 7–12 ($p = 0.021$) and 7–18 ($p = 0.01$). The cladribine group also had significantly fewer enhancing lesions on brain MRI compared to placebo ($p = 0.001$). There were no side effects or adverse events reported except for mild segmental herpes zoster in two cladribine-treated patients and one placebo patient.

As cladribine is used as a chemotherapeutic agent, most of the safety data on cladribine are derived largely from patients who have been treated with courses of antineoplastic agents. These studies indicated that the primary toxicity of cladribine is myelosuppression, particularly on lymphocyte counts, which can be dose-limiting and may persist for several months after cessation of treatment (Brousil et al. 2006). Severe aplastic anemia and thrombocytopenia have occurred with cladribine in patients who have previously been treated with other myelo-suppressive drugs (Brousil et al. 2006). Because of these risks of myelosuppression, pretreatment hematologic safety criteria with minimum levels of platelet counts, granulocyte counts, and hemoglobin concentration have been established (Brousil et al. 2006).

Recently, the results of a 2-year, double-blind, placebo-controlled phase 3 trial of cladribine in RRMS patients were reported (Giovannoni et al. 2010). This trial enrolled 1,326 patients with endpoints of clinical relapses, disability scores, and lesions on brain MRI. Cladribine was administered at two different doses (either 3.5 or 5.25 mg/kg) given in two or four courses for the first 48 weeks, and then in two courses at week 48 and at week 52. Treatment with cladribine at either dose resulted in a significantly lower annualized relapse rate compared with placebo (0.14 for the lower dose, 0.15 for the higher dose, and 0.33 for placebo, $p < 0.001$), a lower risk of disability progression compared with placebo ($p = 0.02$ for the lower dose, $p = 0.03$ for the higher dose), and lower brain lesion counts ($p < 0.001$ for all comparisons). Adverse events included lymphocytopenia and herpes zoster.

Cladribine is also being studied as an add-on treatment to interferon β-1a (Rebif) in a 2-year, double-blind, placebo-controlled trial of 260 MS patients who have experienced at least one relapse on interferon β-1a. Data from this add-on trial are expected in early 2010.

2.3 Monoclonal Antibodies

Monoclonal antibodies have the advantage of a well-defined molecular target and thus a seemingly clear mechanism of action. However, despite a presumably singular molecular target, as demonstrated in the case of the only monoclonal that is approved for use in MS, natalizumab (Tysabri), there may be unexpected consequences of a monoclonal antibody approach that must be anticipated. Three monoclonal antibodies in late stage clinical development for MS are highlighted that have demonstrated very promising efficacy but have had variable degrees of concerns about safety.

2.3.1 Rituximab

One of the more promising and unique monoclonal antibodies in development for MS is rituximab (Genentech). Rituximab is a mouse/human chimeric monoclonal antibody and has been approved for the treatment of B-cell non-Hodgkin's lymphoma and for rheumatoid arthritis. Its uniqueness lies in the fact that it targets the B-cell marker CD20. Rituximab is known to deplete CD20+ B cells through a variety of mechanisms. For example, rituximab recognizes the CD20 phosphoprotein expressed on the surface of B cells and has been shown in vitro to bind human C1q and mediate complement-dependent cell lysis of human B cell lines (Reff et al. 1994). Targeting of B cells with an MS therapeutic is a relatively novel concept in the field of MS research.

MS has long been thought to be a T-cell-mediated autoimmune disease given the evidence of myelin antigen-specific T cells that confer disease in the EAE animal model and the identification of such cells in demyelinating lesion sites within the CNS of human MS patients. It is clear now, however, that B cells play key roles in the pathogenesis of MS based on a variety of pieces of evidence. Long before the advent of MRI, cerebral spinal fluid analysis of immunoglobulin production and oligoclonal bands have been the mainstay of MS diagnosis (Antel and Bar-Or 2006; Siden 1979). Within MS brain lesions, antibody-complement deposition and expanded B-cell populations can be found (Baranzini et al. 1999; Genain et al. 1999). Furthermore, an antibody-independent role of B cells in the pathogenesis of MS has been postulated. For example, B cells can function as efficient APCs and can themselves release effector cytokines that could contribute to disease progression (Antel and Bar-Or 2006).

A phase 2 trial of rituximab in RRMS patients was recently completed with very encouraging results (Hauser et al. 2008). This was a 48-week, double-blind, placebo-controlled trial in 104 RRMS patients who were randomized to 1,000 mg of intravenous rituximab or placebo on days 1 and 15. Patients who received rituximab had fewer numbers of gadolinium-enhancing lesions on brain MRI at weeks 12, 16, 20, and 24 as compared to patients who received placebo ($p < 0.001$). The significant reductions in lesions were sustained for 48 weeks ($p < 0.001$).

Furthermore, the proportion of patients in the rituximab group with relapses was significantly lower than placebo at week 24 (14.5% vs. 34.3% respectively, $p = 0.02$) and week 48 (20.3% vs. 40.0% respectively, $p = 0.04$). With respect to adverse events, within 24 h of the first infusion, a high number of patients in the rituximab group (78.3%) experienced adverse events than in the placebo group (40%), most of which were graded as mild to moderate. After the second infusion, the numbers of adverse events were similar in the two groups. Three patients who were treated with rituximab had grade 4 adverse events: one with ischemic coronary artery syndrome, one with a malignant thyroid neoplasm, and one with worsening MS.

Although in this trial, the drug was overall well tolerated with relatively few side effects, rituximab does carry a black box warning for potentially fatal infusion reactions, as well as other serious effects such as cardiac arrhythmias and renal failure. Therefore, although there is considerable enthusiasm and interest for rituximab in MS, there is significant concern about its safety especially if used on a chronic basis. Reports of the potentially fatal disease progressive multifocal leukoencephalopathy have also been reported in trials of rituximab in lupus patients.

Currently, rituximab is being tested in another phase 2 trial in RRMS patients. It was also recently announced that rituximab failed to slow disease progression in a trial of primary progressive MS.

2.3.2 Daclizumab

Daclizumab is a humanized monoclonal antibody against the alpha subunit of the IL-2 receptor (CD25) on activated T cells. IL-2 is a cytokine that causes the proliferation of lymphocytes and the secretion of proinflammatory cytokines. It has been approved for use in controlling kidney transplant rejection. Rationale for its use in MS comes from both preclinical and clinical studies. Daclizumab has shown efficacy in the EAE model of MS and in a phase I/II clinical trial of autoimmune uveitis (Nussenblatt et al. 2003). Subsequently, daclizumab has been tested in several clinical trials of MS.

The first clinical trials of daclizumab in MS were small open label trials. In the first of these trials, it was tested as an add-on therapy in MS patients who failed to respond to interferon β (Bielekova et al. 2004). This was a phase II open label trial in six RRMS and five SPMS patients in whom daclizumab was administered intravenously at 1 mg/kg/dose 2 weeks apart for the first two doses and every 4 weeks thereafter for a total of seven doses. The treatment resulted in a 78% decrease in new contrast-enhancing lesion on brain MRI on study compared to baseline ($p = 0.004$). There was a significant reduction in the exacerbation rate during treatment compared to baseline of 81% ($p = 0.047$). There were increases in the number of infections during treatment, which consisted of urinary and upper respiratory tract infections, as well as two cases of transient elevation of liver function tests and bilirubin.

In another trial, daclizumab was tested both as monotherapy and as add-on to other therapy (Rose et al. 2004). A total of 19 RRMS and SPMS patients were enrolled in this open label study, and daclizumab was administered intravenously at 1 mg/kg/dose in the same manner as the previous study, with the first two doses 2 weeks apart and subsequent doses every 4 weeks but for anywhere from 5 to 25 months. Another difference is that daclizumab was administered as monotherapy in 16 patients, while it was given in combination with interferon β in two patients and with methylprednisolone in one other patient. The results demonstrate that the patients segregated within two to four treatments into either responders (ten patients) or those whose disease stabilized (nine patients). The responders had an average reduction in EDSS of approximately 2.5 points on treatment compared to baseline ($p < 0.001$). Those whose disease stabilized had either no change or ≤0.5 point decline in their EDSS on treatment. There was also a decrease in relapse rates for all patients from an annualized relapse rate of 1.23 at baseline compared to 0.32 on treatment. A significant decrease in the proportion of active MRI scans (defined as gadolinium-enhancing lesions, new T2 lesions, or both) was observed from 7 of 18 scans at baseline to 6 of 46 scans with treatment ($p < 0.05$). Side effects were mild, although one patient discontinued treatment because of paraesthesias, and another because of anemia and upper respiratory tract infection. Transient liver function test level elevation was noted in one patient.

More recently, the results of a much larger 230-patient phase 2 trial of daclizumab was reported at the 60th Annual Meeting of the AAN in April, 2008. This trial enrolled a total of 230 relapsing MS patients on interferon β who were randomized to either daclizumab 1 mg/kg/dose, daclizumab 2 mg/kg/dose, or placebo. Doses were administered every 4 weeks for 24 weeks, and then patients were observed for a total of 48 weeks. There was a 25% reduction in active lesions on MRI in the lower daclizumab dose group and a 72% reduction in the higher dose group compared to the interferon β and placebo group.

Daclizumab shows promise as an add-on therapy in these trials, and it appears to have the best safety profile of all of the monoclonal antibodies being tested in MS trials. However, daclizumab suffers from modest efficacy and lack of robust data as a monotherapy in RRMS.

2.3.3 Alemtuzumab

Alemtuzumab is a depleting monoclonal antibody against the CD52 antigen, which is a glycoprotein of unknown function present on all mature lymphocytes and monocytes. Alemtuzumab is more commonly known by its trade name "Campath-1H" (Genzyme), whose name derives from the university department in which the antibody was first developed, "Cambridge Pathology" (Coles et al. 2004). The mechanism of alemtuzumab's cell depletion is through complement-mediated cell lysis, and induction of apoptosis (Cree 2006). Alemtuzumab is currently licensed for use in the treatment of chronic lymphocytic leukemia, and it is being tested in trials of transplantation and autoimmunity, including graft versus host disease (GVHD), autoimmune thrombocytopenic purpura, and cutaneous scleroderma (Coles et al. 2004).

Because of the potential severe side effects of complete lymphocyte depletion, treatment of MS patients with alemtuzumab proceeded with caution. These trials in MS began with the treatment of one patient in 1992, six additional during 1993, and to a cumulative total of 36 in 1999, all of whom had SPMS (Coles et al. 2004). Just one or two pulses of alemtuzumab were enough to suppress inflammation on brain MRI for at least 6 years, but disease progression was not halted. It was decided that patients should be treated earlier in their course of disease, and, thus, a total of 22 patients with RRMS were then treated with alemtuzumab. In this group of patients, it was found that their relapse rates declined by 94% (from 2.21 to 0.14 per patient per year). There was, however, a mean annual increase in the disability scores of 2.2 points on the EDSS.

In these 58 patients (36 SPMS and 22 RRMS), profound CD4 and CD8 depletion that persisted for a median of 61 and 30 months, respectively, was noted. Seven incidences of infections occurred that were considered mild. More significantly, however, 15 new cases of Graves' disease were observed in these patients, with one additional case of autoimmune hypothyroidism. These patients were treated with standard medical therapy, but nine of them required radioactive thyroid ablation. One of the patients who was diagnosed with Graves' disease also developed a permanent and cosmetically unpleasant ophthalmopathy.

Despite these adverse effects, alemtuzumab was more recently tested in a phase 2 trial of 334 RRMS patients in a direct comparison with interferon β-1a (Rebif). This trial compared high (24 mg/day) or low (12 mg/day) dose of alemtuzumab, given intravenously with interferon β-1a in treatment-naïve RRMS patients. Alemtuzumab was given as five daily infusions at month 0 to 216 patients, then as three daily infusions at month 12 to 207 patients, and then as three daily infusions at month 24 to 46 patients prior to voluntary suspension of dosing in September 2005. It had previously been reported that several patients developed idiopathic thrombocytopenic purpura and that one proved to be fatal, and thus the trial was suspended. Although dosing was stopped, the patients who had already been dosed were observed for 3 years, and these results were reported at the 60th annual meeting of the AAN in April 2008. Pooled data on patients taking either dose of alemtuzumab demonstrated a 73% reduction in the risk of relapse compared to interferon β-1a ($p < 0.0001$), and a 71% reduction in the risk of sustained accumulation of disability ($p < 0.0001$). Furthermore, there was an increase in the 3-year mean disability change from baseline by 0.39 points in the EDSS in patients treated with interferon β-1a, but there was an improvement by 0.39 points in the alemtuzumab treated patients ($p < 0.0001$).

Despite the concerns about its safety profile, because of the impressive efficacy of alemtuzumab, especially with regards to reduction in disability, phase 3 trials in MS with this drug have begun. This is a 2-year study of 12 mg/day alemtuzumab given as two annual cycles compared to interferon β-1a in treatment-naïve RRMS patients, with a target enrollment of 525 patients. The trial is estimated to be completed in March 2011. In another similar phase 3 trial, two different doses of alemtuzumab (12 mg/day or 24 mg/day) will be compared to interferon β-1a in a 2-year study of 1,200 RRMS patients. This study is also expected to be completed by March 2011.

3 Antigen-Specific Therapies

Antigen-specific therapies offer the opportunity for the most directed method of treatment of autoimmune diseases, as the target of the therapy is the autoimmune T- or B-cell against a specific antigen. Whether such a focused method of treatment of MS will result in sufficient efficacy to change clinical course remains to be determined, but such an approach should certainly provide the least amount of untoward effects. Two antigen-specific methods in late stage clinical trials are discussed and contrasted.

3.1 MBP8298

An antigen-specific approach in late stage clinical testing employs a peptide of MBP, and is called MBP8298 (BioMS). The hypothesis behind this peptide-based approach is the phenomenon of "high zone tolerance," which is observed when administration of high doses of soluble antigen results in immunological tolerance (Liblau et al. 1994; Mitchison 1964). It has been demonstrated that strong and repeated stimulation by antigen can result in programmed cell death or unresponsiveness of T cells and B cells. The reason for the choice of this particular peptide region is that MBP 82-98 has been shown to be immunodominant in MBP-specific T cells as well as in autoantibodies, especially from patients with particular HLAs such as DR2 (Martin et al. 1990; Ota et al. 1990; Warren et al. 1995).

In a phase 1 trial of the MBP8298 peptide, the safety and attainment of immunological tolerance were demonstrated in a small trial in which different routes of administration were tested (Warren et al. 1997). In this trial of 53 chronic progressive MS patients, the peptide was administered via the intrathecal, subcutaneous, or intravenous routes. It was determined that tolerance, as measured by quantification of antibodies to MBP in the CSF, was achieved only via the intravenous route. Also, the duration of tolerance was most prolonged in patients with the HLA-DR2 haplotype, and it persisted in this group of patients for up to 1 year after only two doses of the peptide.

The MBP8298 peptide was then tested in a larger phase 2 trial of 32 patients with progressive MS (Warren et al. 2006). This was a placebo-controlled trial of 500 mg of intravenous MBP8298 administered every 6 months for 2 years in either primary progressive or in SPMS patients. Although, overall, there were no differences in the rate of progression in the EDSS score between MBP8298-treated and placebo patients, patients who were of a defined HLA haplotype demonstrated a significant benefit to the peptide. Within patients who were either HLA-DR2 or DR4, none of the ten MBP8298-treated patients had EDSS progression, whereas six of ten placebo patients progressed in disability ($p=0.01$). All patients were offered the option of continued treatment with the peptide in an open label fashion for up to 5 years. In this long-term follow-up period, HLA-DR2 or DR4 patients who received the MBP8298 peptide had a median time to progression of disability of 78 months compared to 18 months on placebo ($p = 0.004$). Interestingly, CSF anti-MBP autoantibody levels were reduced with MBP8298 in the majority of patients regardless of HLA type. The peptide was found to be very well tolerated and safe with no serious adverse events, and only mild local injection site reactions and blood pressure decreases were observed. MRI with gadolinium was performed to assess safety, and no differences were seen in lesion characteristics between the peptide and placebo.

MBP8298 was then tested in a phase 3 clinical trial in SPMS patients as well as in a phase 2 trial in RRMS patients. The SPMS trial was a placebo-controlled trial in 510 patients dosed with 500 mg of MBP8298 intravenously every 6 months for 24 months. Unfortunately, the trial failed to meet its primary outcome of time to progression as measured by EDSS. The RRMS trial was a placebo-controlled, 15-month trial with a 12-month open label extension in 218 patients. This trial also failed to meet its primary efficacy outcome as measured by relapse rate, MRI activity, and disease progression.

3.1.1 BHT-3009

Another similar yet quite distinct antigen-specific treatment in clinical testing for RRM is BHT-3009, a DNA vaccine encoding full-length MBP (Bar-Or et al. 2007; Garren et al. 2008). The similarity between BHT-3009 and MBP8298 is that they both act to cause immune tolerance toward the MBP autoantigen. However, there are far more differences than similarities. The BHT-3009 DNA vaccine encodes the entire MBP molecule, thus having the potential to tolerize to all possible MBP epitopes rather than a single peptide epitope that is HLA restricted. The DNA vaccine is administered intramuscularly rather than intravenously. The DNA backbone of BHT-3009 has been modified to further enhance immune tolerance by the incorporation of inhibitory dinucleotide sequences, whereas the peptide is simply the antigen with no additional enhancers of tolerance.

BHT-3009 was tested in a placebo-controlled phase 1 trial of 30 SPMS or RRMS patients (Bar-Or et al. 2007). Four intramuscular administrations of the DNA vaccine were given at weeks 1, 3, 5, and 9 after enrollment. Three doses were tested: 0.5, 1.5, and 3 mg. The trial demonstrated that BHT-3009 was safe and well tolerated with no significant adverse events greater than placebo, including injection site reactions. Trends in benefit of lesion reduction in brain MRI with gadolinium were observed with BHT-3009 at all doses. Furthermore, evidence of antigen-specific tolerance in both peripheral T cells and CSF myelin autoantibody titers was observed. This tolerance was observed not only in the MBP reactive T cells and autoantibodies, but also in other myelin-associated autoantigens. Thus, BHT-3009 appears to cause "bystander suppression" of the autoimmune response to several myelin autoantigens, and could benefit a larger proportion of patients in a heterogeneous autoimmune disease such as MS than a single epitope approach.

Based on these findings, a larger 289-patient phase 2 placebo-controlled trial in RRMS patients was recently completed with BHT-3009 (Garren et al. 2008). The DNA vaccine was administered intramuscularly at weeks 0, 2, 4, and then every 4 weeks, thereafter until 44 weeks after enrollment. Two doses were tested and compared against placebo: 0.5 and 1.5 mg. In addition, more than 80 patients provided CSF and serum at baseline and at the end of the study to examine the autoimmune response in CSF. Although the primary endpoint of lesion rate reduction on brain MRI was not observed in the overall trial, in all patients who provided serum at baseline, it was found that 0.5 mg of BHT-3009 caused a significant reduction in brain lesion activity that was directionally proportional to the level of anti-MBP antibodies in serum. In other words, the magnitude of lesion activity reduction increased as the serum anti-MBP levels increased. A direct correlation was also observed in the baseline serum anti-MBP level with the magnitude of relapse rate reduction in patients treated with 0.5 mg of BHT-3009 compared to placebo. Stated another way, the more immunologically active patients at baseline as measured by the level of anti-MBP antibodies present in serum were the best responders to BHT-3009 both based on MRI and clinical relapse rate parameters.

Bystander suppression was again demonstrated in this phase 2 trial when examining the changes in the CSF autoantibody levels to BHT-3009 treatments (Garren et al. 2008). Significant decreases in anti-MBP but also to other myelin-specific autoantibodies were observed in patients who received 0.5 mg BHT-3009, but not in those who were given placebo. The higher 1.5 mg dose neither caused immune tolerance in CSF autoantibodies nor demonstrated any efficacy in MRI or clinical parameters.

4 Conclusions

An effective yet safe therapy for MS remains a clinical challenge. The therapies that are either approved or in clinical development for MS span the range from highly effective but potentially quite harmful compounds on the one hand, to less effective but much safer compounds on the other. The therapies in late stage clinical trials reviewed here can be classified as falling within various points along that spectrum. It is likely that the optimal balance between safety and efficacy will be achieved by one or more or a combination of the approaches reviewed here. What is clear is that whether that optimal balance is achieved with an oral agent, monoclonal antibody, an antigen-specific therapy, or any of the other 130 or so recent clinical trials listed by the National MS Society, given the profusion of clinical trial activity in MS, there remains a great deal of interest in new treatments. The results of these trials are eagerly anticipated by the MS community at large and will in due course give the physician a number of new options with which to manage this debilitating disease.

References

Andersen O, Lycke J, Tollesson PO, Svenningsson A, Runmarker B, Linde AS, Astrom M, Gjorstrup P, Ekholm S (1996) Linomide reduces the rate of active lesions in relapsing-remitting multiple sclerosis. Neurology 47:895–900

Antel J, Bar-Or A (2006) Roles of immunoglobulins and B cells in multiple sclerosis: from pathogenesis to treatment. J Neuroimmunol 180:3–8

Arnett HA, Fancy SP, Alberta JA, Zhao C, Plant SR, Kaing S, Raine CS, Rowitch DH, Franklin RJ, Stiles CD (2004) bHLH transcription factor Olig1 is required to repair demyelinated lesions in the CNS. Science 306:2111–2115

Baranzini SE, Jeong MC, Butunoi C, Murray RS, Bernard CC, Oksenberg JR (1999) B cell repertoire diversity and clonal expansion in multiple sclerosis brain lesions. J Immunol 163:5133–5144

Barnett MH, Prineas JW (2004) Relapsing and remitting multiple sclerosis: pathology of the newly forming lesion. Ann Neurol 55:458–468

Barnett MH, Henderson AP, Prineas JW (2006) The macrophage in MS: just a scavenger after all? Pathology and pathogenesis of the acute MS lesion. Mult Scler 12:121–132

Bar-Or A (2008) The immunology of multiple sclerosis. Semin Neurol 28:29–45

Bar-Or A, Vollmer T, Antel J, Arnold DL, Bodner CA, Campagnolo D, Gianettoni J, Jalili F, Kachuck N, Lapierre Y et al (2007) Induction of antigen-specific tolerance in multiple sclerosis after immunization with DNA encoding myelin basic protein in a randomized, placebo-controlled phase 1/2 trial. Arch Neurol 64:1407–1415

Becher B, Bechmann I, Greter M (2006) Antigen presentation in autoimmunity and CNS inflammation: how T lymphocytes recognize the brain. J Mol Med 84:532–543

Beutler E (1992) Cladribine (2-chlorodeoxyadenosine). Lancet 340:952–956

Bielekova B, Goodwin B, Richert N, Cortese I, Kondo T, Afshar G, Gran B, Eaton J, Antel J, Frank JA et al (2000) Encephalitogenic potential of the myelin basic protein peptide (amino acids 83-99) in multiple sclerosis: results of a phase II clinical trial with an altered peptide ligand. Nat Med 6:1167–1175

Bielekova B, Richert N, Howard T, Blevins G, Markovic-Plese S, McCartin J, Frank JA, Wurfel J, Ohayon J, Waldmann TA et al (2004) Humanized anti-CD25 (daclizumab) inhibits disease activity in multiple sclerosis patients failing to respond to interferon beta. Proc Natl Acad Sci USA 101:8705–8708

Black JA, Liu S, Carrithers M, Carrithers LM, Waxman SG (2007) Exacerbation of experimental autoimmune encephalomyelitis after withdrawal of phenytoin and carbamazepine. Ann Neurol 62:21–33

Brinkmann V (2007) Sphingosine 1-phosphate receptors in health and disease: mechanistic insights from gene deletion studies and reverse pharmacology. Pharmacol Ther 115:84–105

Brousil JA, Roberts RJ, Schlein AL (2006) Cladribine: an investigational immunomodulatory agent for multiple sclerosis. Ann Pharmacother 40:1814–1821

Buchli AD, Schwab ME (2005) Inhibition of Nogo: a key strategy to increase regeneration, plasticity and functional recovery of the lesioned central nervous system. Ann Med 37:556–567

Cepok S, Zhou D, Srivastava R, Nessler S, Stei S, Bussow K, Sommer N, Hemmer B (2005) Identification of Epstein-Barr virus proteins as putative targets of the immune response in multiple sclerosis. J Clin Invest 115:1352–1360

Chang A, Tourtellotte WW, Rudick R, Trapp BD (2002) Premyelinating oligodendrocytes in chronic lesions of multiple sclerosis. N Engl J Med 346:165–173

Cherwinski HM, McCarley D, Schatzman R, Devens B, Ransom JT (1995) The immunosuppressant leflunomide inhibits lymphocyte progression through cell cycle by a novel mechanism. J Pharmacol Exp Ther 272:460–468

Chora AA, Fontoura P, Cunha A, Pais TF, Cardoso S, Ho PP, Lee LY, Sobel RA, Steinman L, Soares MP (2007) Heme oxygenase-1 and carbon monoxide suppress autoimmune neuroinflammation. J Clin Invest 117:438–447

Cohen JA, Barkhof F, Comi G, Hartung HP, Khatri BO, Montalban X, Pelletier J, Capra R, Gallo P, Izquierdo G, Tiel-Wilck K, de Vera A, Jin J, Stites T, Wu S, Aradhye S, Kappos L, TRANSFORMS Study Group (2010) Oral fingolimod or intramuscular interferon for relapsing multiple sclerosis. N Engl J Med 362:402–415

Coles A, Deans J, Compston A (2004) Campath-1H treatment of multiple sclerosis: lessons from the bedside for the bench. Clin Neurol Neurosurg 106:270–274

Cree B (2006) Emerging monoclonal antibody therapies for multiple sclerosis. Neurologist 12:171–178

de Jong R, Bezemer AC, Zomerdijk TP, van de Pouw-Kraan T, Ottenhoff TH, Nibbering PH (1996) Selective stimulation of T helper 2 cytokine responses by the anti-psoriasis agent monomethylfumarate. Eur J Immunol 26:2067–2074

Engelhardt B, Ransohoff RM (2005) The ins and outs of T-lymphocyte trafficking to the CNS: anatomical sites and molecular mechanisms. Trends Immunol 26:485–495

Filbin MT (2003) Myelin-associated inhibitors of axonal regeneration in the adult mammalian CNS. Nat Rev Neurosci 4:703–713

Fontoura P, Steinman L, Miller A (2006) Emerging therapeutic targets in multiple sclerosis. Curr Opin Neurol 19:260–266

Freund P, Schmidlin E, Wannier T, Bloch J, Mir A, Schwab ME, Rouiller EM (2006) Nogo-A-specific antibody treatment enhances sprouting and functional recovery after cervical lesion in adult primates. Nat Med 12:790–792

Frohman EM, Racke MK, Raine CS (2006) Multiple sclerosis–the plaque and its pathogenesis. N Engl J Med 354:942–955

Fukushima R, Kanamori S, Hirashiba M, Hishikawa A, Muranaka RI, Kaneto M, Nakamura K, Kato I (2007) Teratogenicity study of the dihydroorotate-dehydrogenase inhibitor and protein tyrosine kinase inhibitor Leflunomide in mice. Reprod Toxicol 24:310–316

Garren H, Robinson WH, Krasulova E, Havrdova E, Nadj C, Selmaj K, Losy J, Nadj I, Radue EW, Kidd BA et al (2008) Phase 2 trial of a DNA vaccine encoding myelin basic protein for multiple sclerosis. Ann Neurol 63:611–620

Genain CP, Cannella B, Hauser SL, Raine CS (1999) Identification of autoantibodies associated with myelin damage in multiple sclerosis. Nat Med 5:170–175

Giovannoni G, Comi G, Cook S, Rammohan K, Rieckmann P, Soelberg Sørensen P, Vermersch P, Chang P, Hamlett A, Musch B, Greenberg SJ, CLARITY Study Group (2010) A placebo-controlled trial of oral cladribine for relapsing multiple sclerosis. N Engl J Med 362:416–426

Glass CK, Ogawa S (2006) Combinatorial roles of nuclear receptors in inflammation and immunity. Nat Rev Immunol 6:44–55

Grieb P, Ryba M, Stelmasiak Z, Nowicki J, Solski J, Jakubowska B (1994) Cladribine treatment of multiple sclerosis. Lancet 344:538

Hafler DA, Slavik JM, Anderson DE, O'Connor KC, De Jager P, Baecher-Allan C (2005) Multiple sclerosis. Immunol Rev 204:208–231

Hafler DA, Compston A, Sawcer S, Lander ES, Daly MJ, De Jager PL, de Bakker PI, Gabriel SB, Mirel DB, Ivinson AJ et al (2007) Risk alleles for multiple sclerosis identified by a genomewide study. N Engl J Med 357:851–862

Hauser SL, Waubant E, Arnold DL, Vollmer T, Antel J, Fox RJ, Bar-Or A, Panzara M, Sarkar N, Agarwal S et al (2008) B-cell depletion with rituximab in relapsing-remitting multiple sclerosis. N Engl J Med 358:676–688

Hemmer B, Archelos JJ, Hartung HP (2002) New concepts in the immunopathogenesis of multiple sclerosis. Nat Rev Neurosci 3:291–301

Jiang H, Chess L (2004) An integrated view of suppressor T cell subsets in immunoregulation. J Clin Invest 114:1198–1208

John GR, Shankar SL, Shafit-Zagardo B, Massimi A, Lee SC, Raine CS, Brosnan CF (2002) Multiple sclerosis: re-expression of a developmental pathway that restricts oligodendrocyte maturation. Nat Med 8:1115–1121

Jonsson S, Andersson G, Fex T, Fristedt T, Hedlund G, Jansson K, Abramo L, Fritzson I, Pekarski O, Runstrom A et al (2004) Synthesis and biological evaluation of new 1, 2-dihydro-4-hydroxy-2-oxo-3-quinolinecarboxamides for treatment of autoimmune disorders: structure-activity relationship. J Med Chem 47:2075–2088

Kanter JL, Narayana S, Ho PP, Catz I, Warren KG, Sobel RA, Steinman L, Robinson WH (2006) Lipid microarrays identify key mediators of autoimmune brain inflammation. Nat Med 12:138–143

Kappos L, Antel J, Comi G, Montalban X, O'Connor P, Polman CH, Haas T, Korn AA, Karlsson G, Radue EW (2006a) Oral fingolimod (FTY720) for relapsing multiple sclerosis. N Engl J Med 355:1124–1140

Kappos L, Miller DH, MacManus DG, Gold R, Havrdova E, Limmroth V, Polman C, Schmierer K, Yousry T, Yang M et al (2006b) Efficacy of a novel oral single-agent fumarate, BG00012, in patients with relapsing-remitting multiple sclerosis: results of a phase 2 study. J Neurol 253:27

Kappos L, Radue EW, O'Connor P, Polman C, Hohlfeld R, Calabresi P, Selmaj K, Agoropoulou C, Leyk M, Zhang-Auberson L, Burtin P, FREEDOMS Study Group (2010) A placebo-controlled trial of oral fingolimod in relapsing multiple sclerosis. N Engl J Med 362:387–401

Karadottir R, Cavelier P, Bergersen LH, Attwell D (2005) NMDA receptors are expressed in oligodendrocytes and activated in ischaemia. Nature 438:1162–1166

Karussis DM, Meiner Z, Lehmann D, Gomori JM, Schwarz A, Linde A, Abramsky O (1996) Treatment of secondary progressive multiple sclerosis with the immunomodulator linomide: a double-blind, placebo-controlled pilot study with monthly magnetic resonance imaging evaluation. Neurology 47:341–346

Kebir H, Kreymborg K, Ifergan I, Dodelet-Devillers A, Cayrol R, Bernard M, Giuliani F, Arbour N, Becher B, Prat A (2007) Human TH17 lymphocytes promote blood-brain barrier disruption and central nervous system inflammation. Nat Med 13:1173–1175

Kielian T (2006) Toll-like receptors in central nervous system glial inflammation and homeostasis. J Neurosci Res 83:711–730

Korn T, Toyka K, Hartung HP, Jung S (2001) Suppression of experimental autoimmune neuritis by leflunomide. Brain 124:1791–1802

Krogsgaard M, Wucherpfennig KW, Cannella B, Hansen BE, Svejgaard A, Pyrdol J, Ditzel H, Raine C, Engberg J, Fugger L (2000) Visualization of myelin basic protein (MBP) T cell epitopes in multiple sclerosis lesions using a monoclonal antibody specific for the human histocompatibility leukocyte antigen (HLA)-DR2-MBP 85-99 complex. J Exp Med 191:1395–1412

Kutzelnigg A, Lucchinetti CF, Stadelmann C, Bruck W, Rauschka H, Bergmann M, Schmidbauer M, Parisi JE, Lassmann H (2005) Cortical demyelination and diffuse white matter injury in multiple sclerosis. Brain 128:2705–2712

Lassmann H (2008) Models of multiple sclerosis: new insights into pathophysiology and repair. Curr Opin Neurol 21:242–247

Lassmann H, Bruck W, Lucchinetti C (2001) Heterogeneity of multiple sclerosis pathogenesis: implications for diagnosis and therapy. Trends Mol Med 7:115–121

Liblau RS, Pearson CI, Shokat K, Tisch R, Yang XD, McDevitt HO (1994) High-dose soluble antigen: peripheral T-cell proliferation or apoptosis. Immunol Rev 142:193–208

Lucchinetti C, Bruck W, Parisi J, Scheithauer B, Rodriguez M, Lassmann H (2000) Heterogeneity of multiple sclerosis lesions: implications for the pathogenesis of demyelination. Ann Neurol 47:707–717

Lundmark F, Duvefelt K, Iacobaeus E, Kockum I, Wallstrom E, Khademi M, Oturai A, Ryder LP, Saarela J, Harbo HF et al (2007) Variation in interleukin 7 receptor alpha chain (IL7R) influences risk of multiple sclerosis. Nat Genet 39:1108–1113

Magliozzi R, Howell O, Vora A, Serafini B, Nicholas R, Puopolo M, Reynolds R, Aloisi F (2007) Meningeal B-cell follicles in secondary progressive multiple sclerosis associate with early onset of disease and severe cortical pathology. Brain 130:1089–1104

Marik C, Felts PA, Bauer J, Lassmann H, Smith KJ (2007) Lesion genesis in a subset of patients with multiple sclerosis: a role for innate immunity? Brain 130:2800–2815

Martin R, Jaraquemada D, Flerlage M, Richert J, Whitaker J, Long EO, McFarlin DE, McFarland HF (1990) Fine specificity and HLA restriction of myelin basic protein-specific cytotoxic T cell lines from multiple sclerosis patients and healthy individuals. J Immunol 145:540–548

Martino G, Pluchino S (2006) The therapeutic potential of neural stem cells. Nat Rev Neurosci 7:395–406

Matysiak M, Stasiolek M, Orlowski W, Jurewicz A, Janczar S, Raine CS, Selmaj K (2008) Stem cells ameliorate EAE via an indoleamine 2, 3-dioxygenase (IDO) mechanism. J Neuroimmunol 193:12–23

McDonald WI, Compston A, Edan G, Goodkin D, Hartung HP, Lublin FD, McFarland HF, Paty DW, Polman CH, Reingold SC et al (2001) Recommended diagnostic criteria for multiple sclerosis: guidelines from the International Panel on the diagnosis of multiple sclerosis. Ann Neurol 50:121–127

McFarland HF, Martin R (2007) Multiple sclerosis: a complicated picture of autoimmunity. Nat Immunol 8:913–919

McGeachy MJ, Cua DJ (2008) Th17 cell differentiation: the long and winding road. Immunity 28:445–453

Mi S, Miller RH, Lee X, Scott ML, Shulag-Morskaya S, Shao Z, Chang J, Thill G, Levesque M, Zhang M et al (2005) LINGO-1 negatively regulates myelination by oligodendrocytes. Nat Neurosci 8:745–751

Mi S, Hu B, Hahm K, Luo Y, Kam Hui ES, Yuan Q, Wong WM, Wang L, Su H, Chu TH et al (2007) LINGO-1 antagonist promotes spinal cord remyelination and axonal integrity in MOG-induced experimental autoimmune encephalomyelitis. Nat Med 13:1228–1233

Mitchison NA (1964) Induction of immunological paralysis in two zones of dosage. Proc R Soc Lond B Biol Sci 161:275–292

Neumann H, Medana IM, Bauer J, Lassmann H (2002) Cytotoxic T lymphocytes in autoimmune and degenerative CNS diseases. Trends Neurosci 25:313–319

Noseworthy JH, Wolinsky JS, Lublin FD, Whitaker JN, Linde A, Gjorstrup P, Sullivan HC (2000) Linomide in relapsing and secondary progressive MS: part I: trial design and clinical results. North American Linomide Investigators. Neurology 54:1726–1733

Nussenblatt RB, Thompson DJ, Li Z, Chan CC, Peterson JS, Robinson RR, Shames RS, Nagarajan S, Tang MT, Mailman M et al (2003) Humanized anti-interleukin-2 (IL-2) receptor alpha therapy: long-term results in uveitis patients and preliminary safety and activity data for establishing parameters for subcutaneous administration. J Autoimmun 21:283–293

Ockenfels HM, Schultewolter T, Ockenfels G, Funk R, Goos M (1998) The antipsoriatic agent dimethylfumarate immunomodulates T-cell cytokine secretion and inhibits cytokines of the psoriatic cytokine network. Br J Dermatol 139:390–395

O'Connor PW, Li D, Freedman MS, Bar-Or A, Rice GP, Confavreux C, Paty DW, Stewart JA, Scheyer R (2006) A Phase II study of the safety and efficacy of teriflunomide in multiple sclerosis with relapses. Neurology 66:894–900

Ota K, Matsui M, Milford EL, Mackin GA, Weiner HL, Hafler DA (1990) T-cell recognition of an immunodominant myelin basic protein epitope in multiple sclerosis. Nature 346:183–187

Patrikios P, Stadelmann C, Kutzelnigg A, Rauschka H, Schmidbauer M, Laursen H, Sorensen PS, Bruck W, Lucchinetti C, Lassmann H (2006) Remyelination is extensive in a subset of multiple sclerosis patients. Brain 129:3165–3172

Pirko I, Lucchinetti CF, Sriram S, Bakshi R (2007) Gray matter involvement in multiple sclerosis. Neurology 68:634–642

Platten M, Ho PP, Youssef S, Fontoura P, Garren H, Hur EM, Gupta R, Lee LY, Kidd BA, Robinson WH et al (2005) Treatment of autoimmune neuroinflammation with a synthetic tryptophan metabolite. Science 310:850–855

Pluchino S, Zanotti L, Rossi B, Brambilla E, Ottoboni L, Salani G, Martinello M, Cattalini A, Bergami A, Furlan R et al (2005) Neurosphere-derived multipotent precursors promote neuroprotection by an immunomodulatory mechanism. Nature 436:266–271

Polman C, Barkhof F, Sandberg-Wollheim M, Linde A, Nordle O, Nederman T (2005) Treatment with laquinimod reduces development of active MRI lesions in relapsing MS. Neurology 64:987–991

Prinz M, Schmidt H, Mildner A, Knobeloch K-P, Hanisch U-K, Raasch J, Merkler D, Detje C, Gutcher I, Mages J et al (2008) Distinct and nonredundant in vivo functions of IFNAR on myeloid cells limit autoimmunity in the central nervous system. Immunity 28:1–12

Reff ME, Carner K, Chambers KS, Chinn PC, Leonard JE, Raab R, Newman RA, Hanna N, Anderson DR (1994) Depletion of B cells in vivo by a chimeric mouse human monoclonal antibody to CD20. Blood 83:435–445

Rice GP, Filippi M, Comi G (2000) Cladribine and progressive MS: clinical and MRI outcomes of a multicenter controlled trial. Cladribine MRI Study Group. Neurology 54:1145–1155

Romine JS, Sipe JC, Koziol JA, Zyroff J, Beutler E (1999) A double-blind, placebo-controlled, randomized trial of cladribine in relapsing-remitting multiple sclerosis. Proc Assoc Am Physicians 111:35–44

Rose JW, Watt HE, White AT, Carlson NG (2004) Treatment of multiple sclerosis with an anti-interleukin-2 receptor monoclonal antibody. Ann Neurol 56:864–867

Schilling S, Goelz S, Linker R, Luehder F, Gold R (2006) Fumaric acid esters are effective in chronic experimental autoimmune encephalomyelitis and suppress macrophage infiltration. Clin Exp Immunol 145:101–107

Schimrigk S, Brune N, Hellwig K, Lukas C, Bellenberg B, Rieks M, Hoffmann V, Pohlau D, Przuntek H (2006) Oral fumaric acid esters for the treatment of active multiple sclerosis: an open-label, baseline-controlled pilot study. Eur J Neurol 13:604–610

Schreibelt G, van Horssen J, van Rossum S, Dijkstra CD, Drukarch B, de Vries HE (2007) Therapeutic potential and biological role of endogenous antioxidant enzymes in multiple sclerosis pathology. Brain Res Rev 56:322–330

Serafini B, Rosicarelli B, Franciotta D, Magliozzi R, Reynolds R, Cinque P, Andreoni L, Trivedi P, Salvetti M, Faggioni A, Aloisi F (2007) Dysregulated Epstein-Barr virus infection in the multiple sclerosis brain. J Exp Med 204:2899–2912

Seto S, Carrera CJ, Kubota M, Wasson DB, Carson DA (1985) Mechanism of deoxyadenosine and 2-chlorodeoxyadenosine toxicity to nondividing human lymphocytes. J Clin Invest 75:377–383

Siden A (1979) Isoelectric focusing and crossed immunoelectrofocusing of CSF immunoglobulins in MS. J Neurol 221:39–51

Sipe JC, Romine JS, Koziol JA, McMillan R, Zyroff J, Beutler E (1994) Cladribine in treatment of chronic progressive multiple sclerosis. Lancet 344:9–13

Smith T, Groom A, Zhu B, Turski L (2000) Autoimmune encephalomyelitis ameliorated by AMPA antagonists. Nat Med 6:62–66

Sospedra M, Martin R (2005) Immunology of multiple sclerosis. Annu Rev Immunol 23:683–747

Steinman L (1996) Multiple sclerosis: a coordinated immunological attack against myelin in the central nervous system. Cell 85:299–302

Steinman L (2001) Multiple sclerosis: a two-stage disease. Nat Immunol 2:762–764

Steinman L (2005) Blocking adhesion molecules as therapy for multiple sclerosis: natalizumab. Nat Rev Drug Discov 4:510–518

Steinman L, Zamvil SS (2005) Virtues and pitfalls of EAE for the development of therapies for multiple sclerosis. Trends Immunol 26:565–571

Szczucinski A, Losy J (2007) Chemokines and chemokine receptors in multiple sclerosis. Potential targets for new therapies. Acta Neurol Scand 115:137–146

Tan IL, Lycklama a Nijeholt GJ, Polman CH, Ader HJ, Barkhof F (2000) Linomide in the treatment of multiple sclerosis: MRI results from prematurely terminated phase-III trials. Mult Scler 6:99–104

Vandermeeren M, Janssens S, Borgers M, Geysen J (1997) Dimethylfumarate is an inhibitor of cytokine-induced E-selectin, VCAM-1, and ICAM-1 expression in human endothelial cells. Biochem Biophys Res Commun 234:19–23

Viglietta V, Baecher-Allan C, Weiner HL, Hafler DA (2004) Loss of functional suppression by CD4+CD25+ regulatory T cells in patients with multiple sclerosis. J Exp Med 199:971–979

Wakkee M, Thio HB (2007) Drug evaluation: BG-12, an immunomodulatory dimethylfumarate. Curr Opin Investig Drugs 8:955–962

Warren KG, Catz I, Steinman L (1995) Fine specificity of the antibody response to myelin basic protein in the central nervous system in multiple sclerosis: the minimal B-cell epitope and a model of its features. Proc Natl Acad Sci USA 92:11061–11065

Warren KG, Catz I, Wucherpfennig KW (1997) Tolerance induction to myelin basic protein by intravenous synthetic peptides containing epitope P85 VVHFFKNIVTP96 in chronic progressive multiple sclerosis. J Neurol Sci 152:31–38

Warren KG, Catz I, Ferenczi LZ, Krantz MJ (2006) Intravenous synthetic peptide MBP8298 delayed disease progression in an HLA Class II-defined cohort of patients with progressive multiple sclerosis: results of a 24-month double-blind placebo-controlled clinical trial and 5 years of follow-up treatment. Eur J Neurol 13:887–895

Waxman SG (2006) Axonal conduction and injury in multiple sclerosis: the role of sodium channels. Nat Rev Neurosci 7:932–941

Wolinsky JS, Narayana PA, Noseworthy JH, Lublin FD, Whitaker JN, Linde A, Gjorstrup P, Sullivan HC (2000) Linomide in relapsing and secondary progressive MS: part II: MRI results. MRI Analysis Center of the University of Texas-Houston, Health Science Center, and the North American Linomide Investigators. Neurology 54:1734–1741

Yang JS, Xu LY, Xiao BG, Hedlund G, Link H (2004) Laquinimod (ABR-215062) suppresses the development of experimental autoimmune encephalomyelitis, modulates the Th1/Th2 balance and induces the Th3 cytokine TGF-beta in Lewis rats. J Neuroimmunol 156:3–9

Zamvil SS, Steinman L (1990) The T lymphocyte in experimental allergic encephalomyelitis. Annu Rev Immunol 8:579–621

Examination of the Role of MRI in Multiple Sclerosis: A Problem Orientated Approach

Henry F. McFarland

Abstract Current multiple sclerosis (MS) is generally thought to consist of two general pathological processes; acute inflammation and degeneration. The relationship between these two components is not understood. What is clear, however, is that the measures of acute inflammation are a poor predictor of long-term disability. Although some have suggested that inflammation may not contribute directly to the essential pathology in MS or that it is secondary to tissue degeneration, most students of the disease believe that the two processes are linked. Therefore, applications of MRI to measure both components of the disease are important. As most readers know, considerable success has been achieved in measuring acute inflammation and very little success has been obtained in identifying measures that correlate with disability and the prediction of future disability has not been achieved.

In this review, we will examine the successes and failures of MRI in measuring these two components of the disease process. Consequently, we will not attempt to provide a detailed review of each MRI technique or sequence that has been applied to MS (a number of excellent reviews are available) but rather discuss how these techniques have been applied to answer specific questions. We will provide some comments on the use of MRI in clinical trials as well as in clinical practice. Finally, we will end with a brief discussion of future challenges.

1 Measuring Acute Disease Using MRI

As mentioned above, a common hypothesis in MS is that the disease process is initiated by an acute inflammatory process. While many believe that the inflammation reflects an autoimmune process, direct proof is limited; acute inflammation could be initiated by other mechanisms such as an immune response to an infectious agent. Regardless of the cause, a focus of MRI research has been to measure

H.F. McFarland (✉)
Neuroimmunology Branch, NINDS, NIH, Bethesda, MD, USA
e-mail: mcfarlandh@ninds.nih.gov

Results Probl Cell Differ, DOI 10.1007/400_2009_33

the inflammatory component of the disease. Early in the application of MRI to MS, evidence emerged showing that contrast enhancement was a precursor of new lesions (Harris et al. 1991; Miller et al. 1993) and the contrast enhancement was related to acute inflammation (Katz et al. 1993). Subsequently, considerable success has been achieved using responses seen on postcontrast T1-weighted MRI to study the natural history of MS as well as to measure the effectiveness of therapies targeting the early, inflammatory aspect of MS (Barkhof et al. 1992; Frank et al. 1994; McFarland et al. 1992). The use of contrast-enhancing lesions (CELs) as an outcome has proved invaluable in phase I and early phase II clinical trials to establish preliminary evidence of the effect of the therapy and to justify advancing testing of a therapy to phase III trials that are costly both in dollars as well as in human resources. A major concern exists, however, in regard to the understanding of the significance of CELs. Multiple studies have examined the relationship between enhancing lesions and relapses and all have found that while a relationship exists it is weak (Albert et al. 1994; Kappos et al. 1999). A recent study has explored the extent to which CELs could meet the criteria for a validated surrogate for relapses (Petkau et al. 2008). The ability to use CELs as a surrogate would be valuable in shortening clinical trials but, unfortunately, CELs failed to meet the formal criteria for validating a biomarker as a surrogate. Some studies have suggested a modest relationship between CELs and disability over short periods of time (Smith et al. 1993; Losseff et al. 2001) but an analysis of the relationship of the average number of CELs on three serial MRIs and disability on average 8 years later failed to show any relationship with disability (Stone et al. submitted). In contrast, a long-term follow-up study of a cohort of patients presenting with a clinically isolated syndrome (CIS) has shown a relationship between the number of lesions seen on a T2-weighted image and disability 20 years later (Fisniku et al. 2008a). Since it is generally held that most if not all CELs persist as lesions on T2-weighted images and that essentially all lesions seen on T2-weighted images begin with the blood brain barrier (BBB) breakdown, the lack of a relationship between CELs and disability becomes more curious. The modest relationship between CELs and short term disability or conversion to CDMS (discussed later in this review) is most likely related to the relationship of CELs to relapses. The disconnect between enhancing lesions and longer-term disability suggests that processes other than just evidence of disruption of the BBB are contributing to disability. Several explanations may contribute to the lack of correlation between CELs and disability. Location of the lesion is certainly critical and it is well understood that some lesions such as those in the spinal cord contribute disproportionally to disability especially as measured by the EDSS. Yet it would be expected that lesions in the cord would be more frequent in patients with a higher frequency of enhancing lesions and that a relationship between the number of CELS and disability would exist. A second possibility is that not all enhancing lesions are equal and, in fact, considerable evidence for heterogeneity exists. Also it is now well documented that the damage in white matter distal to the lesions, the so-called normal appearing white matter (NAWM) contributes to disability. Finally, disease activity involving the GM no doubt contributes to disability. Each of these possibilities will be discussed below.

2   Evidence for Lesion Heterogeneity

It is now clear that all CELs are not the same; some are associated with greater destruction than others. While essentially all lesions seen on a T2-weighted image begin as a contrast-enhancing lesion, variability is evident using other imaging techniques such as magnetization transfer, diffusion weighted imaging, or appearance on noncontrast T1-weighted images. Some but not all acute lesions will remain or become hypointense on precontrast T1-weighted images (black holes) (van den Elskamp et al. 2008). Acute CELs that enhance for longer periods of time when followed using serial monthly MRIs are more likely to become persistent black holes (Bagnato et al. 2003) suggesting that either a qualitative or quantitative difference exists in the inflammatory response between those lesions that become black holes versus those that do not. A second technique used to demonstrate lesion heterogeneity is magnetization transfer imaging. The results are expressed as a ratio (MTR). Magnetization transfer ratio (MTR) imaging depends on the exchange of protons between macromolecular structure and water. The macromolecular content of tissue can be indirectly measured by using two images; an off-resonance pulse to saturate bound protons associated with macromolecular structure will affect the magnetic exchange with free water protons and alter the signal from the free water pool. The results are expressed as the signal intensity with and without the saturation pulse. A decrease in MTR indicates a probable decrease in macromolecular structure. Since macromolecular structure rich in protons in white matter is predominantly myelin, MTR is thought to represent predominantly a measure of myelin integrity (Schmierer et al. 2004). Clearly, however, MTR is sensitive to loss of axons. A strong correlation has been shown between black holes, the decrease in MTR, and axonal loss in tissue samples (van Walderveen et al. 1998). A number of investigators have reported differing evolutions of MTR in new CELs (Filippi et al. 1999; Richert et al. 2001). Figure 1 shows serial MTR values for three regions of interest (ROIs) that will become CELs at one time point (Figure 1).

The actual pathological substrate for the recovery of MTR in lesions is uncertain and could represent decreasing inflammation, remyelination, or both. The interpretation of the heterogeneity of MTR recovery of lesions is complex. Most likely both the magnitude of the inflammatory response and level of tissue destruction are reflected. The important point is that those lesions which persist with reduced MTR values are lesions which probably have the greatest levels of myelin and axonal loss. The association between decreased MTR and axonal loss has been demonstrated in a study examining the pathological changes seen in lesions that persist as black holes (van Walderveen et al. 1998). Using tissue samples from patients who have died from MS, a relationship between the degree of hypointensity on T1weighted images and MTRs within these lesions have been shown to correlate with axonal density.

Similar evidence has evolved from studies using diffusion tensor imaging. Two primary diffusion measurements can be made. The first represents the diffusion of protons along a confined longitudinal space (fractional anisotropy [FA]) and the

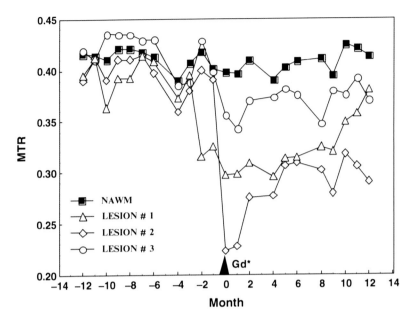

Fig. 1 Richert et al. (2001)

Table 1 Relationship between MTR values and axonal density using postmortum tissue

Lesion characteristic	MTR	Axon density (%)
No T2	0.32	90
T2 Isotropic on T1	0.30	80
Mild T1 hypointensity	0.24	50
Marked t1 hypointensity	0.15	30

Adapted from van Walderveen et al. (1998)

second is radial diffusion as would occur in free water (diffusivity). Again, uncertainty exists as to the relative contributions of myelin and the axon to FA; undoubtedly both contribute. Regardless, the results indicate that as the degree of tissue destruction increases, FA will decrease and diffusivity will increase (Schmierer et al. 2007). Again using this technique, results indicate differences in the level of tissue destruction (Kolind et al. 2008; Werring et al. 1999). Recently, a technique that combines selective imaging of the three tissue compartments in the brain: WM, GM, and CSF has been applied to studying heterogeneity of lesions (Riva et al. 2009). Using this approach, a small number of lesions that have signal intensity identical to CSF can be identified. Characterization of these lesions using MTR and DTI indicate that these lesions represent those with the greatest degree of tissue destruction. The differences in FA and MD between that are seen as hypointense on

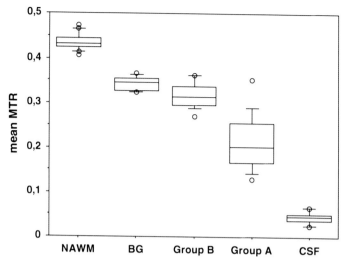

Fig. 2 Differences in MTR values between type 1 and type 2 lesions using tissue specific imaging (Riva et al. 2009)

a T1-weighted image and on the TSI image (group B) and those that are seen only on the t1-weighted image (group A) are shown in Fig. 2.

It is also worth noting that new contrast agents have been applied to the study of MS and results indicate that varying levels of lesion development are being identified (Vellinga et al. 2008; Bendszus et al. 2008; Linker et al. 2006). (Kolind et al. 2008; Werring et al. 1999).

All of the evidence described above indicates that considerable differences exist with respect to the level of tissue damage following the appearance of a CEL. Thus, it is likely that a part of the disconnect between the frequency of CELs and future disability is related to the level of tissue destruction that remains following the initial lesion.

3   Involvement of NAWM

In addition to the damage associated with the acute lesions recognized by a CEL or increased signal on a T2-weighted image, strong evidence exists for abnormalities in NAWM. Abnormalities have been demonstrated using a number of imaging techniques including MTR, MRS, DTI, and T2 relaxation (Griffin et al. 2002; Richert and Frank 1999; Rocca et al. 2008; Roosendaal et al. 2009; Vrenken et al. 2005; Yu et al. 2008). Evidence supports significant changes in NAWM beginning at an early stage of disease in MS; patients with CIS tend to have identifiable changes in MTR ratios in NAWM as compared to control individuals (Rocca et al. 2008). Of note, when correlations were sought between MRI measures and conversion from CIS to MS, abnormities in NAWM did not contribute while the number of lesions seen on T2- weighted images did. The findings are consistent with the idea that focal lesions have the strongest relationship to relapse which is necessary for clinical conversion to MS from CIS. What remains unclear is whether the abnormalities in NAWM are a result of acute focal lesion or represent a partially independent process. Studies of the pathology of MS have described the presence of immune cells in NAWM (Allen et al. 1981; Kutzelnigg et al. 2005), and diffuse inflammation could represent the substrate for the development of acute lesions. Consistent with this hypothesis is the finding that abnormalities in MTR may predict the development of an acute lesion many months in advance (Goodkin et al. 1998; Filippi et al. 1998). An example of decreased MTR preceding the development of acute lesions can be seen in Fig. 1 (Richert et al. 2001). In contrast, analysis of the magnitude of damage in NAWM-surrounding lesions indicates that the damage is greatest close to the lesion and diminishes with distance from the lesion (Guo et al. 2002). Consequently, the processes explaining abnormalities in NAWM are not well understood, but are doubtlessly critical to our understanding of the disease. As will be discussed below, considerable evidence indicating the involvement of gray matter is available now. It is possible that neuronal damage resulting in structural or metabolic impairment of axons in the white matter could contribute to focal alterations in axon-myelin integrity.

4   Involvement of Gray Matter

Numerous studies have focused on the damage to gray matter in MS. Lesions in the cortex are seen on pathological examination and can be seen on MRI. The cortical lesions have been classified based on appearance and the extent of cortex involved and if the lesion involves white matter as well as gray matter (Kidd et al. 1999; Peterson et al. 2001). Included in the classification are subpial lesions involving the first two layers of the cortex, lesions that are entirely confined to the cortex but may involve all layers of the cortex and lesions that involve both gray and white matter. Several characteristics of the pathology of cortical lesions are important to note in considering the ability to identify these lesions using MRI. Lesions within the cortex generally have relatively small amounts of inflammation but are demyelinating. Of note lesions that involve both gray and white matter have been shown to have the characteristic acute inflammatory characteristics of MS in the white matter portion but only limited inflammation within the gray matter portion of the lesion (Peterson et al. 2001). The results suggest that processes within the brain that contribute to amplification of inflammation may differ between gray and white matter. Only occasionally are CELs seen in the cortex. The demonstration of cortical lesions by MRI has been an area of considerable interest in the past few years and has met with limited success (reviewed in (Geurts and Barkhof 2008; Geurts 2008)). A variety of imaging techniques have been used including DIR, SPGR,and FLAIR (Bagnato et al. 2009; Bagnato et al. 2006; Nelson et al. 2008; Nelson et al. 2007).

In addition to lesions within the cortex, atrophy of both cortical and deep gray matter is well documented (Geurts 2008). Again a variety of techniques have been used to measure atrophy. Atrophy of the cortex can be seen early in the disease course and has been reported to be more marked than atrophy of WM (Fisniku et al. 2008b). Of note, atrophy of gray matter appears to be more marked in progressive disease than in RRMS (Fisher et al. 2008). The finding suggests that independent processes may be involved and this hypothesis is consistent with a lack of correlation between the pathological changes in WM compared to those in GM (Bo et al. 2007). Atrophy of deep gray matter structures is also observed in MS (Neema et al. 2009). Changes are more evident in the thalamus than in other components of the basal ganglion. Correlations with atrophy of deep GM including the thalamus and hippocampus and some measures of cognitive function have been reported (Benedict et al. 2004; Sicotte et al. 2008). Still uncertain is the relationship between gray matter lesions, gray matter atrophy, and disease in WM.

Currently a widely accepted concept of MS is that it begins with an inflammatory process but then evolves into a neurodegenerative disease. The relationships between acute lesions and other components of the disease process such as damage in NAWM and in both cortical and deep GM have been discussed above. Again it should be stressed that the relationship between acute inflammation and measures of more diffuse disease remains uncertain. In contrast to measurements of damage to specific tissue compartments a number of global measures have been used to access the overall degenerative component of the disease. The most commonly used metric is brain atrophy. Various techniques have been used to measure atrophy and an analysis of the relative merits of these various techniques is beyond the scope of this review. Generally all have reported about a 1% loss of tissue per year compared to a loss of about 0.1% in healthy individuals. Brain atrophy is currently the most extensively studied of all the global measures of tissue damage and is currently being used as an outcome in trials examining neuroprotective strategies (Altmann et al. 2009).

5   Use of MRI to Monitor Clinical Trials

In addition to being an important biomarker for MS, MRI has also had an important role in advancing new therapies (McFarland et al. 2002). Since CELs are recognized as reflecting an acute inflammatory stage of disease, CELs have been used as an outcome for various types of clinical trials. As discussed above CELs do not meet the criteria for a validated surrogate and therefore cannot be used in place of a clinical outcome in clinical trials which will be used to request registration of the treatment by either the American or European health authorities. However, CELs have been used to access safety in phase I studies and as an outcome in early phase II studies designed as proof-or-principle studies. Finally, imaging outcomes represent important secondary outcomes in phase III clinical trials and are useful in helping to understand the mechanisms of the therapy.

5.1 *Phase I Studies*

Many of the new therapies in development over the past several years have had mechanisms of action that are poorly understood. Consequently the initial exposure of a patient with MS to the therapy is done with the possibility that the therapy may increase rather than decrease disease activity. Since MRI is a very sensitive measure of new inflammatory disease activity, MRI outcomes are effective in monitoring safety. An example is a recent study of a PDE4 inhibitor (Bielekova et al. 2000a). Studies in the mouse EAE model indicated that PDE4 inhibition was associated with a decrease in disease activity both clinically and pathologically. The therapy was shown to decrease Th1 activity and promote Th2 activity. The initial clinical trial of a PDE4 inhibitor, rolipram, was done to examine safety. An initial cohort of patients with no ongoing activity was studied to see if new activity occurred on therapy. Next a second cohort with ongoing new activity was studied to access the effect of the therapy on patients with active disease. The results indicated that the therapy most likely increased disease activity and the study was stopped.

5.2 *Phase IIa Studies (Proof-of-Principle)*

Probably the most effective use of imaging as an outcome has come in early phase II studies. An interesting example is the study of IFNb1b that used a simple cross-over design of a to examine the effect of IFNb1b on CELs after the drug had completed phase III testing and had been approved for use (Stone et al. 1995). The study demonstrated clearly that IFNb1b targeted an early inflammatory stage of disease. Since then studies with similar designs have been used in proof-or-principle studies of a number of therapies. MRI represents an important outcome in early testing of essentially all of the therapies now in phase III testing.

A recent study of a monoclonal antibody, daclizumab, that blocks the alpha chain of the high affinity IL-2 receptor, CD25, has been studied in a series of small clinical trials using MRI as the primary outcome (Bielekova et al. 2009; Bielekova et al. 2004). The results of these studies have shown that daclizumab substantially reduces the frequency of CELs. The design used in these studies was similar to that described above for rolipram. In addition to accessing the effect of therapy on the imaging outcome, the effect on the patient's immune response was also interrogated. The immunological studies demonstrated that therapy was associated with an increase in the population of NK cells that appear to have regulatory function. The combination of the detailed immunology with the imaging outcomes demonstrated a very strong association between the expansion of the NK population and the reduction in the number of CELs while on therapy (Fig. 3).

A second example with very different results is a study of an altered peptide ligand (APL) in the treatment of MS (Bielekova et al. 2000b). APLs based on an immunodominant portion of myelin basic protein (MBP) had been tested in the

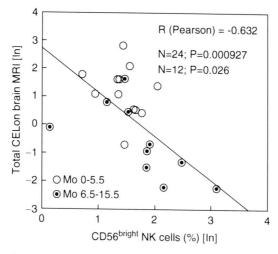

Fig. 3 Correlation between expansion of CD56bright NK cells and inhibition of brain inflammatory activity during treatment with daclizumab

mouse and had been shown to be effective in reducing disease. The clinical trial in patients again followed a design described for rolipram. Early in the study, a patient was identified as having a substantial increase in CELs. Two additional patients had increases in disease activity that seemed atypical, based on their prior levels of disease activity. Since these patients had also been carefully studied immunologically, it was possible to identify a relationship between the substantial increase in disease activity in the patient and the frequency of T cells reacting with MBP (Muraro et al. 2003). In distinction to generating just a regulatory phenotype, the therapy had resulted in a substantial increase in Th1 cells thought to promote disease. The results of this study demonstrate the power of combining careful MRI monitoring of proof-of-principle trials with detailed biological analysis of the effect of the therapy.

A final example is campath 1H or alemtuzumab (Coles et al. 2004). Alemtuzumab is a monoclonal antibody directed at CD52 which is expressed on a number of immunologically active cells. Alemtuzumab therapy results in significant immuno-suppression and had been used to treat B-cell chronic lymphocytic leukemia. The initial study of alemtuzumab in MS consisted of treating a relatively small cohort of patients including both RRMS and SPMS clinical courses. The results demonstrated substantial reduction in CELs and T2 lesion load. There was also reduction in relapse frequency. However many patients, particularly those already in the SPMS phase of the disease, continued to have progression of disability and progression of brain atrophy. The results of this study provided one of the clearest demonstrations that progression of disability and brain atrophy can become relatively independent of acute inflammatory activity once the disease was well established (Coles et al. 2006). The results support the hypothesis that while MS begins as an acute inflammatory process, disease progression is related to degenerative processes relatively independent of new inflammation. The hypothesis has been further validated by a subsequent study of alemtuzumab in patients with early MS; in this cohort a substantial reduction and even improvement in disability was reported (Coles et al. 2008).

5.3   Phase III Studies

MRI represents a secondary outcome in most phase III clinical trials in MS. Since all therapies examined in definitive phase III clinical trials to date have been therapies that target an inflammatory component of disease, evidence that the therapies decrease CELs or accumulate disease burden have been useful in confirming the effectiveness of the therapy. More important, however, will be the use of advance imaging techniques to better understand the disease process and the effect of the new therapy on that process. Since many of the new therapies are very effective in reducing inflammation, using advanced imaging techniques that should provide better information on the effect of the therapy on tissue damage will be valuable.

Unfortunately the ability to fund the application of advanced imaging and immunological evaluations as part of clinical trials is often problematic. Also problematic is the ability to get sponsors of clinical trials to invest in imaging that goes beyond the standard measures of lesion load and inflammation.

6   Use of Imaging in the Diagnosis of MS

Since the initial demonstration of the ability of MRI to image areas of damage in MS, MRI has been attractive as a tool in assisting in the diagnosis of the disease. Prior to the use of MRI in helping to establish dissemination in space and time, the diagnosis of MS was often difficult in the early stages of the disease. Several paradigms have been used in applying MRI criteria to assist with the diagnosis. In 2000 an international panel proposed a set of unified diagnostic criteria incorporating MRI parameters that would establish dissemination in space and time (McDonald et al. 2001). The criterion was based on a series of natural history studies that had examined MRI findings in patients with CIS that predict definite MS (Barkhof et al. 1997; Tintore et al. 2000). In 2007 the diagnostic criteria was revised and is outlined in Table 2 (Polman et al. 2005). Readers are advised to review the entire diagnostic criteria; only the MRI criteria for dissemination in space or time are shown in Table 2.

Recently, some investigators have proposed simplified criteria that would allow diagnosis based on a single MRI done at the time of presentation with an initial clinical event. Hopefully, a consensus will be reached quickly on these differences in approach. It is important to note that the criteria emerging from the international panel placed emphasis on the specificity of the diagnostic criteria. A goal was to propose criteria that would have the greatest chance of correctly establishing the

Table 2 MRI criteria for dissemination in time and space

Dissemination in space	Three of the following
	At least 1 CEL or 9 T2 lesions
	At least 1 juxtacortical lesion
	At least 1 infratentorial lesion
	At least 3 periventricula lesions
Dissemination in time	One of the following
	Demonstration of a CEL on an MRI done at least 3 months after the clinical onset and in a location not corresponding to the site of the initial event
	Detection of a new T2 lesion on any scan compared to a reference scan done 3 months after initial event

diagnosis, since the diagnosis of MS carries with it may potential difficulties such as employment and insurance. Proposals for more liberal diagnostic criteria have tended to place emphasis on sensitivity or the ability to diagnose the illness at the earliest stage in order to allow individuals to be started on therapy. Although the studies that have supported the more liberal criteria have shown reasonable specificity in addition to sensitivity, the studies have focused only on patients who have been followed in tertiary MS centers. Consequently, the rate of misdiagnosis in a general neurology clinic is unknown and is of concern.

7 Use of Imaging for Patient Management

Extrapolation of the successes in using MRI in clinical trials and the study of the natural history of MS to the care of patients in the clinic has been complicated. Identification of the effectiveness of a therapy in a clinical trial requires either knowledge of the level of disease in an individual patient prior to beginning therapy or the ability to compare the response of a group of patients to matched patients on placebo. Even the crossover design discussed previously has limitations in interpreting the effect of therapy in an individual patient. Levels of disease activity are extremely variable and the frequency of new lesions tends to decrease with time. So, the use of MRI in the clinic to judge the effectiveness of treatment in an individual patient has considerable risk of error. For example, examining the data in Fig. 4 it can be seen that if only one MRI was available pretreatment, the conclusions could be very different regarding the effect of therapy. The potential error is compounded when the single MRI is repeated while on therapy; the timing of the MRI could again lead to different conclusions.

Consequently, the use of MRI to measure actual disease as an assessment of the effectiveness of therapy must be done with caution. Clearly patients having high levels of disease on a therapy known to reduce acute disease activity can be considered to be nonresponders. Low levels of disease or even lack of acute disease are not easily interpreted on a single MRI.

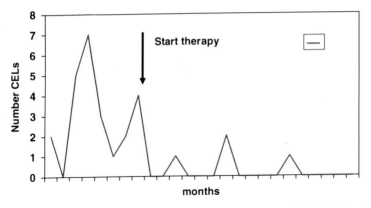

Fig. 4 Difficulties in using MRI to monitor treatment response in individual patients

Are MRI measures other than those that identify acute disease activity valuable in monitoring patients? A number of global measures of disease are often monitored including T2 lesion load, atrophy, and volume of T1 hypointensities. All of these measures require some level computer based image analysis in order to be accurate. For global measures to be helpful, careful attention must be paid to the technical aspects of the images. Essentially pretreatment and treatment images should be done using similar techniques which is often not the case when imaging is done outside of a research setting. Finally there is concern over the interpretation of changes in global measures since the effect of therapy is less well understood. In summary any effort to use MRI to monitor the effect of a particular therapy in the clinic must be done with considerable care and the realization that findings may be difficult to interpret.

8   Conclusions

MRI has provided a powerful tool for studying MS. It has advanced our understanding of the natural history of the disease and shown that the disease is usually active at the time of initial clinical presentation. This finding has contributed to the understanding of the need for early therapy. MRI has also contributed to our ability to bring new therapies to the clinic. Finally, MRI has helped in establishing an accurate diagnosis early in the disease process which is important now that disease modifying therapies exist. Despite these advances many questions remain with respect to the relationship between what we see on MRI and the disease process. Hopefully, the use of high field imaging along with the use of MRI together with therapies that target various phases of the disease will lead to a more complete understanding of the disease.

References

Albert PS, McFarland HF, Smith ME, Frank JA (1994) Time series for modelling counts from a relapsing-remitting disease: application to modelling disease activity in multiple sclerosis. Stat Med 13:453–466

Allen IV, Glover G, Anderson R (1981) Abnormalities in the macroscopically normal white matter in cases of mild or spinal multiple sclerosis (MS). Acta Neuropathol Suppl 7:176–178

Altmann DR, Jasperse B, Barkhof F et al (2009) Sample sizes for brain atrophy outcomes in trials for secondary progressive multiple sclerosis. Neurology 72:595–601

Bagnato F, Jeffries N, Richert ND et al (2003) Evolution of T1 black holes in patients with multiple sclerosis imaged monthly for 4 years. Brain 126:1782–1789

Bagnato F, Butman JA, Gupta S et al (2006) In vivo detection of cortical plaques by MR imaging in patients with multiple sclerosis. AJNR Am J Neuroradiol 27:2161–2167

Bagnato F, Yao B, Cantor F et al (2009) Multisequence-imaging protocols to detect cortical lesions of patients with multiple sclerosis: Observations from a post-mortem 3 Tesla imaging study. J Neurol Sci

Barkhof F, Valk J, Hommes OR et al (1992) Gadopentetate dimeglumine enhancement of multiple sclerosis lesions on long TR spin-echo images at 0.6 T. AJNR Am J Neuroradiol 13:1257–1259

Barkhof F, Filippi M, Miller DH et al (1997) Comparison of MRI criteria at first presentation to predict conversion to clinically definite multiple sclerosis. Brain 120(Pt 11):2059–2069

Bendszus M, Ladewig G, Jestaedt L et al (2008) Gadofluorine M enhancement allows more sensitive detection of inflammatory CNS lesions than T2-w imaging: a quantitative MRI study. Brain 131:2341–2352

Benedict RH, Weinstock-Guttman B, Fishman I et al (2004) Prediction of neuropsychological impairment in multiple sclerosis: comparison of conventional magnetic resonance imaging measures of atrophy and lesion burden. Arch Neurol 61:226–230

Bielekova B, Lincoln A, McFarland H, Martin R (2000a) Therapeutic potential of phosphodiesterase-4 and -3 inhibitors in Th1-mediated autoimmune diseases. J Immunol 164:1117–1124

Bielekova B, Goodwin B, Richert N et al (2000b) Encephalitogenic potential of the myelin basic protein peptide (amino acids 83–99) in multiple sclerosis: results of a phase II clinical trial with an altered peptide ligand. Nat Med 6:1167–1175

Bielekova B, Richert N, Howard T et al (2004) Humanized anti-CD25 (daclizumab) inhibits disease activity in multiple sclerosis patients failing to respond to interferon beta. Proc Natl Acad Sci USA 101:8705–8708

Bielekova B, Howard T, Packer AN et al (2009) Effect of anti-CD25 antibody daclizumab in the inhibition of inflammation and stabilization of disease progression in multiple sclerosis. Arch Neurol 66:483–489

Bo L, Geurts JJ, van der Valk P et al (2007) Lack of correlation between cortical demyelination and white matter pathologic changes in multiple sclerosis. Arch Neurol 64:76–80

Coles A, Deans J, Compston A (2004) Campath-1H treatment of multiple sclerosis: lessons from the bedside for the bench. Clin Neurol Neurosurg 106:270–274

Coles AJ, Cox A, Le Page E et al (2006) The window of therapeutic opportunity in multiple sclerosis: evidence from monoclonal antibody therapy. J Neurol 253:98–108

Coles AJ, Compston DA, Selmaj KW et al (2008) Alemtuzumab vs. interferon beta-1a in early multiple sclerosis. N Engl J Med 359:1786–1801

Filippi M, Rocca MA, Martino G et al (1998) Magnetization transfer changes in the normal appearing white matter precede the appearance of enhancing lesions in patients with multiple sclerosis. Ann Neurol 43:809–814

Filippi M, Rocca MA, Sormani MP et al (1999) Short-term evolution of individual enhancing MS lesions studied with magnetization transfer imaging. Magn Reson Imaging 17:979–984

Fisher E, Lee JC, Nakamura K, Rudick RA (2008) Gray matter atrophy in multiple sclerosis: a longitudinal study. Ann Neurol 64:255–265

Fisniku LK, Brex PA, Altmann DR et al (2008a) Disability and T2 MRI lesions: a 20-year follow-up of patients with relapse onset of multiple sclerosis. Brain 131:808–817

Fisniku LK, Chard DT, Jackson JS et al (2008b) Gray matter atrophy is related to long-term disability in multiple sclerosis. Ann Neurol 64:247–254

Frank JA, Stone LA, Smith ME et al (1994) Serial contrast-enhanced magnetic resonance imaging in patients with early relapsing-remitting multiple sclerosis: implications for treatment trials. Ann Neurol 36 Suppl:S86–90

Geurts JJ (2008) Is progressive multiple sclerosis a gray matter disease? Ann Neurol 64:230–232

Geurts JJ, Barkhof F (2008) Grey matter pathology in multiple sclerosis. Lancet Neurol 7:841–851

Goodkin DE, Rooney WD, Sloan R et al (1998) A serial study of new MS lesions and the white matter from which they arise. Neurology 51:1689–1697

Griffin CM, Dehmeshki J, Chard DT et al (2002) T1 histograms of normal-appearing brain tissue are abnormal in early relapsing-remitting multiple sclerosis. Mult Scler 8:211–216

Guo AC, MacFall JR, Provenzale JM (2002) Multiple sclerosis: diffusion tensor MR imaging for evaluation of normal-appearing white matter. Radiology 222:729–736

Harris JO, Frank JA, Patronas N et al (1991) Serial gadolinium-enhanced magnetic resonance imaging scans in patients with early, relapsing-remitting multiple sclerosis: implications for clinical trials and natural history. Ann Neurol 29:548–555

Kappos L, Moeri D, Radue EW et al (1999) Predictive value of gadolinium-enhanced magnetic resonance imaging for relapse rate and changes in disability or impairment in multiple sclerosis: a meta-analysis. Gadolinium MRI Meta-analysis Group. Lancet 353:964–969

Katz D, Taubenberger JK, Cannella B et al (1993) Correlation between magnetic resonance imaging findings and lesion development in chronic, active multiple sclerosis. Ann Neurol 34:661–669

Kidd D, Barkhof F, McConnell R et al (1999) Cortical lesions in multiple sclerosis. Brain 122(Pt 1):17–26

Kolind SH, Laule C, Vavasour IM et al (2008) Complementary information from multi-exponential T2 relaxation and diffusion tensor imaging reveals differences between multiple sclerosis lesions. Neuroimage 40:77–85

Kutzelnigg A, Lucchinetti CF, Stadelmann C et al (2005) Cortical demyelination and diffuse white matter injury in multiple sclerosis. Brain 128:2705–2712

Linker RA, Kroner A, Horn T et al (2006) Iron particle-enhanced visualization of inflammatory central nervous system lesions by high resolution: preliminary data in an animal model. AJNR Am J Neuroradiol 27:1225–1229

Losseff NA, Miller DH, Kidd D, Thompson AJ (2001) The predictive value of gadolinium enhancement for long term disability in relapsing-remitting multiple sclerosis–preliminary results. Mult Scler 7:23–25

McDonald WI, Compston A, Edan G et al (2001) Recommended diagnostic criteria for multiple sclerosis: guidelines from the International Panel on the diagnosis of multiple sclerosis. Ann Neurol 50:121–127

McFarland HF, Frank JA, Albert PS et al (1992) Using gadolinium-enhanced magnetic resonance imaging lesions to monitor disease activity in multiple sclerosis. Ann Neurol 32:758–766

McFarland HF, Barkhof F, Antel J, Miller DH (2002) The role of MRI as a surrogate outcome measure in multiple sclerosis. Mult Scler 8:40–51

Miller DH, Barkhof F, Nauta JJ (1993) Gadolinium enhancement increases the sensitivity of MRI in detecting disease activity in multiple sclerosis. Brain 116(Pt 5):1077–1094

Muraro PA, Wandinger KP, Bielekova B et al (2003) Molecular tracking of antigen-specific T cell clones in neurological immune-mediated disorders. Brain 126:20–31

Neema M, Arora A, Healy BC et al (2009) Deep gray matter involvement on brain MRI scans is associated with clinical progression in multiple sclerosis. J Neuroimaging 19:3–8

Nelson F, Poonawalla AH, Hou P et al (2007) Improved identification of intracortical lesions in multiple sclerosis with phase-sensitive inversion recovery in combination with fast double inversion recovery MR imaging. AJNR Am J Neuroradiol 28:1645–1649

Nelson F, Poonawalla A, Hou P et al (2008) 3D MPRAGE improves classification of cortical lesions in multiple sclerosis. Mult Scler 14:1214–1219

Peterson JW, Bo L, Mork S et al (2001) Transected neurites, apoptotic neurons, and reduced inflammation in cortical multiple sclerosis lesions. Ann Neurol 50:389–400

Petkau J, Reingold SC, Held U et al (2008) Magnetic resonance imaging as a surrogate outcome for multiple sclerosis relapses. Mult Scler 14:770–778

Polman CH, Reingold SC, Edan G et al (2005) Diagnostic criteria for multiple sclerosis: 2005 revisions to the "McDonald Criteria". Ann Neurol 58:840–846

Richert ND, Frank JA (1999) Magnetization transfer imaging to monitor clinical trials in multiple sclerosis. Neurology 53:S29–32

Richert ND, Ostuni JL, Bash CN et al (2001) Interferon beta-1b and intravenous methylprednisolone promote lesion recovery in multiple sclerosis. Mult Scler 7:49–58

Riva M, Ikonomidou VN, Ostuni JJ et al (2009) Tissue-specific imaging is a Robust methodology to differentiate in vivo T1 black holes with advanced multiple sclerosis-induced damage. AJNR Am J Neuroradiol 30(7):1394–401

Rocca MA, Agosta F, Sormani MP et al (2008) A three-year, multi-parametric MRI study in patients at presentation with CIS. J Neurol 255:683–691

Roosendaal SD, Geurts JJ, Vrenken H et al (2009) Regional DTI differences in multiple sclerosis patients. Neuroimage 44:1397–1403

Schmierer K, Scaravilli F, Altmann DR et al (2004) Magnetization transfer ratio and myelin in postmortem multiple sclerosis brain. Ann Neurol 56:407–415

Schmierer K, Wheeler-Kingshott CA, Boulby PA et al (2007) Diffusion tensor imaging of post mortem multiple sclerosis brain. Neuroimage 35:467–477

Sicotte NL, Kern KC, Giesser BS et al (2008) Regional hippocampal atrophy in multiple sclerosis. Brain 131:1134–1141

Smith ME, Stone LA, Albert PS et al (1993) Clinical worsening in multiple sclerosis is associated with increased frequency and area of gadopentetate dimeglumine-enhancing magnetic resonance imaging lesions. Ann Neurol 33:480–489

Stone LA, Frank JA, Albert PS et al (1995) The effect of interferon-beta on blood-brain barrier disruptions demonstrated by contrast-enhanced magnetic resonance imaging in relapsing-remitting multiple sclerosis. Ann Neurol 37:611–619

Tintore M, Rovira A, Martinez MJ et al (2000) Isolated demyelinating syndromes: comparison of different MR imaging criteria to predict conversion to clinically definite multiple sclerosis. AJNR Am J Neuroradiol 21:702–706

van den Elskamp IJ, Lembcke J, Dattola V et al (2008) Persistent T1 hypointensity as an MRI marker for treatment efficacy in multiple sclerosis. Mult Scler 14:764–769

van Walderveen MA, Kamphorst W, Scheltens P et al (1998) Histopathologic correlate of hypointense lesions on T1-weighted spin-echo MRI in multiple sclerosis. Neurology 50:1282–1288

Vellinga MM, Oude Engberink RD, Seewann A et al (2008) Pluriformity of inflammation in multiple sclerosis shown by ultra-small iron oxide particle enhancement. Brain 131:800–807

Vrenken H, Barkhof F, Uitdehaag BM et al (2005) MR spectroscopic evidence for glial increase but not for neuro-axonal damage in MS normal-appearing white matter. Magn Reson Med 53:256–266

Werring DJ, Clark CA, Barker GJ et al (1999) Diffusion tensor imaging of lesions and normal-appearing white matter in multiple sclerosis. Neurology 52:1626–1632

Yu CS, Lin FC, Liu Y et al (2008) Histogram analysis of diffusion measures in clinically isolated syndromes and relapsing-remitting multiple sclerosis. Eur J Radiol 68:328–334

Index